Microbes for Plant Stress Management

Microbes for Plant Stress Management

D.J. Bagyaraj
Centre for Natural Biological Resources and Community Development
University of Agricultural Sciences (UAS)
Bangalore, India

Jamaluddin
Department of Biological Science
Rani Durgavati University
Jabalpur, Madhya Pradesh, India

CRC Press
Taylor & Francis Group

NEW INDIA PUBLISHING AGENCY
New Delhi – 110 034

CRC Press
Taylor & Francis Group
6000 Broken Sound Parkway NW, Suite 300
Boca Raton, FL 33487-2742

© 2019 by New India Publishing Agency

CRC Press is an imprint of the Taylor & Francis Group, an informa business

No claim to original U.S. Government works

International Standard Book Number-13: 978-0-367-14071-7 (Hardback)

Print edition not for sale in South Asia (India, Sri Lanka, Nepal, Bangladesh, Pakistan or Bhutan)

Library of Congress Cataloging in Publication Data
A catalog record has been requested

Visit the Taylor & Francis Web site at
http://www.taylorandfrancis.com

and the CRC Press Web site at
http://www.crcpress.com

National Academy of Agricultural Sciences

June 22, 2016

Foreword

Dr. S. Ayyappan
President

As the world marches towards the year 2050 with the predicted population of 9.7 billion, the daunting task faced by the agricultural scientists is to find out methods to feed the growing population with shrinking natural resources. It is also predicted that the demand for food has to be raised by about 60% compared to the current food production. This challenge gets more complicated with the current threat of global climate change resulting in increased temperature, drought, salinity, etc. These abiotic stresses will vigorously affect the agricultural productivity. It has already been observed that yield of crops is reduced due to various stresses. Abiotic stresses predominantly affect the genetic potential of food crops to the tune of nearly 69%. Drought is the most serious environmental factor limiting the production of agricultural crops with serious economical and social impacts. Salinity is another severe environmental stress decreasing crop productivity, mainly in irrigated land, worldwide. Hence there is an urgent need to address the abiotic stresses in order to achieve the goal of increased food production. Certain agronomic and breeding strategies are recommended for mitigating abiotic stresses. Microorganisms could play an important role in adaptation strategies and increase tolerance to abiotic stresses in agricultural crops. The mechanisms adopted by microorganisms to mitigate abiotic stresses in crop plants are many. The information available on the utilization of microbes for plant stress management is limited.

NASC
DPS Marg, P.O. Pusa
New Delhi-110 012

Tel.: 91-11-25846055
91-11-25846051
91-11-25846052
Fax : 91-11-25846054
Email : naas@vsnl.com
Web.: www.naasindia.org

Viewing from the above stated perspective, the book on "Microbes for Plant Stress Management" by Dr. D.J. Bagyaraj and Dr. Jamaluddin is very timely. The book will be a valuable companion to students, teachers, researchers, administrators and policy makers who foresee the need for managing stresses affecting crop productivity using microbes for feeding mankind in the years to come.

S. Ayyappan

Preface

The world population is predicted to reach about 9.7 billion by the end of the year 2050 and to feed this population the current food production levels have to be raised by nearly 60% with limited land resource. This challenge gets compounded by the threat of global climate change leading to erratic rainfall, drought, salinity, increased temperature, etc. These abiotic stress consequences will seriously threaten the sustainable agricultural production. Hence there is a need to address this issue by all available means in order to achieve the goal of enhanced food production. Certain agronomic strategies like change in sowing dates, alley cropping, zero tillage, etc. have been proposed to mitigate stress. Breeding strategies recommend mitigating abiotic stresses by choice of resistant cultivars, use of hyper-accumulator plants, etc. A lesser explored potential option for abiotic stress alleviation is in the utilization of stress tolerant microbial resources, which have the ability to promote and sustain crop growth during adverse environmental conditions. This approach has gained popularity in the recent years and seems to be a potential option for the future. Thus the present book brings out the role of different groups of microorganisms like plant growth promoting rhizomicroorganisms (PGPR), arbuscular mycorrhizal fungi (AMF), endophytes, etc. in alleviating abiotic stress in crop plants.

The book contains 14 chapters written by distinguished scientists of the country having expertise in dealing with microorganisms and exploiting them for the benefits of mankind. Chapters 1 and 2 deal with exploring microbes from extreme environments and rhizomicrobiome respectively, for use in sustainable agriculture. Chapter 3 describes the recent developments in utilizing microorganisms for abiotic stress management in crop plants. Chapter 4 helps to understand the innate stress management mechanisms in plants. Chapter 5 deals with nanoparticles synthesized by microorganisms for use in agricultural ecosystem. Chapter 6 covers the role of plant growth promoting rhizomicroorganisms in supporting the growth of plants under stressed environment. Chapters 7 and 8 discuss the role of endophytic microorganisms in alleviation of abiotic stress in crop plants. Chapters 9 to 12 cover the role of arbuscular mycorrhizal fungi in alleviating drought and salt tolerance in different crops important in agriculture and horticulture. Chapter 13 is devoted to bioconversion of municipal solid wastes and its use in soil fertility. Chapter 14 provides an overview of microbial inoculants for quality seedling production in forestry. Thus the book is a comprehensive and detailed analysis of the subject.

We profusely thank all the authors for their keen interest, kind efforts and rich contributions in making this book highly informative and productive. Special thanks are due to Mr. R. Ashwin for his help in bringing out this book. We are also thankful to New India Publishing Agency, New Delhi for taking keen interest in bringing out this book in time.

D. J. Bagyaraj
Jamaluddin

Contents

List of Contributors

Abhishek Bharti, Microbiology Section, ICAR- Indian Institute of Soybean Research, Khandwa Road, Indore 452001, *abhishekmic@gmail.com*

Abhishek Mundaragi, Department of Botany, Karnatak University, Dharwad 580003, *mabhishekphd@gmail.com*

Anil Prakash, Department of Microbiology, Barkatullah University, Bhopal 462026 *dranilprakash98@gmail.com*

Ashwin R, Centre for Natural Biological Resources and Community Development (CNBRCD), 41 RBI Colony, Anand Nagar, Bangalore 560024, *ashwin.bengaluru@gmail.com*

Bagyaraj DJ, Centre for Natural Biological Resources and Community Development (CNBRCD), 41 RBI Colony, Anand Nagar, Bangalore 560024

Baskaran V, M.S. Swaminathan Research Foundation, 3rd Cross Road, Taramani, Chennai 600113, *baskaranvj@gmail.com*

Desai S, Central Research Institute for Dryland Agriculture, Hyderabad 500 059, desai1959@yahoo.com

Devarajan Thangadurai, Department of Botany, Karnatak University, Dharwad 580003, *drthanga.kud@gmail.com*

Ganga V, M.S. Swaminathan Research Foundation, 3rd Cross Road, Taramani, Chennai 600113, *ganga03on@gmail.com*

Hruda Ranjan Sahoo, Division of Plant Pathology and Microbiology, Regional Plant Resource Centre, Bhubaneswar 751015, *hruda26@gmail.com*

Indira Rathore, Central Arid Zone Research Institute, Jodhpur 342003 *rathoreindusamrpan@gmail.com*

Tarafdar JC, Central Arid Zone Research Institute, Jodhpur 342 003, jctarafdar@yahoo.in

Jamaluddin, Department of Biological Science, Rani Durgavati University, Jabalpur, MP-482001

Jegan S, M.S. Swaminathan Research Foundation, 3rd Cross Road, Taramani, Chennai 600113, *jeganmicro1@gmail.com*

Jeyabalan Sangeetha, Department of Environmental Science, Central University of Kerala, Tejaswini Hills, Kasaragod, Kerala 671316, *drtsangeetha@gmail.com*

Kamal K Pal, ICAR-Directorate of Groundnut Research, Ivnagar Road, Junagadh 362001, *kamalk_pal@yahoo.co.in*

Kathiravan R, M.S. Swaminathan Research Foundation, 3rd Cross Road, Taramani, Chennai 600113, *krsnakat@gmail.com*

Krishna Sundari Sattiraju, Department of Biotechnology, Jaypee Institute of Information Technology (JIIT), A-10, Sector 62, Noida 201304, *krishna.sundari@jiit.ac.in, skrishnasundari@gmail.com*

Mahaveer P Sharma, Microbiology Section, ICAR- Indian Institute of Soybean Research, Khandwa Road, Indore 452001, *mahaveer620@gmail.com*

Maheswari M, Central Research Institute for Dryland Agriculture, Hyderabad, 500 059, *mmandapaka59@gmail.com*

Minakshi Grover, Central Research Institute for Dryland Agriculture, Hyderabad, 500 059, *minigt3@yahoo.co.in*

Muniswamy David, Department of Zoology, Karnatak University, Dharwad, 580003, *mdavid.kud@gmail.com*

Muthukumar T, Root and Soil Biology Laboratory, Department of Botany, Bharathiar University, Coimbatore 641046, *tmkum@yahoo.com*

Nibha Gupta, Division of Plant Pathology and Microbiology, Regional Plant Resource Centre, Bhubaneswar-751015, *nguc2003@yahoo.co.in*

Padmavathi Tallapragada, Department of Microbiology, Centre for PG Studies, Jain University, Jayanagar, Bangalore 560011, *vam2010tpraviju@gmail.com*

Parkash Vipin, Mycology and Soil Microbiology Research and Technology Laboratory, Rain Forest Research Institute, ICFRE, Deovan, Jorhat 785001 *bhardwajvpnpark@rediffmail.com*

Poonam Verma, Forest Pathology Division, Tropical Forest Research Institute, Jabalpur 482021, *poonamverma8624@gmail.com*

Prabavathy VR, M.S. Swaminathan Research Foundation, 3rd Cross Road, Taramani, Chennai 600113, *prabavathyvr@mssrf.res.in*

Purushotham Prathima, Department of Botany, Karnatak University, Dharwad, 580003, Karnataka, India, *prathimapacharya3@gmail.com*

Rinku Dey, ICAR-Directorate of Groundnut Research, Ivnagar Road, Junagadh 362001, *rinku_dy@yahoo.co.in*

Saikia AJ, Mycology and Soil Microbiology Research and Technology Laboratory, Rain Forest Research Institute, ICFRE, Deovan, Jorhat 785001, *ankurjyoti05@rediffmail.com*

Saikia M, Mycology and Soil Microbiology Research and Technology Laboratory, Rain Forest Research Institute, ICFRE, Deovan, Jorhat 785001, *saikiamonali@gmail.com*

Sanjay K Singh, Biodiversity and Palaeobiology Group, MACS' ARI, GG Agarkar Road, Pune 411004, *sksingh@aripune.org*

Shinde BP, Vidya Pratisthans Arts, Science and Commerce College Baramati, Dist. Pune 413133, *scindiab2002@gmail.com*

Shivani Garg, Microbiology Section, ICAR- Indian Institute of Soybean Research, Khandwa Road, Indore 452001, *gargsshivani@gmail.com*

Shrinivas Jadhav, Department of Zoology, Karnatak University, Dharwad, 580003, *sspj123@gmail.com*

Srinivasa Rao Ch, Central Research Institute for Dryland Agriculture, Hyderabad 500059, India, *cherukumalli2011@gmail.com*

Srishti Kotiyal, Department of Biotechnology, Jaypee Institute of Information Technology (JIIT), A-10, Sector 62, Noida 201304, *srishti11101053@gmail.com*

Venkateswarlu B, Vasantrao Naik Marathwada Krishi Vidyapeeth, Parbhani 431402, *vbandi_1953@yahoo.com*

Yadav SK, Central Research Institute for Dryland Agriculture, Hyderabad 500059, *skyadav@crida.in*

1

Exploring Microbes from Extreme Environments for Crop Productivity and Environmental Sustainability

Jeyabalan Sangeetha, Purushotham Prathima, Devarajan Thangadurai, Muniswamy David, Abhishek Mundaragi, Shrinivas Jadhav

Abstract

Ever growing population and associated food scarcity are the major issues of concern globally. Microorganisms are the ultimate solution for addressing this severe problem in association with recent advances made in genomics and molecular biology. Many bacterial and fungal species which are in association with plant rhizosphere soil can immensely contribute in better growth and developments, ultimately leading to better yield. Exploring these novel organisms happens to be the need of an hour. Metagenomics and next generation sequencing technologies can contribute in identifying those microbes which are otherwise impossible to cultivate traditionally. Extremophiles make up those classes of microbes which are explored to a lesser extent. This is due to their extreme habitat conditions and impracticable cultivation methods. However, recent sophisticated technologies make isolation and identification of certain traits in these organisms which can be made use in agronomic applications. There are number of extremophiles which could be explored and these are divided according to environment where they exist: halophilic, thermophilic, psychrophiles, acidophilic and alkalophilic microorganisms. These extremophiles are better measures to counter play toxic infiltration in cultivation soil due to excessive use of chemical fertilizers. Since their extreme tolerance capabilities exploiting them in agriculture can replenish essential vital components to plants by means of rapid and efficient nutrient recycling. In this scenario, current chapter describes an insight on various aspects of exploring extremophiles for sustainable agriculture and provides an understanding on their practical applications in food, agriculture and sustainability.

Keywords: Ammonia-oxidizing archaea, Ectomycorrhizal fungi, Metagenomics, Mycorrhizae, Plant growth promoting rhizobacteria

1. Introduction

Across globe the demand for food is greatly increasing due to the rise in population. Several types of stress factors including drought, nutrient deficiencies, diseases and pests have been serious issues during these days (Jaggard et al. 2010). Also agriculture practices are changing drastically to meet that demand. Though agriculture occupies major portion of Earth's surface (approximately 38%) the current agricultural practices includes extensive use of chemical fertilizers and pesticides which have resulted in devastation of soil fertility, soil microbial diversity and regulation of biogeochemical cycles (Horrigan et al. 2002). Microbes play a significant role in regulation of biogeochemical cycles such as carbon, nitrogen and phosphorus. Considering these facts, concerns have been raised regarding the agricultural expansion that will have greater impact on environment. The concept of low-input agriculture is one of the popular approaches. According to Edwards (1989) the controlled use of fertilizers and pesticides and cultivation methods play a key role in sustainable agriculture. Inorganic fertilizers and harmful pesticides should be minimized and abandoned. Organic farming and integrated pest management systems should be promoted.

The microbial communities found in the soil have crucial roles in overall development of plant, for instance, enhancing nitrogen fixation, promoting plant growth and protection against pathogens. Analysis of soil microbial communities appears as a promising approach to meet sustainable environment and agronomic issues (Gupta et al. 2015). Furthermore, microbes play a vital and often inimitable role in the proper functioning of ecosystems and in preserving a sustainable biosphere (Elmqvist et al. 2010). Currently, an increased knowledge of microbial diversity and innovations in high throughput technologies of genomics has ensured dynamic progress in achieving sustainable agriculture with a minimal harm to the nature and environment. Detailed studies on microbial interactions and microbe-plant interaction are very useful in agriculture and significant for the improvement of soil fertility and crop production. The symbiotic association of plant and beneficial microbes such as diazotrophs, plant growth promoting rhizobacteria (PGPRs), arbuscular mycorrhizal fungi (AMF) are essential for the development of sustainable agriculture. Greater emphasis is being now laid on biological techniques like vermicomposting, legume biological nitrogen fixation (BNF), biofertilizers, biopesticides, bioherbicides and bioinsecticides

(Russo *et al.* 2012; Gupta *et al.* 2015). Thus, bio-inoculants play crucial functions in preserving soil health, sustained crop production and soil microbial communities (Trabelsi and Mhamdi 2013).

Microorganisms are the largest, unexplored and most divergent group of organisms found in biosphere that contains approximately $4\text{-}6 \times 10^{30}$ prokaryotic cells (Whitman *et al.* 1998) constituting nearly 60% of the Earth's biomass. Till date only 1% of microorganisms have been studied and distinguished. Till today, several microbes are unidentified and unculturable because nutritional requirement of each bacterium is different from one another. The conventional method of culture dependent identification is tedious and hectic. Nevertheless, recent advancements in genomics and next generation sequencing technologies, analysis of microbial community from various environmental niche using culture-independent techniques such as metagenomics have paved the way for understanding microbial ecology better (Simon and Daniel 2011). Wherein, now the technologies have unleashed several microbial genomes that were unknown earlier and have provided deeper insights in understanding complexity involved in the functioning of the ecosystem.

Exploring microbial diversity from various extreme environmental niches will aid to find the novel strains and species of microbes and enzymes that can be truly benefit in near future (Hibbing *et al.* 2010). For instance, the discovery of novel bacterium *Thermus aquaticus* by Thomus Brock in 1967 shed the light on the study of extremophiles, later a thermostable enzyme of this bacterium was found useful and is extensively being used in the polymerase chain reaction (PCR) technique for DNA amplification (Brock 1967). Microbial diversity analysis of extreme environments has made remarkable progress because of their unique properties and opportunities that they offer in identification of novel functionalities and genes (Panada *et al.* 2015). Thus, conservation of biodiversity is a primary requisite for the overall development of ecosystem (Goyal and Arora 2009). This chapter presents an overview on the application of extremophiles in sustainable agriculture, advantages and potentialities of currently used molecular techniques to investigate and identify these microbial communities and future challenges and perspectives for applying microbial biotechnology for sustainable agriculture.

2. Exploring microbes using classical and molecular approaches

Microbial communities are the earliest predictors for change in soil quality, and also the critical components of soil (Scow *et al.* 1998). Recently, soil quality has been defined as the ability of soil to sustain biological functions and to promote plant growth and animal health (Doran and Parkin 1994). The maintenance of microbial diversity in soil is a promising task, hence microorganisms are involved in several functions such as carbon and nitrogen fixing, soil structure, tilth and other organic matter transformation (Parkinson and Coleman 1991; Tisdall 1991). Several traditional approaches have been used in studying microbial diversity of agricultural soils (Turco *et al.* 1994; Jordan *et al.* 1995). Molecular biology and biochemical approaches are new tools for analysis of soil microbial components without utilization of any traditional approaches.

Culturable methods, viable, CLPP and flow cytometry are several methods used in analysis of soil microbial diversity. Torsvik *et al.* (1996) discussed the limitations of various culturable methods which show the strong evidence that the bacteria which are observed under microscope are unable to form visible colonies on agar plates though they are viable and active when observed under microscope (Staley and Konopka 1985; Torsvik *et al.* 1990; Amann *et al.* 1990). Molecular techniques are useful in determining both the culturable and unculturable microorganisms based on several criteria like extraction, purification and characterization of nucleic acids from environmental samples, which provides accurate measurement of extent of microbial diversity in soil. Also the nucleic acid information is used for investigation and compares the diversity at different organization levels, the variability within the species of communities (Torsvik 1980; Johnsen *et al.* 2001). This section deals with some of molecular methods which are currently used to study microbial communities.

2.1 Cultivation based methods

Traditionally, the soil microorganisms are analyzed based on the cultivation and isolation (van Elsas *et al.* 1998); to maximize the diverse microbial group a wide array of culture media has been designed. Community-Level Physiological Profiling (CLPP), denotes the biolog-based method which directly analyzes the potential activity of soil microbial communities (Garland 1996), has also been introduced. Due to assessment of small fraction of microbial cells in soil, the study of cultivation-based methods has been limited, a recent study has been evolved which substantially raises the percentage by using special cultivation techniques (Janssen 2002).

2.2 Cultivation independent methods

Molecular technology has developed cutting-edges for the studies of microbial communities; the study basically deals with PCR or RT-PCR of specific DNA or RNA. For prokaryotes and eukaryotes 16S and 18S ribosomal RNA (rRNA) or ribosomal DNA (rDNA) represents as useful ecological markers. Janssen (2002) and Von Wintzingerode *et al.* (1997) has documented some major drawbacks of these respective techniques. The generated primers of PCR products is mainly based on the regions of 16S or 18S rDNA that are conserved, a mixture of DNA fragments is yielded which represents the overall PCR-accessible species which are present in the soil. Those mixed PCR products can be used for (a) preparing clone libraries (Borneman and Triplett 1997; McCaig *et al.* 2001), and (b) a range of microbial community fingerprinting techniques.

2.3 Clone libraries

The characterization and identification of dominant bacterial or fungal type in soil is carried through clone libraries, which provides clear picture for diversification. The accurate description of microbial diversity in soil usually requires quite large clone libraries. Several techniques, rarefaction analysis, calculation of coverage values, or other statistical techniques has been carried out to estimate, whether the number of screened clones is sufficient to realistically examine the true diversity (Ravenschlag *et al.* 1999; Curtiss *et al.* 2002).

2.4 Microbial community fingerprinting techniques

To fingerprint soil microbial communities a wide array of techniques has been developed which includes denaturing or temperature gradient gel electrophoresis (DGGE/TGGE) (Muyzer *et al.* 1993; Heuer *et al.* 1997; Muyzer *et al.* 1998), amplified rDNA restriction analysis (ARDRA) (Massol-Deya *et al.* 1995), terminal restriction fragment length polymorphism (T-RFLP) (Liu *et al.* 1997), single-strand conformational polymorphism (SSCP) (Schmalenberger and Tebbe 2002), and ribosomal intergenic spacer length polymorphism (RISA) (Ranjard 2001). Most widely used methods in studying microbial communities in environmental samples were PCR-DGGE, thus help to understand specific subgroups of complex environmental communities. It is often tedious approach when analyzed with universal (bacterial) primers; less abundant organisms may escape detection (Gelsomino *et al.* 1999; Boon *et al.* 2002). Prokaryotic groups such as actinomycetes (Heuer *et al.* 1997), *Bacillus* (Garbeva *et al.* 2003), *Paenibacillus* (Da Silva *et al.* 2003), *Pseudomonas* (Gyamfi *et al.* 2002; Garbeva *et al.* 2004), the α- and β-Proteobacteria (Gomes *et al.* 2001), methanotrophic bacteria of the α- and β-Proteobacteria (Henckel *et al.* 1999), ammonia-oxidizing bacteria (Kowalchuk *et al.* 1998), and N_2-fixing bacteria (Lovell *et al.* 2001) uses group-specific PCR-DGGE systems. DNA array is one of the promising techniques for the analysis of microbial communities and selected genes from the environmental samples (Hurt *et al.* 2001; Tiedje *et al.* 2001).

2.5 Metabolically Active Communities and Other Molecular Approaches

Felske and Akkermans (1998) has examined that by using rRNA rather than use of rDNA, it helps to obtain information about metabolically active members of microbial communities in the soil. The incorporation of 5-bromo-2-deoxyuridine (BrdU) into DNA, is one of the most evident method, for linking the microbial activity of phylogenetic information, which has been followed by fingerprinting of active communities, addition to this, the incorporation of ^{13}C was followed by separation and fingerprinting for the assessment of metabolically active fractions (Urbach et al. 1999; Yin et al. 2000; Bailey and McGill 2002). This advancement helps us to identify organisms' activity in soil, microbial community structure and with that of how the system functions.

Several other methods which are equally potentially applicable as that of above mentioned methods are fluorescent in situ hybridization (FISH) (Dokic et al. 2010) and pyrosequencing (one of the DNA sequencing based analysis for community analysis) (Lauber et al. 2009; Fakruddin et al. 2012), Illumina-based lights for high throughput microbial community analysis (Caporaso et al. 2012; Degnan and Ochman 2012) without cultivated strains, it has been possible to discover a new groups of microorganism in complex environments by pyrosequencing and illumina-based sequencing which sheds a light on the complexities of microbial communities and it also helps in resolving highly complex microbiota, linking community diversity with niche function (Bartram et al. 2011; Fakruddin and Mannan 2012).

2.6 Metagenomic analysis of microbial communities

The functional and sequence-based analysis of collective microbial genomes in the environmental samples is metagenomics (Zeyaullah et al. 2009). Metagenomics has the potential to give a comprehensive view of genetic diversity, species composition, evolution and interactions with environment of natural microbial communities by direct assay of collective genome of co-occurring microbes (Simon and Daniel 2011). Ultimately, the genomics reveals how individually the species and strains contribute to the net activity of the microbial community (Allen and Banfield 2005).

The natural microbial phenomena analysis which includes biogeochemical activities, population ecology, evolutionary processes like lateral gene transfer (LGT) events and interactions of microbes are collectively called as community genomics analyzing methods (Allen and Banfield 2005). It mainly focuses on the linkages between genes and gene families, and also the distribution of metabolic functions among the communities. To analyze community genomics data various assemblers are currently available: ARACHNE, CAP, CELERA,

EULER, JAZZ, PHRAP and TIGR (Tyson *et al*. 2004). The characterization and sequencing of metatranscriptomes was employed for the identification of RNA-based regulation, which expresses biological signatures even in complex ecosystems (Zeyaullah *et al*. 2009). For the assessment of immediate catalytic potential of microbial community the proteomic analysis of mixed microbial communities was emerged. The identification of proteins in complex mixture is done rapidly by mass-spectroscopy based proteomic methods (Schloss and Handelsman 2003).

Metagenomics has certain limitations in processing of huge amount of data obtained by the various community to overcome from the limitation, the statistical methods have been incorporated, where various mixture of organisms are assigned to phylogenetic group based on the taxonomic origins (Tyson *et al*. 2004). Metagenome Analyzer is the best example for bioinformatics tool based on similarity binning, whereas, CARMA is the ortholog-based approach in improving taxonomic estimation for metagenomic sequences (Simon and Daniel 2011).

3. Extremophiles and their sustainable uses

Today's, food production is a main concern in different parts of world. Ecological stresses typically divided into biotic stress and abiotic stress, which comprises plant diseases, salinity, nutrient deficiency and temperature, which setbacks the agriculture productivity globally (Atkinson and Urwin 2012; Maheshwari 2013). One of the great contributions towards sustainability and agricultural productivities was the implication of chemical fertilizers for high yield, due to overusing or misusing which ultimately lead to the development of resistant pests and also microbes developed against chemical pesticides. Genetically modified crops do not withstand the great versatility in several crop plants; it remains with some controversy about possible ecological and health consequences.

The functional role of microbial diversity in a variety of environments, and different microorganisms, their genes and biomolecules with a special, but not exclusive, focus on extremophiles were investigated by a group of microbial diversity on extreme environments. The majority form of life on earth is microorganisms which are visible only through microscope. Microbes having vast diversity are present everywhere in the environment, which adopt to extreme cold (below 0°C) to extreme hot (above 100°C) conditions, from high acidic concentrations to higher salinity, high pressure, or else anywhere which might not be suitable for the survival of organisms. Microorganisms thriving under extreme conditions are the extremophiles. The variability in the selection of systems is reflected by the huge diversity existing in the microbial world. Due to the level of microbial diversity it is focused as the main topic of debate in current microbial ecology. Many microorganisms are easily diversified across the planet, hence the so called extreme environments represent one of the selective factor for the living beings to develop in them. While extreme environments lead to an ideal model for studying ecological properties of microorganisms, their physiology, adaptive properties and many other characteristics which are related to microbial communities and specific microbial cells. Microorganisms are always found to be in a community, often composed of large number of different types of cells. The major key aspects to understand the role and functions of microorganisms in nature, and also the microbial life implications at local and global scales in our planet, require studying the interaction of microbial communities within the environment (Tapilatu *et al.* 2010).

Some microorganisms have adopted few strategies which allow them to develop in an extremely unique environment. Hence the uniqueness in the organisms, biomolecules, implies interest in biotechnology due to its high stability to withstand, used in potential applications of industrial processes or for their commercial purpose. Wide arrays of techniques are required to search the unique nature of microorganisms and molecules. Extremophiles are the novel approach for biotechnological interest; hence they produce an enzyme, extremozyme, which certainly functions under extreme environmental conditions. The ability to remain active under severe conditions typically employs the enzyme, extremozymes for extensive use of industrial production processes and research applications. Normally, extremophiles are actively involved in the production of enzymes and secondary metabolites in extreme environment in industries and pharmaceutical companies. There are limited extremophiles such as halophiles and halotolerants, thermophiles and thermotolerants, psychrophiles, acidophiles and alkaliphiles are involved in agricultural practices. Some of the common extremophiles employed in the agricultural field is given in table 1.

3.1 Halophilic and halotrophic microbes for producing plant growth promoters

The slowdown in production of world's agricultural land has depleted to 5% (1.5 billion hectares) due to high salinity level in the soil content, about 20% (308 million hectares) of cultivated land is salt affected, and the area of salt affected land has been increasing (Hamedi *et al.* 2015). For instance, by recruiting agriculturally beneficial microorganisms it is possible to substitute chemically enhanced agricultural productivity and also decreasing detrimental effects of salt stress (Strap 2011). Increase in soil salinity drastically affecting certain physiological process in plants including seed germination (Raafat *et al.* 2011), several enzyme activities (Almansouri *et al.* 1999), bioenergetic processes

Table 1: List of extremophiles and their application in sustainable agricultural practices.

Extremophiles	Microbial species	Application	References
Halophiles	*Brevibacterium epidermidis* *Brevibacterium iodinum* *Micrococcus yannanensis* *Zhihengliuela alba*	Plant growth promoters	Hamedi *et al.* (2015)
Thermophiles	*Bacillus oceanisediminis, Bacillus safensis* *Anureibacillus thermoaerophilus* *Bacillus licheniformis* *Bacillus pallidus* *Bacillus smithii* *Bacillus stearothermophillus* *Bacillus thermodenitrificans* *Bacillus thermodurans* *Geobacillus thermodenitrificans* *Staphylococcus similans* *Ureibacillus suwonensis*	Biocontrol agents Composting	Berrada *et al.* (2012) Ishii *et al.* (2000); Hassen *et al.* (2001); Charbonneau *et al.* (2012); Rawat and Johri (2014)
Psychrophiles	*Acinetobacter* sp. *Aspergillus sydowii* *Paecilomyces hepialid* *Penecilium raistrickii* *Pseudomonas fluorescence* *Pseudomonas putida* *Pseudomonas striata*	Phosphate solubilizers	Trivedi and Sa 2007; Rinu *et al.* (2012); Pallavi and Gupta (2013); Joshi *et al.* (2014)
Acidophiles	*Bradyrhizobium* sp. *Burkholderia* sp. *Cupriavidus asilensis* *Elusimicrobium* sp. *Mesorhizobium* sp.	Ammonia oxidizers	Lehtovirta-Morley *et al.* (2011)

(Contd.)

Extremophiles	Microbial species	Application	References
Alkaliphiles	*Nitrosotalea devanaterra* *Rhizobium* sp. *Aspergills niger* *Aspergillus caespitosus* *Bacillus cereus* *Bacillus flexus* *Bacillus licheniformis* *Bacillus megaterium* *Bacillus subtilis* *Cenococcum graniforme* *Colletotrichum lindemuthianum* *Escherichia coli* *Micrococcus sodonensis* *Mucor rouxii* *Rhizobium microsporus* *Rhizobium stolonifer*	Phosphate solubilizers	Bae and Barton (1989); Guimarães *et al.* (2006); Falguni and Sharma (2012); Priya *et al.* (2013)

affects photosynthesis (Sudhir and Murthy 2004), the length, shoot and root weights (Jamil *et al*. 2006). Several halophilic or halotolerant microbial species help in promoting the plant growth under saline conditions.

The market share of certain microorganisms is comparatively small, for instance it shows developments in several biological application of agronomically important microorganisms pertaining to some conditions in past few years. Currently many applications were undertaken for the enhancement of agricultural productivity in saline soil, which typically includes halophilic and halotolerant actinobacteria. Those groups were applied for biocontrol of phytopathogens and promoting plant growth by solubilization of essential elements (Berrada *et al*. 2012). Sadfi-Zouaoui *et al*. (2008) demonstrated that halophilic bacteria (e.g. *Bacillus* spp., *Halomonas* spp., *Planococcus* spp., *Salinicoccus* spp., *Halobacillus* spp. and *Marinococcus* spp.) inhibiting *B. cinerea* could be considered as potential biocontrol agents.

3.2 Thermophilic and thermotrophic microbes for increasing soil fertility

The natural process of composting involves the decomposition of large amounts of organic material based on its habitual nature which are naturally occurring microflora, it results for the formation of humus and it is an excellent route of nutrient turnover in the natural ecosystem. It is possible to obtain the ideal growth and activity of microbial decomposers by accelerating certain conditions even though the humus formation is literally a slow process. Change in microbial community structure and composition strongly influenced by changes in soil physicochemical condition. In the process of composting it defines thermophilic microflora, which constitutes the predominant component and provides selectively to compost. The inside temperature of a compost pile reaches up to 75-80°C. Most of the moderate thermophiles were isolated which belongs to the genera of *Geobacillus*, *Bacillus* and *Clostridium*, and thermophilic methanogens (Bhattacharya and Pletschke 2014). The conversion of waste material into important byproducts involves the thermophilic composting method, which extensively uses three different phases, such as mesophilic, thermophilic and curing or maturation phase. During the entire process of composting the thermophilic *Bacillus* spp. remarkably shows the ability to utilize the diverse compounds in which carbon as major source and also the spectrum of their thermostability, as most of the *Bacillus* spp. typically represents to grow at both 30 and 50°C, which allows them to survive and remain metabolically active over all the three phases (Rawat and Johri 2014).

3.3 Alkaliphiles and psychrophiles for phosphate solubilization

Soil ecosystem is generally composed of various microhabitats, each with its own environmental factors like pH, temperature, chemicals, etc. Alkalinity of soil is highly localized and not stable, microbial diversity in this pH is suffering with challenges. Around the world alkaline soils are very common, the ranges of pH would be 10 or higher (Kristjansson and Hreggvidsson 1995). In these alkaline conditions, several microbial species are normally abundant and provide organic matter for diverse groups. High concentration of ammonia leads to protein decaying and urea hydrolysis which raise the pH and supports the high growth of alkaliphiles (Ulukanli and Digrak 2002). So far, several alkaliphiles have been isolated from soil environment, among spore forming bacterial species of *Bacillus* were found to be predominant in the microhabitats of soil. In normal soil the viable count of the alkaliphiles is lower than the alkaline soils (Horikoshi 1996). Increased *Bacillus* species in this alkaline ecosystem are *B. alcaliphilus*, *B. agaradherens*, *B. clarki*, *B. clasusi*, *B. gibsonii*, *B. halmophilus*, *B. halodurans*, *B. horikoshi*, *B. pseudoalcaliphilus* and *B. pseudofirmus* (Nielson *et al.* 1995).

Enormous species of bacteria and fungi have been identified for their phosphate-solubilizing capabilities. Alkaliphiles are normally involved in the phosphate solubilization in agricultural soil ecosystem. Soil microbes produce phosphate solubilizing enzymes such as acid and alkaline phosphomonoesterase depending on pH optima and environmental conditions (Jorquera *et al.* 2008). Typically, alkaline phosphatase is predominately found in alkaline and neutral soil (Renella *et al.* 2006; Sharma *et al.* 2013). This alkaline phosphatase enzyme is having high thermostability produced by various bacterial and fungal alkaliphiles. Although acid phosphatase is produced by plant roots, they rarely produce alkaline phosphatases, hence it is a suitable niche for phosphate solubilizing microbes (Criquet *et al.* 2004). A few evidence also suggest that microbial phosphatases possess greater affinity for Po compounds than plant root produced phosphatases (Richardson *et al.* 2009).

Commercially available biofertilizers when applied in the mountainous regions was found to be ineffective. During the last two decades numerous microbiological and biotechnological techniques were undertaken to overcome those challenges, which also includes the use of cold-loving (psychrophilic) or cold-tolerant (psychrotrophic) organisms. However, despite their great potentiality, the development of cold-tolerant biofertilizers are still in its infancy (Yarzabal 2014). Biological nitrogen fixation process appreciably depends on the availability of phosphorus. Phosphatic fertilizers in huge portion were applied to enhance the availability of phosphorus which quickly in turn transformed to the insoluble form which decreases the efficiency of fertilizers (Pallavi and

Gupta 2013). The conversion of insoluble phosphorus to soluble forms of HPO_4^{2-} and H_2PO_4 is done by utilizing phosphate-solubilizing microbes which are carried through several chemical reactions such as acidification, chelation, exchange reactions and polymeric substance formation (Delvasto *et al.* 2006; Pallavi and Gupta 2013).

A psychrotolerant strain of *Pseudomonas fragi* showed the perplexing temperature effects onto 'P' solubilization which has been reported earlier (Selvakumar *et al.* 2009). Bacteria respond not only to elevated temperature (heat shock), but also at downshifted temperatures (cold shock) by synthesizing a group of heat and cold shock proteins, respectively. The respective proteins are important for the survival of bacteria at higher or lower temperatures (Negi *et al.* 2009). Some microbes under higher psychrophilic 'P' solubilizers are *Pseudomonas fluorescens*, *Pseudomonas putida*, *Pseudomonas striata*, *Acinetobacter* spp., and *Paecilomyces* spp. In the present decade there has been a surge of interest on P solubilization at cold temperatures by natural and mutant psychrotolerant (Trivedi and Sa 2007; Pallavi and Gupta 2013; Joshi *et al.* 2014).

3.4 Acidophiles as ammonia oxidizers

In general, microorganisms thrive at extreme environments in which few preferably grow in low pH typically called as acidophiles. The level of ammonium in acidic soil is generally too low to support for the growth of all known ammonia-oxidizing bacteria (AOB) isolates because of the ionization of ammonia to ammonium (Burton and Prosser 2001; Gubry-Rangin *et al.* 2011). Until recently, it was a difficult task to identify the microorganisms responsible for the nitrification activity in acidic soils. The discovery of *Nitrosotalea devanaterra* proved the existence of acid-tolerance and the extraordinary substrate affinity was demonstrated by ammonia-oxidizing archaea (AOA) which permits growth at extremely low ammonia concentrations (Lehtovirta-Morley *et al.* 2011; Lu *et al.* 2012). Although isolation of obligate acidophilic *N. devanaterra* revealed the important role of archaeal ammonia oxidation in acidic soils, as previously suggested in acidic forest (Boyle-Yarwood *et al.* 2008; Stopnisek *et al.* 2010) and in some agricultural soils (Nicol *et al.* 2008; Offre *et al.* 2009).

4. Mycorrhizal fungi for sustainable agricultural practices

Now-a-days, sustainable high production agriculture is achieved through conservation agriculture and integrated soil fertility management practices worldwide. However, the knowledge on the influence of these practices on arbuscular mycorrhizal fungi (AMF) and ectomycorrhizal (ECM) fungi is inadequate. About two thirds of agricultural crops are having AMF association and there is equivocal evidence that crop plants benefit from these association (Singh 2006). In general, AMF is forming symbiotic association with more than 80% of land plants, liverworts, ferns, gymnosperms and angiosperms (Smith and Read 2008). AMF is playing a vital role in plant nutrient uptake, production of growth promoting substances, tolerance to various environmental factors, promote soil aggregation, synergistic interactions with other beneficial organisms and use a considerable amount of reduced carbon from the plants. These fungi are also known as extremophiles due to their occurrence in extreme environments such as high or low pH, temperature, salt, drought, low nutrients and so on (Singh 2006). Diversity of functional traits of AMF depends on its composition among and within species (Munyanzizaa et al. 1997; Jansa et al. 2006; Smith and Read 2008; Abbasi et al. 2015). The sustainable agricultural soil ecosystem is likely to be more favorable to AMF than those under conventional agriculture or industrial agriculture (Bethlenfalvay and Schuepp 1994; Smith and Read 2008). Crop rotations, cover crops, controlling weeds, retaining soil nutrients and impeding soil erosion are the key parameters for sustainable agriculture (Barea and Jeffries 1995; Galvez et al. 2001). Soils in low-input agricultural systems have larger populations and more inoculum of AMF due to over wintering cover crops than soils under conventional management (Kabir and Koide 2000).

Spores and mycorrhizal roots are associated with a network of hyphae which propagate AMF (Kabir et al. 1998; Galvez et al. 2001). Soil disturbance through agricultural practices such as tillage regimes, nutrient management and cropping systems may affect the hyphal network and contribute to decreased colonization of roots by mycorrhizal fungi and thus reduction of nutrient uptake by plants (McGonigle et al. 1999; Galvez et al. 2001; Johansson et al. 2004; Oehl et al. 2005; Gosling et al. 2006; Verbruggen and Kiers 2010; Schnoor et al. 2011; Daisog et al. 2012; Yang et al. 2012). Understanding of the ecology of AMF species is imperative in order to ensure, role and sustainability of AMF for maximum benefit in low-input agricultural systems.

Higher AMF colonization is possible in organic farming practices compared to conventional systems, which might be attributed to differences in fertilizer inputs. However, any high P-fertilizer either synthetic or organically derived may reduce AMF function. Organic farming systems are less detrimental to AMF since they use water-soluble fertilizers (Gosling et al. 2006). Correlation of soil characteristics with AMF function such as colonization, effects on plant traits may provide an analytical support for identifying the influence of agricultural management practices on the potential benefits of AMF in crop systems (Verbruggen and Kiers 2010; Barber et al. 2013). Agricultural management practices may alter AMF communities, but detailed study is required for understanding the effects of these changes on trait mediated interactions with other communities (Barber et al. 2013).

The most important management practices affecting AMF colonization and possible consequences for agro-ecosystem functioning are nutrient management, soil disturbance and cropping systems. In general, soluble fertilizers are used by the farmers to maintain the soil nutrients level. However, a continuous use of soluble fertilizers ultimately causes soil rigidity through loss of soil organic matter and reduces in microbial diversity (Plenchette *et al.* 2005). In the application of water soluble fertilizers are providing readily available P and N to the plants which severely alters the structure of the AMF (Oehl *et al.* 2004). Alternatively, low-input systems such as organic farming have been shown to be more favorable to AMF diversity and mycorrhizal root colonization (Gosling *et al.* 2006). In modern intensive agriculture, conventional tillage is modifying the structure of soil ecosystem through changes in physical, chemical and biological properties (Kabir 2005). Conventional tillage eventually affects the composition of soil organic matter and loss of C and N (Six *et al.* 2004). These changes in soil structure affect the soil microbial diversity. Conventional agricultural practices through conservation tillage or no-till systems are likely to improve crop residues near the soil surface, which may increase soil organic content, improved soil physical properties, protected soil erosion, increased microbial diversity (Hobbs 2007). No till system is associated with a higher AMF diversity and enhanced functioning, while conventional tilling system have been shown to have negative impact on AMF diversity and functioning (Jansa *et al.* 2003; Castillo *et al.* 2006). Finally, crop rotation also affects AMF diversity and functioning. Cropping system with highly mycorrhizal dependent plants can enhance the colonization of AMF and consequently improve the functioning of AMF for subsequent crops. Development of mycorrhizae in the field is dependent on cropping systems and subsequent crops which exhibit a range of mycorrhizal dependencies (Plenchette *et al.* 2005). Cultivation of non-host crops declines AMF colonization in the soil resulting in less availability of VAM for succeeding crops and delayed AMF colonization, associated with lower yields (Karasawa *et al.* 2001; Karasawa *et al.* 2002).

In recent years, worldwide significant developments have been achieved in mycorrhizal technology which proved that it has great potential to improve productivity of cereal, fruit and vegetable crops and suppress nematode and fungal infestations. The public demand to reduce environmental problems associated with excessive pesticide usage has prompted research on increasing biological inputs such as AMF. In agroecosystem, soil and microbes are integral parts, with potential to provide both benefits and costs to farmers. Farmers are the decision making authority in the agricultural management (Verbruggen and Kiers 2010). Continued work on these questions may allow us to provide specific management recommendations that can increase yield and feed an increasingly populous world.

5. Applying biotechnology for agricultural sustainability

Microorganism's growth remarkably depends on soil, which is the main source for degradation of variety of complex organic compounds due to its versatile metabolism. Microbes play an immense role in recycling of phosphorus, oxygen, carbon, nitrogen and sulphur elements, which replenish the environment by degrading substrates obtained from dead and decaying plants and animal. It is essential to reuse vital elements in which microbes are the decomposers and scavengers to clean up biosphere. Microorganisms are the indicators for the assessment of sustainable land use; they play a key role for maintenance of soil fertility. Due to the pesticidal effects, several ecological processes, such as sulphur oxidation, nitrification and nitrogen fixation are inhibited. Restoration of degraded semiarid and other ecosystems was carried out through mycorrhizal symbiosis (Requena et al. 2001). AMF show tolerance against contamination of heavy metals (Gadd 2010). In low soil pH mycorrhizae show high accumulation of Cu, Ni, Pb and Zn in grass Ehrhartia calycina.

6. Conclusion and future perspectives

It is estimated that the demand in agricultural production is expected to increase by at least 70% by 2050. In addition to this people are becoming aware of sustainable agricultural practices that are fundamental to meet the future world's agricultural demands (Altieri 2004). This is how modern agriculture is being implemented globally and diverse research approaches are undertaken to overcome the environmental and economical sustainability issues. Extensive use of agrochemicals to protect crops against pathogens has been increasing over last few decades. The beneficial microorganisms are applied as biofertilizers and biopesticides for crop production/protection thus leading to a substantial reduction of chemical fertilizer/pesticide use, which is an important source of environmental pollution.

Moreover, now-a-days, microbial inoculants, some of which the well known rhizobia, for the inoculation of legumes carry its historical record since 1896, to control invertebrate pests, Bacillus thuringiensis applied since 1930s were widely applied in modern agriculture as biofertilizers and biocontrol agents. Some other interesting applications of biotechnology include bioremediation, phosphate solubilization, soil aggregation, sewage treatment, bioleaching, oil recovery, coal scrubbing and biogas production. For the future development, from the contribution of scientific disciplines, the primary importance should be given to promote sustainable practices in plant production as well as the conservation and ecosystem restoration. The novel approach for future development involves the construction of ideal multipartite endo and ectosymbiotic communities based on extended metagenomic analyses.

In conclusion, extremophiles growing under extreme environmental conditions are having either active or passive mechanisms for their survival in these conditions. These organisms are producing acids, enzymes and other biomolecules which are slow and steady but effective and prolonged activities shows implication in nutrient cycling. These organisms therefore need focused attention on their diversity, phylogeny, survival strategies and bioprospecting for sustainable agriculture.

References

Abbasi H, Ambreen A, Rushda S (2015) Vesicular arbuscular mycorrhizal (VAM) fungi: a tool for sustainable agriculture. Am. J. Plant Nutr. Fert. Technol. 5, 40-49.

Allen EE, Banfield JF (2005) Community genomics in microbial ecology and evolution. Natl. Rev. Microbiol. 3, 489-498.

Almansouri M, Kinet J, Lutts S (1999) Compared effects of sudden and progressive impositions of salt stress in three durum wheat (*Triticum durum* Desf.) cultivars. J. Plant Physiol. 154(5), 743-752.

Altieri MA (2004) Linking ecologists and traditional farmers in the search for sustainable agriculture. Front. Ecol. Environ. 2, 35-42.

Amann RI, Krumholz L, Stahl DA (1990) Fluorescent-oligonucleotide probing of whole cells for determinative, phylogenetic, and environmental studies in microbiology. J. Bacteriol. 172, 762-770.

Atkinson NJ, Urwin PE (2012) The interaction of plant biotic and abiotic stresses: from genes to the field. J. Exp. Bot. 63(10), 3523-3543.

Bae KS, Barton LL (1989) Alkaline phosphatase and other hydrolyases produced by *Cenococcum graniforme*, an ectomycorrhizal fungus. Appl. Environ. Microbiol. 55(10), 2511-2516.

Bailey VL, McGill WB (2002) Fate of ^{14}C labeled pyrene in a creosote- and octadecane in an oil-contaminated soil. Soil Biol. Biochem. 34(4), 423-433.

Barber NA, Kiers ET, Theis N, Hazzard RV, Adler LS (2013) Linking agricultural practices, mycorrhizal fungi, and traits mediating plant-insect interactions. Ecol. Appl. 23(7), 1519-1530.

Barea JM, Jeffries P (1995) Arbuscular mycorrhizas in sustainable soil-plant systems. In: Mycorrhiza (Ed) Varma A and Hock B, Springer-Verlag, Berlin, pp 521-560.

Bartram AK, Lynch MDJ, Stearns JC, Moreno-Hagelsieb G, Neufeld JD (2011) Generation of multimillion-sequence 16S rRNA gene libraries from complex microbial communities by assembling paired-end illumina reads. Appl. Environ. Microbiol. 77(11), 3846-3852.

Berrada I, Benkhemmar O, Swings J, Bendaou N, Amar M (2012) Selection of halophilic bacteria for biological control of tomato gray mould caused by *Botrytis cinerea*. Phytopathol. Mediterr. 51(3), 625-630.

Bethlenfalvay GJ, Schuepp H (1994) Arbuscular mycorrhizas and agrosystem stability. In: Impact of Arbuscular Mycorrhizas on Sustainable Agriculture and Natural Ecosystems (Ed) Gianinazzi S and Schuepp H, Birkhauser Verlag, Basel, Switzerland, pp 117-131.

Bhattacharya A, Pletschke BI (2014) Thermophilic bacilli and their enzymes in composting. In: Composting for Sustainable Agriculture (Ed) Maheshwari DK, Springer International Publishing, Switzerland, pp 103-124.

Boon N, De Windt W, Verstraete W, Top EM (2002) Evaluation of nested PCR-DGGE (denaturing gradient gel electrophoresis) with group-specific 16S rRNA primers for the analysis of bacterial communities from different wastewater treatment plants. FEMS Microbiol. Ecol. 39, 101-112.

Borneman J, Triplett EW (1997) Molecular microbial diversity in soil from Eastern Amazonia: evidence for unusual microorganisms and microbial population shifts associated with deforestation. Appl. Environ. Microbiol. 63, 2647-2653.

Boyle-Yarwood SA, Bottomleyt PJ, Myrold DD (2008) Community composition of ammonia-oxidizing bacteria and archaea in soils under stands of red alder and Douglas fir in Oregon. Environ. Microbiol. 10(11), 2956-2965.

Brock TD (1967) Life at high temperatures. Science 158, 1012-1019.

Burton SAQ, Prosser JI (2001) Autotrophic ammonia oxidation at low pH through urea hydrolysis. Appl. Environ. Microbiol. 67, 2952-2957.

Caporaso JG, Lauber CL, Walters WA, Berg-Lyons D, Huntley J, Fierer N, Owens SM, Betley J, Fraser L, Bauer M, Gormley N, Gilbert JA, Smith G, Knight R (2012) Ultra-high throughput microbial community analysis on the Illumina HiSeq and MiSeq platforms. ISME J. 6, 1621-1624.

Castillo CG, Rubio R, Rouanet JL, Borie F (2006) Early effects of tillage and crop rotation on arbuscular mycorrhizal fungal propagules in an Ultisol. Biol Fert Soils 43, 83-92.

Charbonneau DM, Meddeb-Mouelhi F, Boissiont M, Sirois M, Beauregard M (2012) Identification of thermophilic bacterial strains producing thermotolerant hydrolytic enzymes from manure compost. Ind. J. Microbiol. 52(1), 41-47.

Criquet S, Ferre E, Farnet AM, Le Petit J (2004) Annual dynamics of phosphatase activities in an evergreen oak litter: influence of biotic and abiotic factors. Soil. Boil. Biochem. 34, 1111-1118.

Curtiss TP, Sloan WT, Scannell JW (2002) Estimating prokaryotic diversity and its limits. Proc. Natl. Acad. Sci. 99(16), 10234-10236.

Da Silva KRA, Salles JF, Seldin L, van Elsas JD (2003) Application of a novel *Paenibacillus*-specific PCR-DGGE method and sequence analysis to assess the diversity of *Paenibacillus* spp. in the maize rhizosphere. J. Microbiol. Methods 54, 213-231.

Daisog H, Sbrana C, Cristani C, Moonen A-C, Giovannetti M, Bàrberi P (2012) Arbuscular mycorrhizal fungi shift competitive relationships among crop and weed species. Plant Soil 353, 395-408.

Degnan PH, Ochman H (2012) Illumina-based analysis of microbial community diversity. ISME J. 6, 183-194.

Delvasto P, Valverde A, Ballester A, Igual JM, Munoz JA, Gonzalez F, Blázquez ML, Garcia C (2006) Characterization of brushite as a re-crystallization product formed during bacterial solubilization of hydroxyapatite in batch cultures. Soil Biol. Biochem. 38, 2645-2654.

Dokic L, Savic M, Narancic T, Vasiljevic B (2010) Metagenomic analysis of soil microbial communities. Arch. Biol. Sci. Belgrade 62(3), 559-564.

Doran JW, Parkin TB (1994) Defining and assessing soil quality. In: Defining Soil Quality for a Sustainable Environment (Ed) Doran JW, Coleman DC, Bezdicek DF and Steward BA, SSSAJ Special Publication, Madison, WI, pp 3-21.

Edwards CA (1989) The importance of integration in sustainable agricultural systems. Agric. Ecosyst. Environ. 27, 25-35.

Elmqvist T, Maltby E, Barker T, Mortimer M, Perrings C (2010) Biodiversity, ecosystems and ecosystem services. In: TEEB Ecological and Economic Foundations (Ed) Kumar P, Earthscan, London, pp 41-111.

Fakruddin M, Chowdhury A (2012) Pyrosequencing - an alternative to traditional sanger sequencing. Am J Biochem Biotechnol. 8(1), 14-20.

Falguni RP, Sharma MC (2012) Optimization of production of alkaline phosphatase by a facultative alkaliphile *Bacillus flexus* FPB17 isolated from alkaline lake soils. Int. J. Agric. Technol. 8(5), 1605-1612.

Felske A, Akkermans ADL (1998) Spatial homogeneity of abundant bacterial 16S rRNA molecules in grassland soils. Microbial Ecol. 36, 31-36.

Gadd GM (2010) Metals, minerals and microbes: geomicrobiology and bioremediation. Microbiology 156, 609-643.

Galvez L, Douds Jr DD, Drinkwater LE, Wagoner P (2001) Effect of tillage and farming system upon VAM fungus populations and mycorrhizas and nutrient uptake of maize. Plant Soil 228(2), 299-308.

Garbeva P, van Veen JA, van Elsas JD (2003) Predominant *Bacillus* spp. in agricultural soil under different management regimes detected via PCR-DGGE. Microbial Ecol. 45, 302-316.

Garbeva P, van Veen JA, van Elsas JD (2004) Assessment of the diversity, and antagonism toward *Rhizoctonia solani* AG3, of *Pseudomonas* species in soil from different agricultural regimes. FEMS Microbiol. Ecol. 47, 51-64.

Garland JL (1996) Analytical approaches to the characterization of samples of microbial communities using patterns of potential C source utilization. Soil Biol. Biochem. 28, 213-221.

Gelsomino A, Keijzer-Wolters A, Cacco G, van Elsas JD (1999) Assessment of bacterial community structure in soil by polymerase chain reaction and denaturing gradient gel electrophoresis. J. Microbiol. Methods 38, 1-15.

Gomes NCM, Heuer H, Schonfeld J, Costa R, Mendonca-Hagler L, Smalla K (2001) Bacterial diversity of the rhizosphere of maize (*Zea mays*) grown in tropical soil studied by temperature gradient gel electrophoresis. Plant Soil 232, 167-180.

Gosling P, Hodge A, Goodlass G, Bending GD (2006) Arbuscular mycorrhizal fungi and organic farming. Agric. Ecosyst. Environ. 113, 17-35.

Goyal AK, Arora S (2009) India's Fourth National Report to the Convention on Biological Diversity. Ministry of Environment and Forests, Government of India, New Delhi, pp 1-86.

Gubry-Rangin C, Hai B, Quince C, Engel M, Thomson BC, James P, Schloter M, Griffiths RI, Prosser JI, Nicol GW (2011) Niche specialization of terrestrial archaeal ammonia oxidizers. Proc. Natl. Acad. Sci. USA 108, 21206-21211.

Guimarães LHS, Peixoto-Nogueira SC, Michelin M, Rizzatti ACS, Sandrim VC, Zanoelo FF, Aquino ACMM, Junior AB, Polizeli MLTM (2006) Screening of filamentous fungi for production of enzymes of biotechnological interest. Braz. J. Microbiol. 37(4), 474-480.

Gupta G, Parihar SS, Ahirwar NK, Snehi SK, Singh V (2015) Plant growth promoting rhizobacteria (PGPR): current and future prospects for development of sustainable agriculture. J. Microb. Biochem. Technol. 7, 96-102.

Gyamfi S, Pfeifer U, Stierschneider M, Sessitch A (2002) Effects of transgenic glucosinate-tolerant oilseed rape (*Brassica napus*) and the associated herbicide application on eubacterial and *Pseudomonas* communities in the rhizosphere. FEMS Microbiol. Ecol. 41, 181-190.

Hamedi J, Mohammadipanah F, Panahi HKS (2015) Biotechnological exploitation of actinobacterial members. In: Halophiles: Biodiversity and Sustainable Exploitation (Ed) Maheshwari DK and Saraf M, Springer International Publishing, Switzerland, pp 113-144.

Hassen A, Belguith K, Jedidi N, Cherif A, Cherif M, Boudabbous A (2001) Microbial characterization during composting of municipal solid waste. Bioresour. Technol. 80, 185-192.

Henckel T, Friedrich M, Conrad R (1999) Molecular analyses of the methane oxidizing microbial community in rice field soil by targeting the genes of the 16S rRNA, particulate methane monooxigenase, and methanol dehydrogenase. Appl. Environ. Microbiol. 65, 1980-1990.

Heuer H, Krsek M, Baker P, Smalla K, Wellington EMH (1997) Analysis of *Actinomycete* communities by specific amplification of gene encoding 16S rDNA and gel-electrophoretic separation in denaturing gradient. Appl. Environ. Microbiol. 63, 3233-3241.

Hibbing ME, Fuqua C, Parsek MR, Peterson SB (2010) Bacterial competition: surviving and thriving in the microbial jungle. Natl. Rev. Microbiol. 8(1), 15-25.

Hobbs PR (2007) Conservation agriculture: what is it and why is it important for future sustainable food production? J. Agric. Sci. 145, 127-137.

Horikoshi K (1996) Alkaliphiles from an industrial point of view. FEMS Microbiol. Rev. 18, 259-270.

Horrigan L, Lawrence RS, Walker P (2002) How sustainable agriculture can address the environmental and human health harms of industrial agriculture. Environ. Health Persp. 110, 445-456.

Hurt RA, Qiu X, Wu L, Roh Y, Palumbo AV (2001) Simultaneous recovery of RNA and DNA from soils and sediments. Appl. Environ. Microbiol. 67(10), 4495-4503.

Ishii K, Fukui M, Takii S (2000) Microbial succession during a composting process as evaluated by denaturing gradient gel electrophoresis analysis. J. Appl. Microbiol. 89, 768-777.

Jaggard KW, Qi A, Ober ES (2010) Possible changes to arable crop yields by 2050. Philos. Trans. R. Soc. Biol. Sci. 365, 2835-2851.

Jamil M, Deog BL, Kwang YJ, Ashraf M, Sheong CL, Eui SR (2006) Effect of Salt (NaCl) stress on germination and early seedling growth of four vegetables species. J. Cent. Eur. Agric. 7(2), 273-282.

Jansa J, Mozafar A, Kuhn G, Anken T, Ruh R, Sanders IR, Frossard E (2003) Soil tillage affects the community structure of mycorrhizal fungi in maize roots. Ecol. Appl. 13, 1164-1176.

Jansa J, Wiemken A, Frossard E (2006) The effects of agricultural practices on arbuscular mycorrhizal fungi. Geological Society, London, pp 89-115.

Janssen PH, Yates PS, Grinton BE, Taylor PM, Sait M (2002) Improved culturability of soil bacteria and isolation in pure culture of novel members of the divisions *Acidobacteria*, *Actinobacteria*, *Proteobacteria*, and *Verrucomicrobia*. Appl. Environ. Microbiol. 68, 2391-2396.

Johansson JF, Paul LR, Finlay RD (2004) Microbial interactions in the mycorrhizosphere and their significance for sustainable agriculture. FEMS Microbiol Ecol. 48, 1-13.

Johnsen K, Jacobsen CS, Torsvik V, Sørensen J (2001) Pesticide effects on bacterial diversity in agricultural soils - a review. Biol. Fertil. Soils 33, 443-453.

Jordan D, Kremer RJ, Berg¢eld WA, Kim KY, Cacnio VN (1995) Evaluation of microbial methods as potential indicators of soil quality in historical agricultural fields. Biol. Fertil. Soils 19, 297-302.

Jorquera MA, Hernandez MT, Rengel Z, Marschner P, Mora MD (2008) Isolation of culturable phosphor bacteria with both phytate-mineralization and phosphate-solubilization activity from the rhizosphere of plants grown in a volcanic soil. Biol. Fertil. Soils 44, 1025-1034.

Joshi P, Joshi GK, Tanuja, Mishra PK, Bisht JK, Bhatt JC (2014) Diversity of cold tolerant phosphate solubilizing microorganisms from North Western Himalayas. In: Bacterial Diversity in Sustainable Agriculture (Ed) Maheshwar DK, Springer International Publishing, New York, pp 227-264.

Kabir Z (2005) Tillage or no-tillage: impact on mycorrhizae. Can. J. Plant Sci. 85, 23-29.

Kabir Z, Koide RT (2000) The effect of dandelion or a cover crop on mycorrhiza inoculum potential, soil aggregation and yield of maize. Agric. Ecosyst. Environ. 78, 167-174.

Kabir Z, O'Halloran IP, Fyles JW, Hamel C (1998) Dynamics of mycorrhizal symbiosis of corn (*Zea mays* L.): effects of host physiology, tillage practice and fertilization on spatial distribution of extra-radical mycorrhizal hyphae in the field. Agr. Eccosys. Environ. 68, 151-163.

Karasawa T, Kasahara Y, Takebe M (2001) Variable response of growth and arbuscular mycorrhizal colonization of maize plants to preceding crops in various types of soil. Biol. Fertil. Soils 33, 286-293.

Karasawa T, Kasahara Y, Takebe M (2002) Differences in growth responses of maize to preceding cropping caused by fluctuation in the population of indigenous arbuscular mycorrhizal fungi. Soil Biol. Biochem. 34, 851-857.

Kowalchuk GA, Bodelier PL, Heilig GHJ, Stephen JR, Laanbroek HL (1998) Community analysis of ammonia-oxidising bacteria, in relation to oxygen availability in soil and root-oxygenated sediments, using PCR, DGGE and oligonucleotide probe hybridisation. FEMS Microbiol. Ecol. 27, 339-350.

Kristjansson JK, Hreggvidsson GO (1995) Ecology and habitats of extremophiles. World J. Microbiol. Biotechnol. 11, 17-25.

Lauber CL, Hamady M, Knight R, Fierer N (2009) Pyrosequencing-based assessment of soil pH as a predictor of soil bacterial community structure at the continental scale. Appl. Environ. Microbiol. 75, 5111-5120.

Lehtovirta-Morley LE, Stoecker K, Vilcinskas A, Prosser JI, Nicol GW (2011) Cultivation of an obligate acidophilic ammonia oxidizer from a nitrifying acid soil. Proc. Natl. Acad. Sci. 108(38), 15892-15897.

Liu WT, Marsh TL, Cheng H, Forney LJ (1997) Characterization of microbial diversity by terminal restriction fragment length polymorphisms of genes encoding 16S rRNA. Appl. Environ. Microbiol. 63, 4516-4522.

Lovell CR, Friez MJ, Longshore JW, Bagwell CE (2001) Recovery and phylogenetic analysis of nif sequences from diazotrophic bacteria associated with dead aboveground biomass of *Spartina alterniflora*. Appl. Environ. Microbiol. 67, 5308-5314.

Lu L, Han W, Zhang J, Wu Y, Wang B, Lin X, Zhu J, Cai Z, Jia Z (2012) Nitrification of archaeal ammonia oxidizers in acid soils is supported by hydrolysis of urea. ISME J. 6, 1978-1984.

Maheshwari DK (2013) Bacteria in Agrobiology: Disease Management, Springer, Heidelberg.

Massol-Deya AA, Odelson DA, Hickey RF, Tiedje JM (1995) Bacterial community fingerprinting of amplified 16S and 16-23S ribosomal RNA gene sequences and restriction endonuclease analysis (ARDRA). In: Molecular Microbial Ecology Manual (Ed) Akkermans ADL, van Elsas JD and de Bruijn FJ, Kluwer Academic Publishers, Dordrecht, The Netherlands, pp 1-8.

McCaig AE, Glover LA, Prosser JJ (2001) Numerical analysis of grassland bacterial community structure under different land management regimens by using 16S ribosomal DNA sequence data and denaturing gradient gel electrophoresis banding patterns. Appl. Environ. Microbiol. 67, 4554-4559.

McGonigle TP, Miller MH, Young D (1999) Mycorrhizae, crop growth and crop phosphorus nutrition in maize-soybean rotations given various tillage treatments. Plant Soil 210, 33-42.

Munyanziza E, Kehri HK, Bagyaraj DJ (1997) Agricultural intensification, soil biodiversity and agro-ecosystem function in the tropics: the role of mycorrhiza in crops and trees. Appl. Soil Ecol. 6(1), 77-85.

Muyzer G, de Waal EC, Uitterlinden AG (1993) Profiling of complex microbial populations by denaturing gradient gel electrophoresis analysis of polymerase chain reaction-amplified genes coding for 16S rRNA. Appl. Environ. Microbiol. 59, 695-700.

Muyzer G, Smalla K (1998) Application of denaturing gradient gel electrophoresis (DGGE) and temperature gradient gel electrophoresis (TGGE) in microbial ecology. Antonie Leeuwenhoek 73, 127-141.

Negi H, Das K, Kapri A, Goel R (2009) Phosphate solubilization by psychrophilic and psychrotolerant microorganisms: An asset for sustainable agriculture at low temperatures. In: Phosphate Solubilizing Microbes for Crop Improvement (Ed) Khan MS and Zaidi A, Nova Science, New York, pp 145-160.

Nicol GW, Leininger S, Schleper C, Prosser JI (2008) The influence of soil pH on the diversity, abundance and transcriptional activity of ammonia oxidizing archaea and bacteria. Environ. Microbiol. 10, 2966-2978.

Nielsen P, Fritze O, Priest FG (1995) Phylogenetic diversity of alkaliphilic *Bacillus* strains: proposal for nine species. Microbiol. 141, 1745-1761.

Oehl F, Sieverding E, Ineichen K, Ris EA, Boller T, Wiemken A (2005) Community structure of arbuscular mycorrhizal fungi at different soil depths in extensively and intensively managed agroecosystems. New Phytol. 165, 273-283.

Oehl F, Sieverding E, Mäder P, Dubois D, Ineichen K, Boller T, Wiemken A (2004) Impact of long-term conventional and organic farming on the diversity of arbuscular mycorrhizal fungi. Oecologia 138, 574-583.

Offre P, Prosser JI, Nicol GW (2009) Growth of ammonia-oxidizing archaea in soil microcosms is inhibited by acetylene. FEMS Microbiol Ecol. 70, 99-108.

Pallavi KP, Gupta PC (2013) A psychrotolerant strain Kluyvera intermedia solubilizes inorganic phosphate at different carbon and nitrogen source. Bioscan 8(4), 1197-1201.

Panda AK, Bisht SS, Kumar NS, De Mandal S (2015) Investigation on microbial diversity of Jakrem hot spring, Meghalaya, India using cultivation-independent approach. Genomics Data 4, 156-157.

Parkinson D, Coleman DC (1991) Microbial communities, activities, and biomass. Agric. Ecosyst. Environ. 34, 3-33.

Plenchette C, Clermont-Dauphin C, Meynard JM, Fortin JA (2005) Managing arbuscular mycorrhizal fungi in cropping systems. Can. J. Plant Sci. 85, 31-40.

Priya S, Paneerselvam T, Sivakumar T (2013) Evaluation of indole-3-acetic acid in phosphate solubilizing microbes isolated from rhizosphere soil. Int. J. Curr. Microbiol. Appl. Sci. 2, 29-36.

Raafat N, Tharwat E, Radwan E (2011) Improving wheat grain yield and its quality under salinity conditions at a newly reclaimed soil by using different organic sources as soil or foliar applications. J. Appl. Sci. Res. 7(1), 42-55.

Ranjard L, Poly F, Lata JC, Mougel C, Thioulouse J, Nazaret S (2001) Characterization of bacterial and fungal soil communities by automated ribosomal intergenic spacer analysis fingerprints: biological and methodological variability. Appl. Environ. Microbiol. 67, 4479-4487.

Ravenschlag K, Sahm K, Pernthaler J, Aman R (1999) High bacterial diversity in permanently cold marine sediments. Appl. Environ. Microbiol. 65, 3982-3989.

Rawat S, Johri BN (2014) Thermophilic fungi: diversity and significance in composting. Kavaka 42, 52-68.

Renella G, Egamberdiyeva D, Landi L, Mench M, Nannipieri P (2006) Microbial activity and hydrolase activities during decomposition of root exudates released by an artificial root surface in Cd-contaminated soils. Soil. Biol. Biochem. 38, 702-708.

Requena N, Pérez-Solis E, Azcón-Aguilar C, Jeffries P, Barea JM (2001) Management of indigenous plant-microbe symbioses aids restoration of desertified ecosystems. Appl. Environ. Microbiol. 67, 495-498.

Richardson AE, Simpson RJ, George TS, Hocking PJ (2009) Plant mechanisms to optimize access to soil phosphorus. Crop Pasture Sci. 60, 124-143.

Rinu K, Pandey A, Palni LMS (2012) Utilization of psychrotolerant phosphate solubilizing fungi under low temperature conditions of the mountain ecosystem. In: Microorganisms in Sustainable Agriculture and Biotechnology (Ed) Satyanarayana T, Johri BN, Prakash A, Springer, New York, pp 77-90.

Russo A, Carrozza GP, Vettori L, Felici C, Cinelli F, Toffanin A (2012) Plant beneficial microbes and their application in plant biotechnology. In: Innovations in Biotechnology (Ed) Agbo EC, InTech, Rijeka, Croatia, pp 57-72.

Sadfi-Zouaoui N, Essghaier B, Hajlaoui MR, Fardeau ML, Cayol JL, Ollivier B (2008) Ability of moderately halophilic bacteria to control grey mould disease on tomato fruits. J Phytopathol. 156, 42-52.
Schloss PD, Handelsman J (2003) Biotechnological prospects from metagenomics. Curr. Opin. Biotechnol. 14, 303-310.
Schmalenberger A, Tebbe CC (2002) Bacterial community composition in the rhizosphere of transgenic, herbicide resistant maize (*Zea mays*) and comparison to its non-transgenic cultivar Bosphore. FEMS Microbiol. Ecol. 40, 29-37.
Schnoor T, Lekberg Y, Rosendahl S, Olsson P (2011) Mechanical soil disturbance as a determinant of arbuscular mycorrhizal fungal communities in semi-natural grassland. Mycorrhiza 21, 211-220.
Scow KM, Bruns MA, Graham K, Bossio D, Schwartz E (1998) Development of indices of microbial community structure for soil quality assessment. In: Soil Quality in the California Environment (Ed) Zabel A and Sposito G, Kearney Foundation of Soil Science, USA, pp 110-123.
Selvakumar G, Joshi P, Mishra PK, Bisht JK, Gupta HS (2009) Mountain aspect influences the genetic glustering of psychrotolerant phosphate solubilizing pseudomonads in the Uttarakhand Himalayas. Curr. Microbiol. 59, 432-443.
Sharma SB, Sayyed RZ, Trivedi MH, Gobi TA (2013) Phosphate solubilizing microbes: sustainable approach for managing phosphorus deficiency in agricultural soils. Springer Plus 2, 587-600.
Simon C, Daniel R (2011) Metagenomic analyses: past and future trends. Appl. Environ. Microbiol. 77(4), 1153-1161.
Singh H (2006) Mycoremediation: Fungal Bioremediation. John Wiley and Sons, Inc., New Jersey, pp 533-535.
Six J, Bossuyt H, Degryze S, Denef K (2004) A history of research on the link between (micro) aggregates, soil biota, and soil organic matter dynamics. Soil Till. Res. 79, 7-31.
Smith SE, Read DJ (2008) Mycorrhizal Symbiosis. Academic Press, London, UK, pp 1-800.
Staley JT, Konopka A (1985) Measurement of *in situ* activities of non photosynthetic microorganisms in aquatic and terrestrial habitats. Annu. Rev. Microbiol. 39, 321-346.
Stopnisek N, Gubry-Rangin C, Höfferle S, Nicol GW, Mandic-Mulec I, Prosser JI (2010) Thaumarchaeal ammonia oxidation in an acidic forest peat soil is not influenced by ammonium amendment. Appl. Environ. Microbiol. 76, 7626-7634.
Strap JL (2011) Actinobacteria-plant interactions: a boon to agriculture. In: Bacteria in Agrobiology: Plant Growth Response (Ed) Maheshwari DK, Springer, Berlin, pp. 285-307.
Sudhir P, Murthy S (2004) Effects of salt stress on basic processes of photosynthesis. Photosynthetica 42(2), 481-486.
Tapilatu YH, Grossi V, Acquaviva M, Militon C, Bertrand JC, Cuny P (2010) Isolation of hydrocarbon-degrading extremely halophilic archaea from an uncontaminated hypersaline pond (Camargue, France). Extremophiles 14, 225-231.
Tiedje JM, Cho JC, Murray A, Treves D, Xia B, Zhou J (2001) Soil teeming with life: new frontiers from soil science. In: Sustainable Management of Soil Organic Matter (Ed) Rees RM, Ball B, Watson C and Campbell C, CAB International, Wallingford, UK, pp 393-412.
Tisdall JM (1991) Fungal hyphae and structural stability of soil. Aust. J. Soil. Res. 29, 729-743.
Torsvik V, Goksøyr J, Daae FL (1990) High diversity in DNA of soil bacteria. Appl. Environ. Microbiol. 56, 782-787.
Torsvik V, Sørheim R, Goksoyr J (1996) Total bacterial diversity in soil and sediment communities - a review. J. Ind. Microbiol. 17, 170-178.
Torsvik VL (1980) Isolation of bacterial DNA from soil. Soil Biol. Biochem. 12, 15-22.

Trabelsi D, Mhamdi R (2013). Microbial inoculants and their impact on soil microbial communities: a review. Biomed. Res. Int. 1, 13.

Trivedi P, Sa T (2007) *Pseudomonas corrugate* (NRRL B-30409) mutants increased phosphate solubilization, organic acid production, and plant growth at low temperatures. Curr. Microbiol. 56, 140-144.

Turco RF, Kennedy AC, Jawson MD (1994) Microbial indicators of soil quality. In: Defining Soil Quality for a Sustainable Environment (Ed) Doran JW, Coleman DC, Bezdicek DF and Steward BA, SSSAJ Special Publication, Madison, WI, pp 73-90.

Tyson GW, Chapman J, Hugenholtz P, Allen EE, Ram RJ, Richardson PM, Solovyev VV, Rubin EM, Rokhsar DS, Banfield JF (2004) Community structure and metabolism through reconstruction of microbial genomes from the environment. Nature 428, 37-43.

Ulukanli Z, Digrak M (2002) Alkaliphilic micro-organisms and habitats. Turk. J. Biol. 26, 181-191.

Urbach E, Vergin KL, Giovannoni SJ (1999) Immunochemical detection and isolation of DNA from metabolically active bacteria. Appl. Environ. Microbiol. 65, 1207-1213.

van Elsas JD, Duarte GE, Rosado AS, Smalla K (1998) Microbiological and molecular methods for monitoring microbial inoculants and their effects in the environment. J. Microbiol. Methods 32, 133-154.

Verbruggen E, Kiers ET (2010) Evolutionary ecology of mycorrhizal functional diversity in agricultural systems. Evol. Appl. 3, 547-560.

Von Wintzingerode F, Gçbel UB, Stackebrandt E (1997) Determination of microbial diversity in environmental samples: pitfalls of PCR-based rRNA analysis. FEMS Microbiol. Rev. 21, 213-229.

Whitman WB, Coleman DC, Wiebe WJ (1998) Prokaryotes: the unseen majority. Proc. Natl. Acad. Sci. USA 95, 6578-6583.

Yang A, Hu J, Lin X, Zhu A, Wang J, Dai J, Wong M (2012) Arbuscular mycorrhizal fungal community structure and diversity in response to 3-year conservation tillage management in a sandy loam soil in North China. J. Soils Sediment 12, 835-843.

Yarzabal LA (2014) Cold-tolerant phosphate solubilizing microorganisms and agriculture development in mountainous regions of the world. In: Phosphate solubilizing microorganisms (Ed) Khan MS, Zaidi A and Musarrat J, Springer, New York, pp 113-135.

Yin B, Crowley D, Sparovek G, De Melo WJ, Borneman J (2000) Bacterial functional redundancy along a soil reclamation gradient. Appl. Environ. Microbiol. 66, 4361-4365.

Zeyaullah M, Kamli MR, Islam B, Atif M, Benkhayal FA, Nehal M, Rizvi MA, Ali A (2009) Metagenomics - an advanced approach for noncultivable micro-organisms. Biotechnol Mol. Biol. Rev. 4(3), 49-54.

2

Rhizomicrobiome – A Biological Software to Augment Soil Fertility and Plant Induced Systemic Tolerance Under Abiotic Stress

Jegan S., Baskaran V., Ganga V., Kathiravan R. and Prabavathy V.R.

Abstract

The rhizomicrobiome plays a vital role in maintaining plant health and is highly influenced by the root exudates of the host plant. Plants select a subset of microbes at different stages of their development, and determine the microbiome community composition in its immediate vicinity presumably for specific functions. The rhizomicrobiome is composed of diversified microbial community with specific functions and is highly influenced by the plant type, soil type and environmental conditions. The rhizomicrobiome communities apart from promoting plant growth also elicit induced systemic tolerance to salinity and drought in stressed plants. Bacterial communities producing 1-aminocyclopropane-1-carboxylate (ACC) deaminase modulate stress ethylene in plants and alleviate the effects of abiotic stress and increases plant adaptability to stressed environment. Inter and intraspecies microbial signals and cross talks modulate plant genes in the rhizobiome and are essential for the functioning and sustainability of plant growth under adverse environmental conditions. With the advance of new technologies such as metatranscriptomics and metaproteomics the structure, function, genomic wealth and complex signaling of the rhizosphere is being explored extensively. In recent years the rhizomicrobiome has received substantial attention as it influences both plant health and productivity under natural and stressed environments. Thus this chapter discusses the role of the root exudates in the selection of the rhizomicrobiome communities during plant-microbe interaction and the mechanism involved in eliciting defense response against abiotic stress, salinity and drought.

Keywords: 1-aminocyclopropane-1-carboxylate (ACC deaminase), Drought, Plant Growth Promoting Rhizobacteria (PGPR), Rhizobiome, Salinity

1. Introduction

The soil microbiome especially the root rhizobiome greatly influences plant health, soil fertility (Philippot *et al.* 2013) and agriculture productivity, and is often referred to as the plant's second genome (Berendsen *et al.* 2012; Chaparro and Sabot 2012). Rhizobacterial colonization in the plant is determined by the root exudates of host plants which act either as an attractant or repellent and is subjective to the developmental stages of plant growth (Peiffer *et al.* 2013; Chaparro *et al.* 2014).. The rhizobacteria respond to the root exudates by chemotaxis (Costa *et al.* 2006b; 2007; Badri and Vivanco 2009; DeAngelis *et al.* 2009) and colonize the root regions. The structure of the rhizosphere microbial community is influenced by soil type and amount of root exudates (Uren 2001; Bais *et al.* 2006; Moe 2013) and the plant microbe association is highly influenced by the host plant which is plant specific or plant part specific (Rout and Southworth 2013). The rhizomicrobiome impart beneficial effects on plants by providing the necessary nutrients and protects the plants from the invading pathogens by producing secondary metabolites (Bais *et al.* 2006; Badri *et al.* 2009; Badri and Vivanco 2009; Shi *et al.* 2011; Philippot *et al.* 2013; Turner *et al.* 2013a). In addition to root exudates plant defense signaling molecules also play a major role in determining root microbiome structure (Doornbos *et al.* 2011). Various factors such as biotic and abiotic stress, climatic conditions, anthropogenic effects and soil health play a major role in determining the plant microbe association and the population dynamics of the rhizomicrobiome (Bulgarelli *et al.* 2012; Lundberg *et al.* 2012).

2. The Rhizosphere and its microbiome

Lorenz Hiltner (1904) first coined the term "rhizosphere" to describe the plant-root interface, a word originated from the Greek word "rhiza", meaning root, the immediate zone of soil that surrounds and is influenced by plant roots (Hartmann *et al.* 2008). "Rhizosphere" is a nutrient-rich zone and (Haas and Defago 2005) acts as a junction for nutrient exchange between plants, soil and microbes (Hinsinger *et al.* 2005; Watt *et al.* 2006). It is the hot spot for the microbial community compared to bulk soil (Sørensen 1997; Bais *et al.* 2006; Raaijmakers *et al.* 2009) and harbours specifically explicit groups e.g., ammonifiers, nitrifiers and denitrifiers which dominates (Vega 2007). Root exudates of the plant genotypes and the varieties determine the microbial community composition of the rhizosphere (Dias *et al.* 2012; Aranda *et al.*

2013) which differs across plant species, among the genotypes at cultivar level, as reported in maize (Peiffer *et al.* 2013), rice (Hardoim *et al.* 2011) and sugar beets (Zachow *et al.* 2008).

The composition of root exudates is reported to vary with plant age and development (Chaparro *et al.* 2014; Huang *et al.* 2014) is composed of proteins and mucilages, and is rich in carbohydrates, organic acids, vitamins, nucleotides, flavonoids, enzymes, hormones, volatile compounds phenolics and secondary metabolites. Root exudates influence rhizobiome microbial community and protects root against pathogenic microbes; alters the chemical properties of rhizosphere soil; and acts as a messenger to mediate both positive and negative interaction in the rhizosphere (Sørensen 1997; Arshad and Frankenberger, 1998; Prescott *et al.* 1999; Nicholas 2007; Bais *et al.* 2008; Turner *et al.* 2013a). Thus the nutrient-rich rhizosphere and the surrounding soil influenced by the host plant and its roots, harbours a plethora of microbes that are of central importance for plant nutrition, health and quality (Nicholas 2007; Hartmann *et al.* 2008; 2009; Berg 2009; Huang *et al.* 2014).

Microbial groups residing in the rhizosphere include bacteria, fungi, archaea, algae, nematodes, protozoa, viruses, oomycetes and microarthropods (Fig. 1) (Lynch 1990; Buée *et al.* 2009; Mendes *et al.* 2013). The bacterial domain leads the microbial population in the rhizosphere soil, followed by fungi, actinomycetes and other groups (Mendes *et al.* 2013; Nunes da Rocha *et al.* 2013). Among bacteria, the phylum proteobacteria represent the dominant

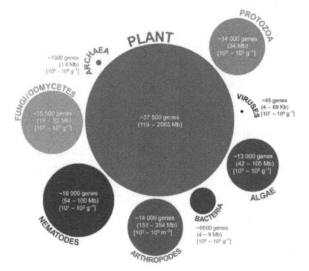

Fig.1: Overview of rhizobiome structure. [Source: Mendes *et al.* (2013)]

members of the rhizosphere microbiome (Uroz *et al*. 2010) followed by other groups like Firmicutes, Actinobacteria, Acidobacteria, Bacteroidetes, Verrucomicrobia and Planctomycetes (Buée *et al*. 2009; Prashar *et al*. 2013; Turner *et al*. 2013a).

Next to bacteria, fungal communities make up a significant group of the rhizosphere microbiome. The phyla include Ascomycota, Basidiomycota, Chytridiomycota and Zygomycota (Vandenkoornhuyse *et al*. 2002; Buée *et al*. 2009; Xu *et al*. 2012). Within the rhizomicrobiome the plant growth promoting rhizobacteria (PGPR) are common inhabitants of the rhizosphere and phyllosphere and survive in a broad range of environmental conditions and influence the soil and plant's health (Costa *et al*. 2006a; Ayyadurai *et al*. 2007; Sekar and Prabavathy, 2014) under stress conditions.The rhizosphere microbiome is involved in a number of processes such as nitrogen fixation, phytohormone production, and biocontrol of phytopathogens under natural or stressed environments. The rhizosphere microbiome with beneficial functions include nitrogen-fixing bacteria (Szeto *et al*. 1987; Kathiravan *et al*. 2013; Sahoo *et al*. 2014), biocontrol (Couillerot *et al*. 2011; Sayyed and Patel, 2011; Singh *et al*. 2011; Santoyo *et al*. 2012; Yin *et al*. 2013; Yokoyama *et al*. 2013; Sekar and Prabavathy, 2014), stress protectant (Zahir *et al*. 2009; Palaniyandi *et al*. 2013; Parihar *et al*. 2015; Shrivastava and Kumar, 2015; Smith *et al*. 2015) and plant growth promoting rhizobacteria (PGPR) (Kloepper 1993; Arshad and Frankenberger, 1998; Picard *et al*. 2000; Saravanakumar *et al*. 2008; Vyas and Gulati, 2009; Farajzadeh *et al*. 2012; Santoyo *et al*. 2012).The microbial community in the rhizosphere harbors members of few groups that adversely affect plant growth and health *viz*., pathogenic fungi, oomycetes, bacteria and nematodes (Raaijmakers *et al*. 2009; Damiani *et al*. 2012; Weller *et al*. 2012; Sekar and Prabavathy, 2014) that influences the plant life cycle either directly or indirectly (Hardoim *et al*. 2011). Though plant microbiome studies have focused on agricultural crops for better protection and productivity enhancement, with an approximate number of 500,000 plant species, the rhizosphere microbiome of plants grown in extreme niches have not been explored. Several studies have proved the impact of human gut microbiome as a source for the alleviation of several human diseases (Heijtz *et al*. 2011; Kinross *et al*. 2011; Engel and Moran 2013). But the core-microbiome in the plants has not been exactly explored as human microbiome, as studies have mostly focused on the rhizobiome community (Berendsen *et al*. 2012; Hirsch and Mauchline, 2012; Gaiero *et al*. 2013; Mendes *et al*. 2013).

3. Role of rhizomicrobiome in soil fertility enhancement and plant growth promotion

The rhizomicrobiome enhances soil fertility by the following functions (i) by releasing plant major and minor nutrients from insoluble inorganic forms, (ii) by decomposing organic residues, (iii) by contributing to soil humus formation, (iv) by improving plant health through symbiotic or mutualistic association, (v) by maintaining soil stability through biogeochemical cycling and (vi) by biological control of plant pathogens, insects and weeds (Weller *et al*. 2002; Gaiero *et al*. 2013; Bhardwaj *et al*. 2014; Gerbore *et al*. 2014).

Rhizobacteria stimulate plant growth under harshening conditions like pest and pathogen infection (Kloepper 1993; Glick and Bashan, 1997); drought and salinity stress. Somers *et al*. (2004) classified PGPR based on their functional activities as (i) biofertilizers (increasing the availability of nutrients to plant), (ii) phytostimulators (plant growth promotion by phytohormones), (iii) rhizoremediators (degrading organic pollutants) and (iv) biopesticides and thus exhibit multiple function. (Lugtenberg and Kamilova 2009).

PGPR strain like *Acinetobacter* (Palaniyandi *et al*. 2013), *Alcaligenes* (Sayyed and Patel, 2011; Yokoyama *et al*. 2013), *Azospirillum* (Combes-Meynet *et al*. 2011; Couillerot *et al*. 2011), *Azotobacter* (Farajzadeh *et al*. 2012; Sahoo *et al*. 2014), *Bacillus* (Joshi and McSpadden Gardener, 2006; Ariffin *et al*. 2008; Santoyo *et al*. 2012) *Rhizobium* (Szeto *et al*. 1987; Schwieger and Tebbe, 2000) *Pseudomonas* (Combes-Meynet *et al*. 2011; Ramette *et al*. 2011; Lanteigne *et al*. 2012; Weller *et al*. 2012; Zhou *et al*. 2012; Yin *et al*. 2013; Sekar and Prabavathy, 2014) exhibit potential biocontrol and PGPR activities. Certain PGPR isolates produce quorum sensing signaling molecules and induce plant immune response (Hartmann and Schikora, 2012; Viswanath *et al*. 2015).

4. Impact of abiotic stress on plant growth and productivity

In the current scenario of climate change agricultural crops are exposed to stress induced by both biotic and abiotic factors. Abiotic stress includes salinity, drought, flooding, high and low temperatures, ultraviolet light, air pollution and heavy metals, all of which affects the yield, quality, quantity, physiology and biochemistry of plants (Sattar *et al*. 2010). The yield losses associated with abiotic stress is predicted as 50–82%, depending on the crop type and stress period (Christensen *et al*. 2007). Currently, over half of the world's irrigated land and 20% of the cultivated land is affected by salinity (Hasegawa *et al*. 2000). Although drought is more pervasive and devastating than salinity; plant responses to both drought and salinity is closely related (Hussain *et al*. 2008).

4.1 Salinity

Salinization is recognized as the main threat to agriculture productivity in many countries, affecting almost 1 billion ha worldwide (Metternicht and Zinck, 2003) with estimated 830 M ha of land or over 6% of the world surfaces affected by salt (Martinez-Beltran and Manzur, 2005) with the scale of 0.25 to 0.5 Mha of agricultural land affected by salinity annually (Peng *et al.* 2008). In India nearly 7.61 Mha of land is under salinity stress (Singh *et al.* 2007). Plants differ greatly in their salinity tolerance level, reflected by growth responses as salts on the roots have an immediate effect on cell growth and associated metabolism; though toxic concentrations of salts take time to accumulate inside plants before they affect plant function (Han *et al.* 2014). Soil salinity makes it harder for roots to extract water leading to osmotic stress, water deficit, stomatal closure and reduced leaf expansion (Rahnama *et al.* 2010; James *et al.* 2011) resulting in osmotic shock and ionic imbalance on plant cells (Zhu 2001). Salt stress negatively affects plant physiology, both the whole plant as well as at cellular levels, it lowers osmotic potential of soil water and consequently the availability of soil water to plants (Ashraf and Harris, 2004). This in turn negatively influence seed germination, growth, flowering, fruit setting, decreasing yield and crop quality (Mahajan and Tuteja, 2005). In addition soil salinity also causes nutrient deficiency of essential ions such as K^+ and elevated Na^+ inside plants which decreases plant photosynthetic rates and biomass accumulation (Mahajan and Tuteja, 2005; Munns and Tester 2008; Zhang and Shi 2013).

4.2 Drought

Drought is another important environmental stress, which decreases crop productivity more than any other environmental stress (Thompson 2008). It is projected that the land area affected by drought will increase by two-fold and water resources will decline by 30% by 2050 (Falkenmark 2013). Drought severely affects the turgor pressure; plant growth and development with substantial reductions in crop growth rate and biomass accumulation (Bartels and Sunkar 2005). The consequences of drought in crop plants are reduced rate of cell division and expansion, resulting in reduction of leaf size, stem elongation and root proliferation, and disturbed stomatal oscillations, plant water and nutrient relations with diminished crop productivity, and water use efficiency (Farooq *et al.* 2009). Drought stress affects morphological, physiological, and biochemical processes in plants and impacts the water relationships on both the cellular and whole plant levels leading to specific and nonspecific phenotype and physiological responses (Beck *et al.* 2007). Different crops respond differently to drought stress, for eg., delayed flowering was observed in maize (Song *et al.* 2010) and rice (Pantuwan *et al.* 2002; Lafitte *et al.* 2007; Kim and

Kim, 2009) on contrary accelerated flowering and physiological maturity was observed in soybean, maize, cotton and rice (Zhu 2002; Brevedan and Egli, 2003; Lafitte *et al.* 2007; Kim and Kim, 2009; Araus *et al.* 2012; Yee-Shan *et al.* 2013), wheat and barley (Samarah 2005). Drought stress modulates plant hormones and other signaling molecules such as salicylic acid, auxins, gibberellins, cytokinins, and abscisic acid which suppress plant growth and yield. Farooq *et al.* (2009) reviewed the impact of drought in several plant species such as barley, maize, rice, wheat and sunflower and reported an average of 50% of growth reduction.

5. Rhizomicrobiome induced systemic tolerance (IST) for alleviating plant abiotic stress

Rhizomicrobiome protects plant from abiotic stress like salinity, drought and temperature, induces physical and chemical changes in plant that results in enhanced tolerance to abiotic stress referred to as Induced Systemic Tolerance (IST) (Paul and Nair, 2008; Yang *et al.* 2009; Shrivastava and Kumar, 2015).

Rhizobacteria are reported to influence growth, development, health and nutrient status of plant under salinity and drought stress through several mechanisms like production or degradation of ethylene, exopolysaccharides (EPS), osmoprotectant, gibberellins (GAs), jasmonic acid (JA), salicylic acid (SA) and to regulate multiple physiological processes in the plant, including root initiation, elongation and root hair formation (Bulgarelli *et al.* 2013; Quiza *et al.* 2015; Smith *et al.* 2015). Generally bacteria belonging to different genera including *Rhizobium, Bacillus, Pseudomonas, Pantoea, Paenibacillus, Burkholderia, Achromobacter, Serratia, Azospirillum, Azotobacter, Microbacterium, Methylobacterium, Variovorax, Enterobacter, Brevibacterium, Brachybacterium, Acinetobacter, Streptomyces,* etc. have been reported to provide tolerance to host plants under different abiotic stress environments (Grover *et al.* 2010).

Dynamic interaction between soil, root, rhizomicrobiome and water modifies the structural and physicochemical properties of the soil (Haynes and Swift, 1990). The EPS produced by bacteria protects it from water stress by enhancing water retention and by regulating the diffusion of organic carbon sources (Chenu and Roberson, 1996). The EPS produced by the colonizing bacteria acts as shield on the root surface and protects the plant from drought stress. In addition, EPS binds to cations including Na^+ thus making it unavailable to plants under saline conditions.

Plants treated with EPS producing bacteria display increased resistance to water and salinity stress due to improved soil structure (Sandhya *et al.* 2009) mediated by the microbial polysaccharides which binds with the soil particles to form micro-aggregates and macro-aggregates and protects plants from abiotic stress and maintains soil moisture. Alami *et al.* (2000) reported increased root adhering soil per root tissue (RAS/RT) ratio under drought conditions in sunflower plants treated with EPS producing *Rhizobium* sp. YAS34. Similarly wheat treated with *Pantoea agglomerans* enhanced growth and yield under salt condition (Amellal *et al.* 1998). Growth hormones cytokinin and antioxidants produced by rhizobacteria leads to enhanced accumulation of abscisic acid (ABA) and degradation of reactive oxygen species in plants as reported in sugar beet when treated with *Azotobacter chroococcum* where enhanced oxidative stress tolerance and increased nitrogen and antioxidant enzymes, superoxide dismutase, peroxidase and catalase was observed (Štajner *et al.* 1997). Proline accumulation enhanced plants osmotic stress tolerance (Chen *et al.* 2007). The *proBA* genes derived from *B. subtilis* transferred to *A. thaliana* resulted in enhanced production of free proline resulting in increased tolerance to osmotic stress in the transgenic plants. Co-inoculation of *Zea mays* with *Rhizobium* and *Pseudomonas* increased proline production, decreased electrolyte leakage, maintained relative water content of leaves and selective uptake of K^+ ions all of which contributed to increased salt tolerance (Bano and Fatima, 2009). Ryu *et al.* (2004) reported induced systemic tolerance in *Arabidopsis* against salt stress by volatile organic compounds (VOCs) from *B. subtilis* GB03.

Bacteria that produce 1- aminocyclo propane 1-carboxylate deaminase (ACC) are key players in induced systemic tolerance (IST) against both salt and drought stress. ACC deaminase producing bacteria use ethylene, the immediate precursor ACC, to produce ketobutyrate and ammonia (Glick 1995). PGPR use ACC as a source of nitrogen and thus lead to decreased ethylene levels in plant and increases root growth (Glick *et al.* 1998; Burd *et al.* 2000; Wang *et al.* 2012; Weller *et al.* 2012; Chang *et al.* 2014; Glick 2014). A number of plant spp. are reported to harbour ACC deaminase producing bacteria in seeds, roots and leaves (Penrose and Glick, 2003). Both rhizospheric and endophytic bacteria produce ACC and these bacteria exert a beneficial effect on abiotically stressed plants (Saleem *et al.* 2007).

PGPR mediated mechanisms reported to influence plant growth under salt stress are enhanced plant nutrient uptake, ACC deaminase and phytohormone production, increased K^+ ion concentration, induced systemic tolerance, ion homeostasis mediation, induced antioxidative enzymes, osmolyte accumulation and production of bacterial extracellular polymeric substance (Yang *et al.* 2009; Nadeem *et al.* 2010; Diby and Harshad 2014). PGPR have been reported to influence plant health under salt stress by impacting several parameters such as increasing biomass, root system surface, improving germination rate, enhancing chlorophyll content and developing resistance to diseases.

Several studies have reported ethylene emission reduction by inoculation of ACC deaminase bacteria that improves plant growth under salt stress ex. *Achromobacter xylosoxidans* in *Catharanthus roseus* (Karthikeyan *et al.* 2012), *Achromobacter piechaudii, Enterobacter* sp. and *Streptomyces* sp. in tomato (Mayak *et al.* 2004; Kim *et al.* 2014; Palaniyandi *et al.* 2014), *Acinetobacter* sp. in barley (Kang *et al.* 2014), *P. pseudoalcaligenes, B. pumilus* and *Alcaligenes faecalis* in rice (Bal *et al.* 2013; Jha *et al.* 2013), *Azospirillum* sp., *Pseudomonas* sp. and *Serratia* sp. in wheat (Zahir *et al.* 2009; Fukami *et al.* 2016), *Brevibacterium casei, Brachybacterium saurashtrense* and *Haererohalobacter* sp. in pea nut (Pushp *et al.* 2012), *Klebsiella oxytoca* in cotton (Yao *et al.* 2010). Thus these ACC deaminase producing organisms could be promising biofertilizer candidates for alleviating abiotic stress in plants. In addition, a *Streptomyces* strain without ACC deaminase was also reported to promote plant growth in wheat under salt stress by the production of Indole acetic acid and auxin (Siddikee *et al.* 2011). Mayank *et al.* (2013) reported improvement in salt stress tolerance in potato treated with *B. pumilus* and *B. firmus*, positive for ACC deaminase production and phosphate solubilisation respectively. ACC deaminase-producing fluorescent pseudomonads from banana rhizosphere have been reported by Naik *et al.* (2008). Cheng *et al.* (2007) have shown the efficiency of ACC deaminase-producing *P. putida* UW4 in facilitating plant growth under salt stress in canola plants. Many research works proved that *Pseudomonas* sp., *Klebsiella* sp., *Burkholderia* sp., *Rahnella* sp. and *Bacillus* sp. are potential stress protectants and promote plant growth under abiotic stress conditions (Wang *et al.* 2000; Saravanakumar and Samiyappan, 2007; Shaharoona *et al.* 2007; Cheng *et al.* 2008; Gamalero *et al.* 2008; Rodriguez *et al.* 2008; Zahir *et al.* 2009; Farajzadeh *et al.* 2010; Kamala-Kannan *et al.* 2010). Pseudomonads exhibiting salinity tolerance up to 1-1.5 M concentration were reported to enhance plant growth in cereals and millets (Rangarajan *et al.* 2002; Sekar and Prabavathy 2014). Use of ACC producing PGPR to alleviate abiotic stress is a promising approach in agricultural practice to maintain productivity under higher salt concentrations and elevated drought conditions (Singh *et al.* 2011; Nadeem *et al.* 2014).

Under salinity stress the availability, transport and mobility of Ca^{2+} and K^+ were affected in growing plant parts. Potassium acts as a cationic solute responsible for stomatal movements as a response to changes in water status on bulk leaf (Caravaca *et al.* 2004). PGPR treatment can influence host physiology by reducing Na^+ and Cl^- ions and accumulating K^+ and Ca^{2+} in foliar tissues. Wheat plants individually inoculated with *P. putida, Enterobacter cloacae, Serratia ficaria* and *P. fluorescens* were reported to increase K^+/Na^+ ratio by increasing

K⁺ and enhancing salinity tolerance (Sajid *et al.* 2013). Inoculation with *Pseudomonas* sp. on eggplant significantly increased K^+ and Ca^{2+} and decreased Na^+ shoot concentrations under saline conditions but not under non-stress conditions (Fu *et al.* 2010). Therefore use of PGPR is a promising agricultural practice to help salt affected crops to maintain an acceptable level of productivity under higher salt concentrations.

ACC deaminase producing *Achromobacter piechaudii* ARV8 conferred IST against drought and salt in pepper and tomato (Mayak *et al.* 2004). Auxin and ACC deaminase producing PGPR *B. licheniformis* K11 enhanced growth of pepper under drought stress (Lim and Kim 2013). *Paenibacillus polymyxa* treated *A. thaliana* enhanced the transcription level of a drought response gene, early responsive to dehydration 15 (ERD15), and of an ABA-responsive gene, RAB18 (Timmusk and Wagner 1999). Another ACC deaminase producing PGPR strain *Variovorax paradoxus* 5C-2 was reported to reduce abscission of mature leaves in *Aquilegia* hybrid experiencing drought stress by lowering ethylene emisision. (Sharp *et al.* 2011).

The production of ectoines, compatible solutes, trehalose, glycinebetaine and osmoregulating proteins by PGPR help plants recover and protect from stress conditions (Mayak *et al.* 2004; He *et al.* 2006). As a response to drought stress plants increase the synthesis of osmolytes thus increasing the osmotic potential within the cells (Farooq *et al.* 2009). Glycine betaine produced by osmotolerant bacteria act synergistically with plant produced glycine betaine in response to stress and increase drought tolerance, especially IAA producing osmotolerant bacteria which improved root proliferation in drought stressed rice and enhanced water uptake (Yuwono *et al.* 2005). In addition, climate models have predicted the increased severity and frequency of drought under the ongoing global climate change scenarios (Walter *et al.* 2001; IPCC 2007). Isolation and identification of potential rhizobacteria which can survive in extreme conditions of salinity and drought and with potential PGPR traits enhance the plant growth and IST under harshening conditions.

6. Endophytes as plant growth promoters under stress

Research focus for the last few decades was mainly on the rhizosphere microbiome, as they play an important role in plant health and productivity. Recent research reveals that apart from rhizobiome other plant microbiome such as the endophytes and the phyllosphere bacteria are involved in the induction of plant growth, disease protection and productivity enhancement (Ryan *et al.* 2008; Delmotte *et al.* 2009; Dias *et al.* 2012; Knief *et al.* 2012; Gopal *et al.* 2013; Berg *et al.* 2014).The endophytes colonizing the internal tissues of their host plant constitute a range of different relationships, including symbiotic, mutualistic, commensalistic and trophobiotic. Endophytic bacteria have been recognized as symbionts with a unique and intimate interaction with the plant (Ryan *et al.* 2008; Reinhold-Hurek and Hurek, 2011) with several beneficial roles in plant growth, soil health and phytoremediation though they were considered as contaminants for a long time.

7. Exploration of rhizomicrobiome community

The rhizosphere associated bacterial community is well characterized by traditional cultivation techniques on different culture media by replicating the natural conditions in lab (Berendsen *et al.* 2012). But the available cultivation dependent methods miss vast majority of the diversity which can be accessed by the modern culture independent molecular techniques to understand the diversity of microbial community in a particular environment (Hugenholtz and Tyson 2008). Diversity analysis based on conserved regions of 16S rRNA (Woese and Fox 1977) was extensively used to study the distribution of microbial community present in an ecosystem. Pinto *et al.* (2014) reported highly dynamic microbiome communities in vineyard by culture independent approaches such as q-PCR and denaturing gradient gel electrophoresis (DGGE). Watanabe *et al.* (2007) analyzed the dynamics of archaeal methanogenic communities in paddy soil by RT-PCR-DGGE analysis. Similarly, Costa *et al.* (2007) analyzed the community structure of *Pseudomonas* sp. with antagonistic activity by PCR-DGGE analysis and the community structure of microbes under elevated carbon dioxide concentration in grasslands was analyzed through DGGE by Drissner *et al.* (2007).

After the first report on microbial community profiling using DGGE (Muyzer *et al.* 1993), many molecular fingerprinting tools like TGGE, SSCP, FISH, clone library construction and advanced sequencing technologies such as next generation sequencing (NGS) provided an in-depth knowledge on microbial community present in an ecosystem (Rincon-Florez *et al.* 2013). Other techniques such as metagenomics which focuses on functional diversity of a microbiome (the abundance of metabolic genes) and metatranscriptomics and metaproteomics which provide information on community-wide gene expression and protein abundance were used to study the community structure.

Metatranscriptomics and metaproteomics are currently employed to understand the microbial community and their function in an ecosystem as these techniques target the specific genes which are expressed by plants and microbes (Knief *et al.* 2012; Rincon-Florez *et al.* 2013; Yergeau *et al.* 2014). Metatranscriptomics sequence analysis of the total RNA from environmental samples provides information on active microbial community and their metabolic pathways (Urich *et al.* 2008), but the significant challenge in metatranscriptomics is the requirement of enriched mRNA (Stewart *et al.* 2010; Yi *et al.* 2011). Turner *et al.* (2013a) showed the predominance of Acidimicrobiales; Actinomycetales and Bacillales in legume and cereal (wheat and oat) rhizospheres through metatranscriptomics.

Metatranscriptome of *A. thaliana* rhizosphere microbiome at different developmental stages of the plant revealed that several microbial genes are involved in carbohydrates, amino acids, secondary metabolites and root exudates metabolism which changed over time (Chaparro *et al.* 2014). Yergeau *et al.* (2014) compared the microbial metatranscriptomic composition of willow rhizosphere along with bulk soil contaminated with organic pollutants and identified genes involved in hydrocarbon degradation in both rhizosphere and bulk soil. Genes related to carbon and amino-acid uptake and utilization were up-regulated in the rhizosphere. The advantage of metatranscriptome analysis is the kingdom level changes in the structure of host plant rhizobiome (Turner *et al.* 2013b) can be determined. Metatranscriptome and metaproteome analysis have been used successfully in determining the microbial community of various systems but not in rhizosphere microbial ecology.

Metagenomic studies can unravel the functional ability of microbes by analyzing the gene abundance that involves in particular metabolic process while metatranscriptomics and metaproteomics can provide information of community level gene expression and protein profiling respectively. Myxococcales and unclassified Deltaproteobacteria as well as Alpha, Beta and Gamma subdivisions namely Rhizobiales, Burkholderiales and Pseudomonadales were reported for their potential plant growth promoting activity (Janssen 2006), from many rhizosphere soils (Lu *et al.* 2006; Yashiro *et al.* 2011) using metagenomic approach. Metagenome analysis of the microbiome of the washed roots and bulk soil of wheat and cucumber showed that Proteobacteria (80%) dominated the root and Actinobacteria (40%) were most abundant in bulk soils. Unno and Shinano (2013) through metagenomics showed enhanced growth of *Lotus japonicus* in the presence of phytic acid and identified genes encoding alkaline phosphatase or citrate synthase related to phytic acid utilization. Likewise, Chhabra *et al.* (2013) by targeted metagenomic approach using a fosmid library in *E. coli*, screened for mineral phosphate solublization activity and identified genes and operons showing similarity to phosphorus uptake, regulatory, and solublization mechanisms through 454 pyrosequencing technology.

Next generation sequencing (NGS) is an impressively accelerated research in microbiome studies in past few years by providing opportunities for a paradigm shift in our ability to resolve the structure and function of rhizosphere microbial communities by generating large volume of sequence datasets and expanding the scope and depth of community distribution studies. Analysis of NGS showed that only a minority (up to 5%) of bacteria have been cultured by current methodologies and that a significant proportion of the bacterial phyla detected by these technologies has no cultured representative yet (Mendes *et al*. 2013).

Currently many reports showed NGS revealed the fundamental and new understanding of the rhizomicrobiome structure in distribution of plant parts and below ground (Metzker 2010; Mendes *et al*. 2013; Knief 2014). Through NGS analysis Sugiyama *et al*. (2014) observed shift of bacterial community in the bulk soils and rhizospheres of soybean at different growth stage at phylum level. Proteobacteria increased, while Acidobacteria and Firmicutes decreased in rhizosphere soil during growth but abundance of potential plant growth promoting rhizobacteria, including *Bacillus*, *Bradyrhizobium* and *Rhizobium* operational taxonomic units were detected. NGS augment the knowledge of microbiome shape and unravel the interactions and functions within the community which determine their effects on plants and microbial community. Similarly, Ofek *et al*. (2014) showed significant reduction in species richness in total roots relative to bulk soil through 16S rRNA gene/transcript sequence analysis.

Atamna-Ismaeel *et al*. (2012) showed the presence of anoxygenic photosynthetic genes in tamarix associated phyllosphere. Ottesen *et al*. (2013) showed the presence of bacterial groups like *Microvirga*, *Xanthomonas*, *Pseudomonas*, *Sphingomonas*, *Brachybacterium*, *Rhizobiales*, *Paracoccus*, *Chryseomonas* and *Microbacterium*. Most of the NGS studied were carried out by targeting the 16S rRNA and ITS variable regions to know the community distribution. The only functional marker gene studied so far in plant associated microorganisms *via* amplicon sequencing is *chiA*, which encodes chitinase (Cretoiu *et al*. 2012); the rhizosphere of *Oxyria dignya* harbours diverse *chiA* gene.

Several quantitative methods widely used in studying rhizosphere microbial community includes MPN-PCR (Rosado *et al*. 1996), competitive PCR (Mauchline *et al*. 2002) and real time PCR (Mavrodi *et al*. 2007). Zancarini *et al*. (2012) analyzed the differential response of rhizosphere microbial community under the influence of nitrogen content and plant genotype and showed that nitrogen content altered the bacterial communities in presence of host plant species. Similarly, Baudoin *et al*. (2009) analyzed the effect of PGPR strain *A. lipoferum* on the structure of microbial communities in field grown maize and showed the differences in microbial community from plant to plant and also the changes in native microbial population. Recently functional microarrays or GeoChip analysis that target specific microbial community functions have been successfully used to analyze the metabolic potential of the microbial communities in soil (He *et al*. 2007; Bai *et al*. 2013; Zhang *et al*. 2013). A major setback of

GeoChip analysis is that it cannot detect novel functions and transcripts (Dugat-Bony *et al.* 2012) and can only detect the functions present in the probe. The various molecular techniques adopted to study the total rhizomicrobiome and its advantage and disadvantage are represented in Table.1

8. Transcriptional Profiling to Identify Novel Stress Protecting Genes from PGPR

Recently, transcriptomic and proteomic analyses are considered as efficient tools to study the microbial expression and to identify the differentially expressed genes of bacteria under different stress condition which will enhance our understanding of plant-microbe interactions under stress conditions. Liu *et al.* (2014) reported the role of three genes in alleviating salt stress in *Mesorhizobium alhagi* CCNWXJ12-2 under high salt condition through transcriptome analysis of which *YadA* domain-containing protein (*yadA*) was involved in salt resistance proved by knockout of other two genes *mttB* encoding trimethylamine methyltransferase and *fhs* gene encoding formate—tetrahydrofolate ligase. Microarray analysis of gene expression during the interaction of *Arabidopsis* with *P. fluorescens* FPT9601-T5 showed up-regulation of putative auxin-regulated genes and nodulin-like genes and down-regulation of some ethylene-responsive genes (Wang *et al.* 2005).

Through transcriptome analysis the effect and mode of interaction of stress protecting *Stenotrophomonas rhizophila* DSM14405^T in oil seed rape revealed that the production of glucosylglycerol (GG) as a remarkable mechanism for the stress protection and spermidine as a novel plant growth regulator to protect against abiotic stress (Alavi *et al.* 2013). Transcriptional responses to salt stress in *Rhodobacter sphaeroides* highly induced several genes encoding putative transcription factors with correlated changes in the compatible solutes and membrane lipids. The *crpO* (RSP1275) gene, a member of the cyclic AMP receptor protein/fumarate and nitrate reduction regulator (CRP/FNR) improved NaCl tolerance in *R. sphaeroides*, when introduced in multiple copies unaltered the expression of the genes involved in the synthesis or transport of compatible solutes but altered the membrane phospholipids composition. These findings set the stage for deciphering the salt stress-responsive regulatory network in *R. sphaeroides* (Tsuzuki *et al.* 2011). Identifying the genetic components for alleviating abiotic stress in plant growth promoting bacteria (PGPB) may be used as potential genes source for developing transgenic host organism (Nadeem *et al.* 2014).

Tolerance level of the transgenic plants to salinity is not high and relatively few mechanisms have been unequivocally demonstrated in explaining the increased

Table 1: List of molecular techniques adopted to study the total rhizomicrobiome

	Methods	Advantages	Disadvantages
Biomass	PLFA	Sensitive detection and accurate quantification of different microbial groups. Rapid and efficient. Useful information on the dynamics of viable bacteria. Reproducible.	Time consuming. Low number of samples can be treated at the same time
	Q-PCR	Quick, accurate and highly sensitive method for sequence quantification that can also be used to quantify microbial groups. Relatively cheap and easy to implement. Specific amplification can be confirmed by melting curve analysis.	Can only be used for targeting known sequences. DNA impurities and artifacts may create false-positives or inhibit amplification.
Diversity	DGGE/TGGE	Sensitive to variation in DNA sequences. Bands can be excised, cloned and sequenced for identification	Time consuming. Multiple bands for a single species can be generated due to micro-heterogeneity. Can be used only for short fragments. Complex communities may appear smeared due to a large number of bands. Difficult to reproduce (gel to gel variation)
	SSCP	Community members can be identified. Screening of potential variations in sequences. Helps to identify new mutations	Short fragments. Lack of reproducibility. Several factors like mutation and size of fragments can affect the sensitivity of the method
	T-RFLP	Enables analyses of a wide array of microbes. Highly reproducible. Convenient way to store data and compare between different samples	Artifacts might appear as false peaks. Distinct sequences sharing a restriction site will result in one peak. Unable to retrieve sequences
Diversity	RISA/ARISA	High resolution when detecting microbial diversity. Quick and sensitive.	More than one peak could be generated for a single organism. Similar spacer length in unrelated organisms may lead to underestimations of community diversity.
	RAPD	Suitable for unknown genomes. Requires low quantities of DNA. Efficient, fast and low cost	Low reproducibility. Sensitive to reaction conditions

(Contd.)

	Methods	Advantages	Disadvantages
	ARDRA	Highly useful for detection of structural changes in simple microbial communities. No special equipment required.	More applicable to environments with low complexity. Several restrictions are needed for adequate resolution. Labour and time intensive. Different bands can belong to the same group.
Diversity	FISH	Allows detection and spatial distribution of more than one sample at the same time	Auto fluorescence of microorganisms. Accuracy and reliability is highly dependent on specificity of probe(s)
	NGS	Rapid method to assess diversity and abundance of microbial taxonomic units simultaneously and at a considerable depth compared to the methods that have been available so far	Relatively expensive. Replication and statistical analysis are essential. Computational intensive and challenging in terms of data analysis
	SIP	High sensitivity. Provides evidence on the function of microorganisms in a controlled experimental setup	Incubation and cycling of the stable isotope might cause bias within the microbial communities
Activity	Meta transcriptomics	Allows rRNA and/or mRNA profiling and quantification without prior knowledge of sequence. Provides a snapshot of microbial transcripts at the time of sampling that may allow detection of microbial ecosystem function. Helps to understand the response of microbial communities to changes in their environment	Many issues with isolation of RNA from soil. mRNA isolation and often amplification are required for gene expression analyses. Current sequencing methods, data bases and computing power are not sufficient yet to cover the high biodiversity in soil.

PLFA- Phospholipid fatty acid; qPCR- quantitative PCR; DGGE/TGGE - Denaturing gradient gel electrophoresis and Temperature gradient gel electrophoresis; SSCP - Single-strand conformation polymorphism; T-RFLP - Terminal restriction fragment length polymorphism; RISA/ARISA - Ribosomal RNA intergenic spacer analysis/Automated ribosomal intergenic spacer analysis; RAPD - Random amplified polymorphic DNA; ARDRA - Amplified ribosomal DNA restriction analysis; FISH - Fluorescence in situ hybridization; NGS - Next-generation sequencing; SIP - Stable isotope probing.

resistance to environmental stress of plants treated with PGPB. Indeed, the majority of genes used for the over expression experiments were sourced from plants and less has been made to identify saline and drought tolerant genes from bacteria.

9. Conclusion

Although plant rhizosphere is recognized as a hot spot for microbial diversity analysis, numerous important crops and their natural relatives have not been studied for their rhizomicrobiome association. The network of microbial interaction occurring in rhizomicrobiome region is highly complex and requires high throughput techniques to analyze the large number of organisms involved in shaping the rhizosphere and influencing plant growth. Previous research evidence the role of rhizosphere bacteria in maintaining soil fertility, plant growth promotion and adaptations of plants under abiotic stress. Despite this, more fundamental and practical studies have to be carried out to understand the processes leading to microbial community assembly and their functions and mechanisms involved in alleviating abiotic stress in plants. As discussed in this chapter, abiotic stress poses serious threat to crop production and hence an in-depth investigation is required to identify potential rhizobacteria that increase soil fertility and enhance plant growth under abiotic stresses like salinity, drought, etc. for sustainable crop production/ protection. The dynamic nature of rhizosphere makes it more challenging in adopting recent molecular tools to provide detailed information on the microbial community. Hence advanced techniques like next generation sequencing techniques, metatranscriptomics and metaproteomics could help us to understand the rhizomicrobiome structure, its activity and ecological behavior under abiotic stress. Discoveries in plant microbiome will advance development of microbial inoculants as biofertilizers, biocontrol agents or stress protectants and to develop relevant bioproducts for stressed soils.

References

Alami Y, Achouak W, Marol C, Heulin T (2000) Rhizosphere soil aggregation and plant growth promotion of sunflowers by an exopolysaccharide-producing *Rhizobium* sp. strain isolated from sunflower roots. Appl. Environ. Microbiol. 66, 3393-3398.

Alavi P, Starcher MR, Zachow C, Muller H, Berg G (2013) Root-microbe systems: the effect and mode of interaction of stress protecting agent (SPA) *Stenotrophomonas rhizophila* DSM14405[T]. Front. Plant Sci. 4, 141.

Amellal N, Burtin G, Bartoli F, Heulin T (1998) Colonization of wheat roots by an exopolysaccharide-producing *Pantoea agglomerans* strain and its effect on rhizosphere soil aggregation. Appl. Environ. Microb. 64, 3740-3747.

Aranda E, Scervino JM, Godoy P, Reina R, Ocampo JA, Wittich RM, Garcia-Romera I (2013) Role of arbuscular mycorrhizal fungus *Rhizophagus custos* in the dissipation of PAHs under root-organ culture conditions. Environ. Pollut. 181, 182-189.

Araus JL, Serret MD, Edmeades GO (2012) Phenotyping maize for adaptation to drought. Front Physiol. 3, 305.

Ariffin H, Hassan MA, Shah UK, Abdullah N, Ghazali FM, Shirai Y (2008) Production of bacterial endoglucanase from pretreated oil palm empty fruit bunch by *Bacillus pumilus* EB3. J. Biosci. Bioeng. 106, 231-236.

Arshad M, Frankenberger WT (1998) Plant growth-regulating substances in the rhizosphere: microbial production and functions. Adv. Agron. 62, 34-151.

Arunachalam Palaniyandi S, Yang SH, Damodharan K, Suh JW (2013) Genetic and functional characterization of culturable plant-beneficial actinobacteria associated with yam rhizosphere. J. Basic Microbiol. 53, 985-995.

Ashraf M, Harris PJC (2004) Potential biochemical indicators of salinity tolerance in plants. Plant Sci. 166, 3-16.

Atamna-Ismaeel N, Finkel O, Glaser F, von Mering C, Vorholt JA, Koblizek M, Belkin S, Beja O (2012) Bacterial anoxygenic photosynthesis on plant leaf surfaces. Environ. Microbiol. Rep. 4, 209-216.

Ayyadurai N, Naik PR, Sakthivel N (2007) Functional characterization of antagonistic fluorescent pseudomonads associated with rhizospheric soil of rice (*Oryza sativa* L.). J. Microbiol. Biotechnol. 17, 919-927.

Badri DV, Vivanco JM (2009) Regulation and function of root exudates. Plant Cell Environ. 32, 666-681.

Badri DV, Quintana N, El Kassis EG, Kim HK, Choi YH, Sugiyama A (2009) An ABC transporter mutation alters root exudation of phytochemicals that provoke an overhaul of natural soil microbiota. Plant Physiol. 151, 2006-2017.

Bai S, Li J, He Z, Van Nostrand JD, Tian Y, Lin G (2013) GeoChip-based analysis of the functional gene diversity and metabolic potential of soil microbial communities of mangroves. Appl. Microbiol. Biot. 97, 7035-7048.

Bais H, Broeckling C, Vivanco J (2008) Root Exudates Modulate Plant—Microbe Interactions in the Rhizosphere. In: Secondary Metabolites in Soil Ecology (Ed) Karlovsky P, Springer Berlin Heidelberg vol. 14, pp. 241-252.

Bais HP, Weir TL, Perry LG, Gilroy S, Vivanco JM (2006) The role of root exudates in rhizosphere interactions with plants and other organisms. Annu. Rev. Plant Biol. 57, 233-266.

Bal HB, Das S, Dangar TK, Adhya TK (2013) ACC deaminase and IAA producing growth promoting bacteria from the rhizosphere soil of tropical rice plants. J. Basic Microb. 53, 972-984.

Bano A, Fatima M (2009) Salt tolerance in *Zea mays* (L). following inoculation with *Rhizobium* and *Pseudomonas*. Biol. Fertil. Soils. 45, 405-413.

Bartels D, Sunkar R (2005) Drought and salt tolerance in plants. Crit. Rev. Plant. Sci. 24, 23-58.

Baudoin E, Nazaret S, Mougel C, Ranjard L, Moënne-Loccoz Y (2009) Impact of inoculation with the phytostimulatory PGPR *Azospirillum lipoferum* CRT1 on the genetic structure of the rhizobacterial community of field-grown maize. Soil Biol. Biochem. 41, 409-413.

Beck EH, Fettig S, Knake C, Hartig K, Bhattarai T (2007) Specific and unspecific responses of plants to cold and drought stress. J. Biosci. 32, 501-510.

Berendsen RL, Pieterse CM, Bakker PA (2012) The rhizosphere microbiome and plant health. Trends Plant Sci. 17, 478-486.

Berg G (2009) Plant-microbe interactions promoting plant growth and health: perspectives for controlled use of microorganisms in agriculture. Appl. Microbiol. Biot. 84, 11-18.

Berg G, Grube M, Schloter M, Smalla K (2014) Unraveling the plant microbiome: looking back and future perspectives. Front. Microbiol. 5, 148.

Bhardwaj D, Ansari M, Sahoo R, Tuteja N (2014) Biofertilizers function as key player in sustainable agriculture by improving soil fertility, plant tolerance and crop productivity. Microbial Cell Fact. 13, 66.

Brevedan RE, Egli DB (2003) Short periods of water stress during seed filling, leaf senescence, and yield of soybean. Crop sci. 43, 2083-2088.

Buée M, Boer Wd, Martin F, Overbeek LSv, Jurkevitch E (2009) The rhizosphere zoo: An overview of plant-associated communities of microorganisms, including phages, bacteria, archaea, and fungi, and some of their structuring factors. Plant Soil. 321, 189-212.

Bulgarelli D, Schlaeppi K, Spaepen S, Ver Loren van Themaat E, Schulze-Lefert P (2013) Structure and functions of the bacterial microbiota of plants. Annu. Rev. Plant Biol. 64, 807-838.

Bulgarelli D, Rott M, Schlaeppi K, Ver Loren van Themaat E, Ahmadinejad N, Assenza F, Rauf P, Huettel B, Reinhardt R, Schmelzer E, Peplies J, Gloeckner FO, Amann R, Eickhorst T, Schulze-Lefert P (2012) Revealing structure and assembly cues for Arabidopsis root-inhabiting bacterial microbiota. Nature 488, 91-95.

Burd GI, Dixon DG, Glick BR (2000) Plant growth-promoting bacteria that decrease heavy metal toxicity in plants. Can J Microbiol. 46, 237-245.

Caravaca F, Figueroa D, Barea JM, Azcón-Aguilar C, Roldán A (2004) Effect of mycorrhizal inoculation on nutrient acquisition, gas exchange, and nitrate reductase activity of two Mediterranean-autochthonous shrub species under drought stress. J. Plant Nutr. 27, 57-74.

Chang P, Gerhardt KE, Huang XD, Yu XM, Glick BR, Gerwing PD, Greenberg BM (2014) Plant growth-promoting bacteria facilitate the growth of barley and oats in salt-impacted soil: implications for phytoremediation of saline soils. Int. J. Phytoremediat. 16, 1133-1147.

Chaparro C, Sabot F (2012) Methods and software in NGS for TE analysis. Methods Mol. Biol. 859, 105-114.

Chaparro JM, Badri DV, Vivanco JM (2014) Rhizosphere microbiome assemblage is affected by plant development. ISME J. 8, 790-803.

Chen M, Wei H, Cao J, Liu R, Wang Y, Zheng C (2007) Expression of *Bacillus subtilis proBA* genes and reduction of feedback inhibition of proline synthesis increases proline production and confers osmotolerance in transgenic *Arabidopsis*. J. Biochem. Mol. Biol. 40, 396-403.

Cheng Z, Park E, Glick BR (2007) 1-Aminocyclopropane-1-carboxylate deaminase from *Pseudomonas putida* UW4 facilitates the growth of canola in the presence of salt. Can. J. Microbiol. 53, 912-918.

Cheng Z, Duncker BP, McConkey BJ, Glick BR (2008) Transcriptional regulation of ACC deaminase gene expression in *Pseudomonas putida* UW4. Can. J. Microbiol. 54, 128-136.

Chenu C, Roberson EB (1996) Diffusion of glucose in microbial extracellular polysaccharide as affected by water potential. Soil Biol. Biochem. 28, 877-884.

Chhabra S, Brazil D, Morrissey J, Burke JI, O'Gara F, D ND (2013) Characterization of mineral phosphate solubilization traits from a barley rhizosphere soil functional metagenome. Microbiology Open. 2, 717-724.

Christensen JH, Hewitson B, Busuioc A, Chen A, Gao X, Held I, Jones R, Kolli RK, Kwon WT, Laprise R (2007) Regional climate projections. In: Climate Change 2007: The Physical Science Basis. Contribution of Working Group I to the Fourth Assessment Report of the Intergovernmental Panel on Climate Change (Ed) Solomon S, Qin D, Manning M, Chen Z, Marquis M, Averyt KB, Tignor M and Miller HL, Cambridge University Press: Cambridge, UK and New York, NY, pp. 848-940.

Combes-Meynet E, Pothier JF, Moenne-Loccoz Y, Prigent-Combaret C (2011) The *Pseudomonas* secondary metabolite 2,4-diacetylphloroglucinol is a signal inducing rhizoplane expression of *Azospirillum* genes involved in plant-growth promotion. Mol. Plant Microbe. Int. 24, 271-284.

Costa R, Salles JF, Berg G, Smalla K (2006a) Cultivation-independent analysis of *Pseudomonas* species in soil and in the rhizosphere of field-grown *Verticillium dahliae* host plants. Environ. Microbiol. 8, 2136-2149.

Costa R, Gotz M, Mrotzek N, Lottmann J, Berg G, Smalla K (2006b) Effects of site and plant species on rhizosphere community structure as revealed by molecular analysis of microbial guilds. FEMS Microbiol. Ecol. 56, 236-249.

Costa R, Gomes NC, Krogerrecklenfort E, Opelt K, Berg G, Smalla K (2007) *Pseudomonas* community structure and antagonistic potential in the rhizosphere: insights gained by combining phylogenetic and functional gene-based analyses. Environ. Microbiol. 9, 2260-2273.

Couillerot O, Combes-Meynet E, Pothier JF, Bellvert F, Challita E, Poirier MA, Rohr R, Comte G, Moenne-Loccoz Y, Prigent-Combaret C (2011) The role of the antimicrobial compound 2,4-diacetylphloroglucinol in the impact of biocontrol *Pseudomonas fluorescens* F113 on *Azospirillum brasilense* phytostimulators. Microbiol. 157, 1694-1705.

Cretoiu MS, Kielak AM, Abu Al-Soud W, Sorensen SJ, van Elsas JD (2012) Mining of unexplored habitats for novel chitinases—chiA as a helper gene proxy in metagenomics. Appl. Microbiol. Biotechnol. 94, 1347-1358.

Damiani I, Baldacci-Cresp F, Hopkins J, Andrio E, Balzergue S, Lecomte P, Puppo A, Abad P, Favery B, Herouart D (2012) Plant genes involved in harbouring symbiotic rhizobia or pathogenic nematodes. New Phytol. 194, 511-522.

DeAngelis KM, Brodie EL, DeSantis TZ, Andersen GL, Lindow SE, Firestone MK (2009) Selective progressive response of soil microbial community to wild oat roots. ISME J. 3, 168-178.

Delmotte N, Knief C, Chaffron S, Innerebner G, Roschitzki B, Schlapbach R, von Mering C, Vorholt JA (2009) Community proteogenomics reveals insights into the physiology of phyllosphere bacteria. Proc Natl Acad Sci U S A. 106, 16428-16433.

Dias ACF, Taketani RG, Andreote FD, Luvizotto DM, da Silva JL, Nascimento RDS, de Melo IS (2012) Interspecific variation of the bacterial community structure in the phyllosphere of the three major plant components of mangrove forests. Braz. J. Microbiol. 43, 653-660.

Diby P, Harshad L (2014) Plant-growth-promoting rhizobacteria to improve crop growth in saline soils: a review. Agron. Sustain. Dev. 34, 737-752.

Doornbos R, Geraats B, Kuramae E, Van Loon L, Bakker P (2011) Effects of jasmonic acid, ethylene, and salicylic acid signaling on the rhizosphere bacterial community of *Arabidopsis thaliana*. Mol. Plant Microbe. Int. 24, 395 - 407.

Drissner D, Blum H, Tscherko D, Kandeler E (2007) Nine years of enriched CO2 changes the function and structural diversity of soil microorganisms in a grassland. Eur. J. Soil Sci. 58, 260-269.

Dugat-Bony E, Peyretaillade E, Parisot N, Biderre-Petit C, Jaziri F, Hill D, Rimour S, Peyret P (2012) Detecting unknown sequences with DNA microarrays: explorative probe design strategies. Environ. Microbiol. 14, 356-371.

Engel P, Moran NA (2013) The gut microbiota of insects - diversity in structure and function. FEMS microbiol. rev. 37, 699-735.

Falkenmark M (2013) Growing water scarcity in agriculture:future challenge to global water security. Philos. Trans. Roy. Soc. A 371, 20120410.

Farajzadeh D, Aliasgharzad N, Sokhandan Bashir N, Yakhchali B (2010) Cloning and characterization of a plasmid encoded ACC deaminase from an indigenous *Pseudomonas fluorescens* FY32. Curr. Microbiol. 61, 37-43.

Farajzadeh D, Yakhchali B, Aliasgharzad N, Sokhandan-Bashir N, Farajzadeh M (2012) Plant growth promoting characterization of indigenous Azotobacteria isolated from soils in Iran. Curr. Microbiol. 64, 397-403.

Farooq M, Wahid A, Kobayashi N (2009) Plant drought stress: effects, mechanisms and management. Agron. Sustain. Dev. 29, 185-212.

Fu Q, Liu C, Ding N, Lin Y, Guo B (2010) Ameliorative effects of inoculation with the plant growth-promoting rhizobacterium *Pseudomonas* sp. DW1 on growth of eggplant (*Solanum melongena* L.) seedlings under salt stress. Agric. Water Manage. 97, 1994-2000.

Fukami J, Nogueira MA, Araujo RS, Hungria M (2016) Accessing inoculation methods of maize and wheat with *Azospirillum brasilense*. AMB Express 6, 3.

Gaiero JR, McCall CA, Thompson KA, Day NJ, Best AS, Dunfield KE (2013) Inside the root microbiome: bacterial root endophytes and plant growth promotion. Am. J. Bot. 100, 1738-1750.

Gamalero E, Berta G, Massa N, Glick BR, Lingua G (2008) Synergistic interactions between the ACC deaminase-producing bacterium *Pseudomonas putida* UW4 and the AM fungus *Gigaspora rosea* positively affect cucumber plant growth. FEMS Microbiol. Ecol. 64, 459-467.

Gerbore J, Benhamou N, Vallance J, Le Floch G, Grizard D, Regnault-Roger C, Rey P (2014) Biological control of plant pathogens: advantages and limitations seen through the case study of *Pythium oligandrum*. Environ. Sci. Pollut. Res. Int. 21, 4847-4860.

Glick BR (1995) The enhancement of plant growth by free-living bacteria. Can. J. Microbiol. 41, 109-117.

Glick BR (2014) Bacteria with ACC deaminase can promote plant growth and help to feed the world. Microbiol. Res. 169, 30-39.

Glick BR, Bashan Y (1997) Genetic manipulation of plant growth-promoting bacteria to enhance biocontrol of phytopathogens. Biotechnol. Adv. 15, 353-378.

Glick BR, Penrose DM, Li J (1998) A model for the lowering of plant ethylene concentrations by plant growth-promoting bacteria. J. Theor. Biol. 190, 63-68.

Gopal M, Gupta A, Thomas GV (2013) Bespoke microbiome therapy to manage plant diseases. Front. Microbiol. 4, 355-355.

Grover M, Ali SZ, Sandhya V, Rasul A, Venkateswarlu B (2010) Role of microorganisms in adaptation of agriculture crops to abiotic stresses. World J. Microbiol. Biotechnol. 27, 1231-1240.

Haas D, Defago G (2005) Biological control of soil-borne pathogens by fluorescent pseudomonads. Nat. Rev. Microbiol. 3, 307-319.

Han QQ, Lu XP, Bai JP, Qiao Y, Pare PW, Wang SM, Zhang JL, Wu YN, Pang XP, Xu WB, Wang ZL (2014) Beneficial soil bacterium *Bacillus subtilis* (GB03) augments salt tolerance of white clover. Front. Plant Sci. 5, 525.

Hardoim PR, Andreote FD, Reinhold-Hurek B, Sessitsch A, van Overbeek LS, van Elsas JD (2011) Rice root-associated bacteria: insights into community structures across 10 cultivars. FEMS Microbiol. Ecol. 154-164.

Hartmann A, Schikora A (2012) Quorum sensing of bacteria and trans-kingdom interactions of N-acyl homoserine lactones with eukaryotes. J. Chem. Ecol. 38, 704-713.

Hartmann A, Rothballer M, Schmid M (2008) Lorenz Hiltner, a pioneer in rhizosphere microbial ecology and soil bacteriology research. Plant Soil 312, 7-14.

Hartmann A, Schmid M, van Tuinen D, Berg G (2009) Plant-driven selection of microbes. Secondary Hartmann A, Schmid M, van Tuinen D, Berg G editors. Vol. 321, pp. 235-257.

Hasegawa PM, Bressan RA, Zhu JK, Bohnert HJ (2000) Plant cellular and molecular responses to high salinity. Annu. Rev. Plant. Physiol. Plant. Mol. Biol. 51, 463-499.

Haynes RJ, Swift RS (1990) Stability of soil aggregates in relation to organic constituents and soil water content. J. Soil. Sci. 41, 73-83.

He J, Jiang JD, Jia KZ, Huang X, Li SP (2006) Glycine betaine supplied exogenously enhance salinity tolerance of *Pseudomonas putida* DLL-1. Wei Sheng Wu Xue Bao. 46, 154-157.

He Z, Gentry TJ, Schadt CW, Wu L, Liebich J, Chong SC (2007) GeoChip: a comprehensive microarray for investigating biogeochemical, ecological and environmental processes. ISME J. 1, 67-77.

Heijtz RD, Wang S, Anuar F, Qian Y, Bjorkholm B, Samuelsson A, Hibberd ML, Forssberg H, Pettersson S (2011) Normal gut microbiota modulates brain development and behavior. Proc. Natl. Acad Sci. USA 108, 3047-3052.

Hinsinger P, Gobran GR, Gregory PJ, Wenzel WW (2005) Rhizosphere geometry and heterogeneity arising from root-mediated physical and chemical processes. New Phytol. 168, 293-303.

Hirsch PR, Mauchline TH (2012) Who's who in the plant root microbiome? Nat. Biotechnol. 30, 961-962.

Huang X-F, Chaparro JM, Reardon KF, Zhang R, Shen Q, Vivanco JM (2014) Rhizosphere interactions: root exudates, microbes, and microbial communities. Botany 92, 267-275.

Hugenholtz P, Tyson GW (2008) Microbiology: metagenomics. Nature 455, 481-483.

Hussain TM, Chandrasekhar T, Hazara M, Sultan Z, Saleh BK, Gopal GR (2008) Recent advances in salt stress biology – a review. Biotechnol. Mol. Biol. Rev. 3, 8-13.

IPCC (2007) Climate Change 2007: The Physical Science Basis. In: Contribution of Working Group I to the Fourth Assessment Report of the Intergovernmental Panel on Climate Change (Ed) Solomon S,Qin D,Manning M,Chen Z,Marquis M,Averyt KB,Tignor M, Miller HL. Cambridge University Press: Cambridge, United Kingdom and USA.

James RA, Blake C, Byrt CS, Munns R (2011) Major genes for Na+ exclusion, Nax1 and Nax2 (wheat HKT1;4 and HKT1;5), decrease Na+ accumulation in bread wheat leaves under saline and waterlogged conditions. J. Exp. Bot. 62, 2939-2947.

Janssen PH (2006) Identifying the dominant soil bacterial taxa in libraries of 16S rRNA and 16S rRNA genes. Appl. Environ. Microbiol. 72, 1719-1728.

Jha M, Chourasia S, Sinha S (2013) Microbial consortium for sustainable rice production. Agroecol. Sus. Food 37, 340-362.

Joshi R, McSpadden Gardener BB (2006) Identification and characterization of novel genetic markers associated with biological control activities in *Bacillus subtilis*. Phytopathology 96, 145-154.

Kamala-Kannan S, Lee KJ, Park SM, Chae JC, Yun BS, Lee YH, Park YJ, Oh BT (2010) Characterization of ACC deaminase gene in *Pseudomonas entomophila* strain PS-PJH isolated from the rhizosphere soil. J. Basic Microbiol. 50, 200-205.

Kang S-M, Khan AL, Waqas M, You Y-H, Kim J-H, Kim J-G, Hamayun M, Lee I-J (2014) Plant growth-promoting rhizobacteria reduce adverse effects of salinity and osmotic stress by regulating phytohormones and antioxidants inCucumis sativus. J. Plant Interact. 9, 673-682.

Karthikeyan B, Joe MM, Islam MR, Sa T (2012) ACC deaminase containing diazotrophic endophytic bacteria ameliorate salt stress in *Catharanthus roseus* through reduced ethylene levels and induction of antioxidative defense systems. Symbiosis. 56, 77-86.

Kathiravan R, Jegan S, Ganga V, Prabavathy VR, Tushar L, Sasikala C, Ramana Ch V (2013) *Ciceribacter lividus* gen. nov., sp. nov., isolated from rhizosphere soil of chick pea (*Cicer arietinum* L.). Int. J. Syst. Evol. Microbiol. 63, 4484-4488.

Kim K, Jang YJ, Lee SM, Oh BT, Chae JC, Lee KJ (2014) Alleviation of salt stress by *Enterobacter* sp. EJ01 in tomato and Arabidopsis is accompanied by up-regulation of conserved salinity responsive factors in plants. Mol. Cells. 37, 109-117.

Kim YS, Kim JK (2009) Rice transcription factor AP37 involved in grain yield increase under drought stress. Plant Signal Behav. 4, 735-736.

Kinross JM, Darzi AW, Nicholson JK (2011) Gut microbiome-host interactions in health and disease. Genome medicine. 3, 14-14.

Kloepper JW (1993) Plant growth-promoting rhizobacteria as biological control agents. In: Soil Microbial Ecology: Applications in Agricultural and Environmental Management in F. B.

Metting. Secondary Kloepper JW editor. Marcel Dekker Inc: New York, USA.

Knief C (2014) Analysis of plant microbe interactions in the era of next generation sequencing technologies. Front. Plant Sci. 5, 216.

Knief C, Delmotte N, Chaffron S, Stark M, Innerebner G, Wassmann R, von Mering C, Vorholt JA (2012) Metaproteogenomic analysis of microbial communities in the phyllosphere and rhizosphere of rice. ISME J. 6, 1378-1390.

Lafitte HR, Yongsheng G, Yan S, Li ZK (2007) Whole plant responses, key processes, and adaptation to drought stress: the case of rice. J. Exp. Bot. 58, 169-175.

Lanteigne C, Gadkar VJ, Wallon T, Novinscak A, Filion M (2012) Production of DAPG and HCN by *Pseudomonas* sp. LBUM300 contributes to the biological control of bacterial canker of tomato. Phytopathology 102, 967-973.

Lim JH, Kim SD (2013) Induction of drought stress sesistance by multi-functional PGPR *Bacillus licheniformis* K11 in pepper. Plant Pathol. J. 29, 201-208.

Liu X, Luo Y, Mohamed OA, Liu D, Wei G (2014) Global transcriptome analysis of *Mesorhizobium alhagi* CCNWXJ12-2 under salt stress. BMC Microbiol. 14, 1.

Lu YH, Rosencrantz D, Liesack W, Conrad R (2006) Structure and activity of bacterial community inhabiting rice roots and the rhizosphere. Environ. Microbiol. 8, 1351-1360.

Lugtenberg B, Kamilova F (2009) Plant-growth-promoting rhizobacteria. Annu. Rev. Microbiol. 63, 541-556.

Lundberg DS, Lebeis SL, Paredes SH, Yourstone S, Gehring J, Malfatti S (2012) Defining the core *Arabidopsis thaliana* root microbiome. Nature 488, 86-90.

Lynch JM (1990) Introduction : some consequences of microbial rhizosphere competence for plant and soil. In: The Rhizosphere. (Ed) Lynch JM. John Wiley & Sons: New York, pp 1-10.

Mahajan S, Tuteja N (2005) Cold, salinity and drought stresses: an overview. Arch. Biochem. Biophys. 444, 139-158.

Martinez-Beltran J, Manzur CL (2005) Overview of salinity problems in the world and FAO strategies to address the problem. Proceedings of the international salinity forum Riverside, pp 311-313.

Mauchline TH, Kerry BR, Hirsch PR (2002) Quantification in soil and the rhizosphere of the nematophagous fungus *Verticillium chlamydosporium* by competitive PCR and comparison with selective plating. Appl. Environ. Microb. 68, 1846-1853.

Mavrodi OV, Mavrodi DV, Thomashow LS, Weller DM (2007) Quantification of 2,4-diacetylphloroglucinol-producing *Pseudomonas fluorescens* strains in the plant rhizosphere by real-time PCR. Appl. Environ. Microbiol. 73, 5531-5538.

Mayak S, Tirosh T, Glick BR (2004) Plant growth-promoting bacteria confer resistance in tomato plants to salt stress. Plant Physiol. Biochem. 42, 565-572.

Mayank AG, Chandrama PU, Venkidasamy B, Jelli V, Akula N, Se WP (2013) Plant growth-promoting rhizobacteria enhance abiotic stress tolerance in *Solanum tuberosum* through inducing changes in the expression of ROS-scavenging enzymes and improved photosynthetic performance. J. Plant Growth Regul. 32, 245-258.

Mendes R, Garbeva P, Raaijmakers JM (2013) The rhizosphere microbiome: significance of plant beneficial, plant pathogenic, and human pathogenic microorganisms. FEMS Microbiol. Rev. 37, 634-663.

Metternicht GI, Zinck JA (2003) Remote sensing of soil salinity: potentials and constraints. Remote Sens. Environ. 85, 1-20.

Metzker ML (2010) Sequencing technologies - the next generation. Nat Rev Genet. 11, 31-46.

Moe LA (2013) Amino acids in the rhizosphere: from plants to microbes. Am. J. Bot. 100, 1692-1705.

Munns R, Tester M (2008) Mechanisms of salinity tolerance. Annu Rev Plant Biol. 59, 651-681.

Muyzer G, de Waal EC, Uitterlinden AG (1993) Profiling of complex microbial populations by denaturing gradient gel electrophoresis analysis of polymerase chain reaction-amplified genes encoding for 16S rRNA. Appl. Environ. Microbiol. 59, 695-700.

Nadeem SM, Zahir ZA, Naveed M, Asghar HN, Arshad M (2010) Rhizobacteria capable of producing ACC-deaminase may mitigate salt stress in wheat. Soil Sci. Soc. Am. J. 74, 533-542.

Nadeem SM, Ahmad M, Zahir ZA, Javaid A, Ashraf M (2014) The role of mycorrhizae and plant growth promoting rhizobacteria (PGPR) in improving crop productivity under stressful environments. Biotechnol. Adv. 32, 429-448.

Naik PR, Sahoo N, Goswami D, Ayyadurai N, Sakthivel N (2008) Genetic and functional diversity among fluorescent pseudomonads isolated from the rhizosphere of banana. Microbial Ecol. 56, 492-504.

Nicholas CU (2007) Types, amounts, and possible functions of compounds released into the rhizosphere by soil-grown plants. In: The Rhizosphere. (Ed) Pinton R, Varanini Z, Nannipieri P, CRC Press, pp 1-21.

Nunes da Rocha U, Plugge CM, George I, van Elsas JD, van Overbeek LS (2013) The rhizosphere selects for particular groups of *Acidobacteria* and *Verrucomicrobia*. PLoS ONE 8, e82443.

Ofek M, Voronov-Goldman M, Hadar Y, Minz D (2014) Host signature effect on plant root-associated microbiomes revealed through analyses of resident vs. active communities. Environ. Microbiol. 16, 2157-2167.

Ottesen AR, Gonzalez Pena A, White JR, Pettengill JB, Li C, Allard S, Rideout S, Allard M, Hill T, Evans P, Strain E, Musser S, Knight R, Brown E (2013) Baseline survey of the anatomical microbial ecology of an important food plant: *Solanum lycopersicum* (tomato). BMC Microbiol. 13, 114.

Palaniyandi SA, Yang SH, Zhang L, Suh JW (2013) Effects of actinobacteria on plant disease suppression and growth promotion. Appl. Microbiol. Biot. 97, 9621-9636.

Palaniyandi SA, Damodharan K, Yang SH, Suh JW (2014) *Streptomyces* sp. strain PGPA39 alleviates salt stress and promotes growth of 'Micro Tom' tomato plants. J. Appl. Microbiol. 117, 766-773.

Pantuwan G, Fukai S, Cooper M, Rajatasereekul S, O'Toole JC (2002) Yield response of rice (*Oryza sativa* L.) genotypes to drought under rainfed lowlands. Field Crops Res. 73, 169-180.

Parihar P, Singh S, Singh R, Singh VP, Prasad SM (2015) Effect of salinity stress on plants and its tolerance strategies: a review. Environ. Sci. Pollut. Res. Int. 22, 4056-4075.

Paul D, Nair S (2008) Stress adaptations in a plant growth promoting rhizobacterium (PGPR) with increasing salinity in the coastal agricultural soils. J. Basic Microbiol. 48, 378-384.

Peiffer JA, Spor A, Koren O, Jin Z, Tringe SG, Dangl JL, Buckler ES, Ley RE (2013) Diversity and heritability of the maize rhizosphere microbiome under field conditions. Proc. Natl. Acad. Sci. USA 110, 6548-6553.

Peng G, Yuan Q, Li H, Zhang W, Tan Z (2008) Rhizobium oryzae sp. nov., isolated from the wild rice Oryza alta. Int. J. Syst. Evol. Microbiol. 58, 2158-2163.

Penrose DM, Glick BR (2003) Methods for isolating and characterizing ACC deaminase-containing plant growth-promoting rhizobacteria. Physiol. Plant. 118, 10-15.

Philippot L, Spor A, Henault C, Bru D, Bizouard F, Jones CM, Sarr A, Maron PA (2013) Loss in microbial diversity affects nitrogen cycling in soil. ISME J. 7, 1609-1619.

Picard C, Di Cello F, Ventura M, Fani R, Guckert A (2000) Frequency and biodiversity of 2,4-diacetylphloroglucinol-producing bacteria isolated from the maize rhizosphere at different stages of plant growth. Appl. Environ. Microbiol. 66, 948-955.

Pinto C, Pinho D, Sousa S, Pinheiro M, Egas C, Gomes AC (2014) Unravelling the diversity of grapevine microbiome. PLoS ONE 9, e85622.

Prashar P, Kapoor N, Sachdeva S (2013) Rhizosphere: its structure, bacterial diversity and significance. Rev. Environ. Sci. Biotechnol. 13, 63-77.

Prescott L, Harley J, Klein DA (1999) Microbiology. Mc-Graw-Hill, Boston.

Pushp S Shukla, Pradeep K Agarwal, Bhavanath J (2012) Improved salinity tolerance of *Arachis hypogaea* (L.) by the interaction of halotolerant plant-growth-promoting rhizobacteria. J. Plant Growth Regul. 31, 195-206.

Quiza L, St-Arnaud M, Yergeau E (2015) Harnessing phytomicrobiome signaling for rhizosphere microbiome engineering. Front. Plant Sci. 6, 507.

Raaijmakers J, Paulitz T, Steinberg C, Alabouvette C, Moënne-Loccoz Y (2009) The rhizosphere: a playground and battlefield for soilborne pathogens and beneficial microorganisms. Plant Soil. 321, 341-361.

Rahnama A, James RA, Poustini K, Munns R (2010) Stomatal conductance as a screen for osmotic stress tolerance in durum wheat growing in saline soil. Func. Plant. Biol. 37, 255-263.

Ramette A, Frapolli M, Fischer-Le Saux M, Gruffaz C, Meyer JM, Defago G, Sutra L, Moenne-Loccoz Y (2011) *Pseudomonas protegens* sp. nov., widespread plant-protecting bacteria producing the biocontrol compounds 2,4-diacetylphloroglucinol and pyoluteorin. Syst. Appl. Microbiol. 34, 180-188.

Rangarajan S, Saleena LM, Nair S (2002) Diversity of *Pseudomonas* spp. isolated from rice rhizosphere populations grown along a salinity gradient. Microbial Ecol. 43, 280-289.

Reinhold-Hurek B, Hurek T (2011) Living inside plants: bacterial endophytes. Curr. Opin. Plant Biol. 14, 435-443.

Rincon-Florez V, Carvalhais L, Schenk P (2013) Culture-independent molecular tools for soil and rhizosphere microbiology. Diversity 5, 581-612.

Rodriguez H, Vessely S, Shah S, Glick BR (2008) Effect of a nickel-tolerant ACC deaminase-producing *Pseudomonas* strain on growth of nontransformed and transgenic canola plants. Curr. Microbiol. 57, 170-4.

Rosado AS, Seldin L, Wolters AC, Elsas JD (1996) Quantitative 16S rDNA-targeted polymerase chain reaction and oligonucleotide hybridization for the detection of *Paenibacillus azotofixans* in soil and the wheat rhizosphere. FEMS Microbiol. Ecol. 19, 153-164.

Rout ME, Southworth D (2013) The root microbiome influences scales from molecules to ecosystems: The unseen majority. Am. J. Bot. 100, 1689-1691.

Ryan RP, Germaine K, Franks A, Ryan DJ, Dowling DN (2008) Bacterial endophytes: recent developments and applications. FEMS Microbiol. Lett. 278, 1-9.

Sahoo RK, Ansari MW, Dangar TK, Mohanty S, Tuteja N (2014) Phenotypic and molecular characterisation of efficient nitrogen-fixing *Azotobacter* strains from rice fields for crop improvement. Protoplasma 251, 511-523.

Sajid MN, Zahir AZ, Muhammad N, Shafqat N (2013) Mitigation of salinity-induced negative impact on the growth and yield of wheat by plant growth-promoting rhizobacteria in naturally saline conditions. Ann. Microbiol. 63, 225-232.

Saleem M, Arshad M, Hussain S, Bhatti AS (2007) Perspective of plant growth promoting rhizobacteria (PGPR) containing ACC deaminase in stress agriculture. J. Ind. Microbiol. Biotech. 34, 635-648.

Samarah NH (2005) Effects of drought stress on growth and yield of barley. Agron. Sustain. Dev. 25, 145–149.

Sandhya V, Sk. Z A, Grover M, Reddy G, Venkateswarlu B (2009) Alleviation of drought stress effects in sunflower seedlings by the exopolysaccharides producing *Pseudomonas putida* strain GAP-P45. Biol. Fertil. Soils. 46, 17-26.

Santoyo G, Orozco-Mosqueda MdC, Govindappa M (2012) Mechanisms of biocontrol and plant growth-promoting activity in soil bacterial species of *Bacillus* and *Pseudomonas*: a review. Biocontrol Sci. Technol. 22, 855-872.

Saravanakumar D, Samiyappan R (2007) ACC deaminase from *Pseudomonas fluorescens* mediated saline resistance in groundnut (*Arachis hypogea*) plants. J. Appl. Microbiol. 102, 1283-1292.

Saravanakumar D, Lavanya N, Muthumeena B, Raguchander T, Suresh S, Samiyappan R (2008) *Pseudomonas fluorescens* enhances resistance and natural enemy population in rice plants against leaffolder pest. J. Appl. Entomol. 132, 469-479.

Sattar S, Hussnain T, Javid A (2010) Effect of NaCl salinity on cotton (*Gossypium arboreum* L.) grown on MS medium and in hydroponic cultures. J. Anim. Plant. Sci. 20, 87-89.

Sayyed RZ, Patel PR (2011) Biocontrol potential of siderophore producing heavy metal resistant *Alcaligenes* sp. and *Pseudomonas aeruginosa* RZS3 vis-a-vis organophosphorus fungicide. Indian J. Microbiol. 51, 266-272.

Schwieger F, Tebbe CC (2000) Effect of field inoculation with *Sinorhizobium meliloti* L33 on the composition of bacterial communities in rhizospheres of a target plant (*Medicago staiva*) and a non-target plant (*Chenopodium album*)—linking of 16S rRNA gene-based single-strand conformation polymorphism community profiles to the diversity of cultivated bacteria. Appl. Environ. Microbiol. 66, 3556-3565.

Sekar J, Prabavathy VR (2014) Novel Phl-producing genotypes of finger millet rhizosphere associated pseudomonads and assessment of their functional and genetic diversity. FEMS Microbiol. Ecol. 89, 32-46.

Shaharoona B, Jamro GM, Zahir ZA, Arshad M, Memon KS (2007) Effectiveness of various *Pseudomonas* spp. and *Burkholderia caryophylli* containing ACC-deaminase for improving growth and yield of wheat (*Triticum aestivum* L.). J. Microbiol. Biotechnol. 17, 1300-1307.

Sharp RG, Chen L, Davies WJ (2011) Inoculation of growing media with the rhizobacterium *Variovorax paradoxus* 5C-2 reduces unwanted stress responses in hardy ornamental species. Sci. Hortic. 129, 804-811.

Shi S, Richardson AE, O'Callaghan M, DeAngelis KM, Jones EE, Stewart A, Firestone MK, Condron LM (2011) Effects of selected root exudate components on soil bacterial communities. FEMS Microbiol. Ecol. 77, 600-610.

Shrivastava P, Kumar R (2015) Soil salinity: A serious environmental issue and plant growth promoting bacteria as one of the tools for its alleviation. Saudi J. Biol. Sci. 22, 123-131.

Siddikee MA, Glick BR, Chauhan PS, Yim W, Sa T (2011) Enhancement of growth and salt tolerance of red pepper seedlings (*Capsicum annuum* L.) by regulating stress ethylene synthesis with halotolerant bacteria containing 1-aminocyclopropane-1-carboxylic acid deaminase activity. Plant Physiol. Biochem. 49, 427-34.

Singh JS, Pandey VC, Singh DP (2011) Efficient soil microorganisms: A new dimension for sustainable agriculture and environmental development. Agric, Ecosys. Environ. 140, 339-353.

Smith DL, Praslickova D, Ilangumaran G (2015) Inter-organismal signaling and management of the phytomicrobiome. Front. Plant Sci. 6, 722.

Somers E, Vanderleyden J, Srinivasan M (2004) Rhizosphere bacterial signalling: a love parade beneath our feet. Crit. Rev. Microbiol. 30, 205-240.

Song Y, Birch C, Qu S, Doherty A, Hanan J (2010) Analysis and modelling of the effects of water stress on maize growth and yield in dryland conditions. Plant Prod. Sci. 13, 199-208.

Sørensen J (1997) The rhizosphere as a habitat for soil microorganisms. In: Modern Soil Microbiology (Ed) Van Elsas JD, Trevors JT, Wellington EMH, Marcel Dekker, Inc: New York, pp 21-45.

Štajner D, Kevrešan S, Gašiæ O, Mimica-Dukiæ N, Zongli H (1997) Nitrogen and *Azotobacter chroococcum* enhance oxidative stress tolerance in sugar beet. Biol. Plantarum. 39, 441-445.

Stewart FJ, Ottesen EA, DeLong EF (2010) Development and quantitative analyses of a universal rRNA-subtraction protocol for microbial metatranscriptomics. ISME J. 4, 896-907.

Sugiyama A, Ueda Y, Zushi T, Takase H, Yazaki K (2014) Changes in the bacterial community of soybean rhizospheres during growth in the field. PLoS ONE. 9, e100709.

Szeto WW, Nixon BT, Ronson CW, Ausubel FM (1987) Identification and characterization of the *Rhizobium meliloti* ntrC gene: *R. meliloti* has separate regulatory pathways for activation of nitrogen fixation genes in free-living and symbiotic cells. J. Bacteriol. 169, 1423-1432.

Thompson K (2008) Plant Physiological Ecology, 2nd edn. Ann. Bot. London. 103, viii-ix.

Timmusk S, Wagner EG (1999) The plant-growth-promoting rhizobacterium *Paenibacillus polymyxa* induces changes in *Arabidopsis thaliana* gene expression: a possible connection between biotic and abiotic stress responses. Mol. Plant Microbe Interact. 12, 951-959.

Tsuzuki M, Moskvin OV, Kuribayashi M, Sato K, Retamal S, Abo M, Zeilstra-Ryalls J, Gomelsky M (2011) Salt stress-induced changes in the transcriptome, compatible solutes, and membrane lipids in the facultatively phototrophic bacterium *Rhodobacter sphaeroides*. Appl. Environ. Microbiol. 77, 7551-7559.

Turner TR, James EK, Poole PS (2013a) The plant microbiome. Genome Biol. 14, 209.

Turner TR, Ramakrishnan K, Walshaw J, Heavens D, Alston M, Swarbreck D, Osbourn A, Grant A, Poole PS (2013b) Comparative metatranscriptomics reveals kingdom level changes in the rhizosphere microbiome of plants. ISME J. 7, 2248-2258.

Unno Y, Shinano T (2013) Metagenomic analysis of the rhizosphere soil microbiome with respect to phytic acid utilization. Microbes Environ. 28, 120-127.

Uren NC (2001) Types, amounts, and possible functions of compounds released into the rhizosphere by soil-grown plants. In: The Rhizosphere - Biochemistry and Organic Substances at the Soil-Plant Interface. (Ed) Pinton R, Varanini Z, Nannipieri P, Marcel Dekker, Inc: New York, pp 19-40.

Urich T, Lanzen A, Qi J, Huson DH, Schleper C, Schuster SC (2008) Simultaneous assessment of soil microbial community structure and function through analysis of the meta-transcriptome. PLoS ONE 3, e2527.

Uroz S, Buee M, Murat C, Frey-Klett P, Martin F (2010) Pyrosequencing reveals a contrasted bacterial diversity between oak rhizosphere and surrounding soil. Environ. Microbiol. Rep. 2, 281-288.

Vandenkoornhuyse P, Baldauf SL, Leyval C, Straczek J, Young JP (2002) Extensive fungal diversity in plant roots. Science 295, 2051.

Vega NWO (2007) A review on beneficial effects of rhizosphere bacteria on soil nutrient availability and plant nutrient uptake. Rev. Fac. Natl. Agr. Medellín 60, 3621-3643.

Viswanath G, Jegan S, Baskaran V, Kathiravan R, Prabavathy VR (2015) Diversity and N-acyl-homoserine lactone production by Gammaproteobacteria associated with Avicennia marina rhizosphere of South Indian mangroves. Syst. Appl. Microbiol. 38, 340-345.

Vyas P, Gulati A (2009) Organic acid production in vitro and plant growth promotion in maize under controlled environment by phosphate-solubilizing fluorescent Pseudomonas. BMC Microbiol. 9, 174.

Walter BP, Heimann M, Matthews E (2001) Modeling modern methane emissions from natural wetlands: 1. Model description and results. J. Geophysical Res. Atmospheres 106, 34189-34206.

Wang C, Knill E, Glick BR, Defago G (2000) Effect of transferring 1-aminocyclopropane-1-carboxylic acid (ACC) deaminase genes into *Pseudomonas fluorescens* strain CHA0 and its gacA derivative CHA96 on their growth-promoting and disease-suppressive capacities. Can. J. Microbiol. 46, 898-907.

Wang CJ, Yang W, Wang C, Gu C, Niu DD, Liu HX, Wang YP, Guo JH (2012) Induction of drought tolerance in cucumber plants by a consortium of three plant growth-promoting rhizobacterium strains. PLoS ONE 7, e52565.

Wang Y, Ohara Y, Nakayashiki H, Tosa Y, Mayama S (2005) Microarray analysis of the gene expression profile induced by the endophytic plant growth-promoting rhizobacteria, Pseudomonas fluorescens FPT9601-T5 in Arabidopsis. Mol. Plant. Microbe Int. 18, 385-396.

Watanabe T, Kimura M, Asakawa S (2007) Dynamics of methanogenic archaeal communities based on rRNA analysis and their relation to methanogenic activity in Japanese paddy field soils. Soil Biol. Biochem. 39, 2877-2887.

Watt M, Silk WK, Passioura JB (2006) Rates of root and organism growth, soil conditions, and temporal and spatial development of the rhizosphere. Ann. Bot. 97, 839-855.

Weller DM, Raaijmakers JM, Gardener BB, Thomashow LS (2002) Microbial populations responsible for specific soil suppressiveness to plant pathogens. Annu. Rev. Phytopathol. 40, 309-348.

Weller DM, Mavrodi DV, van Pelt JA, Pieterse CM, van Loon LC, Bakker PA (2012) Induced systemic resistance in *Arabidopsis thaliana* against *Pseudomonas syringae* pv. tomato by 2,4-diacetylphloroglucinol-producing *Pseudomonas fluorescens*. Phytopathology 102, 403-412.

Woese CR, Fox GE (1977) Phylogenetic structure of the prokaryotic domain: the primary kingdoms. Proc. Natl. Acad. Sci. USA. 74, 5088-90.

Xu L, Ravnskov S, Larsen J, Nicolaisen M (2012) Linking fungal communities in roots, rhizosphere, and soil to the health status of *Pisum sativum*. FEMS Microbiol. Ecol. 82, 736-745.

Yang J, Kloepper JW, Ryu CM (2009) Rhizosphere bacteria help plants tolerate abiotic stress. Trends Plant Sci. 14, 1-4.

Yao L, Wu Z, Zheng Y, Kaleem I, Li C (2010) Growth promotion and protection against salt stress by *Pseudomonas putida* Rs-198 on cotton. Eur. J. Soil Biol. 46, 49-54.

Yashiro E, Spear RN, McManus PS (2011) Culture-dependent and culture-independent assessment of bacteria in the apple phyllosphere. J. Appl. Microbiol. 110, 1284-1296.

Yee-Shan K, Wan-Kin A-Y, Yuk-Lin Y, Man-Wah L, Chao-Qing W, Xueyi L, Hon-Ming L (2013) Drought stress and tolerance in soybean. In: A Comprehensive Survey of International Soybean Research - Genetics, Physiology, Agronomy and Nitrogen Relationships (Ed) James B, INTECH, Croatia, DOI: 10.5772/52945

Yergeau E, Sanschagrin S, Maynard C, St-Arnaud M, Greer CW (2014) Microbial expression profiles in the rhizosphere of willows depend on soil contamination. ISME J. 8, 344-358.

Yi H, Cho YJ, Won S, Lee JE, Jin Yu H, Kim S, Schroth GP, Luo S, Chun J (2011) Duplex-specific nuclease efficiently removes rRNA for prokaryotic RNA-seq. Nucleic Acids Res. 39, e140.

Yin D, Wang N, Xia F, Li Q, Wang W (2013) Impact of biocontrol agents *Pseudomonas fluorescens* 2P24 and CPF10 on the bacterial community in the cucumber rhizosphere. Eur. J. Soil. Biol. 59, 36-42.

Yokoyama S, Adachi Y, Asakura S, Kohyama E (2013) Characterization of *Alcaligenes faecalis* strain AD15 indicating biocontrol activity against plant pathogens. J. Gen. Appl. Microbiol. 59, 89-95.

Yuwono T, Handayani D, Soedarsono J (2005) The role of osmotolerant rhizobacteria in rice growth under different drought conditions. Aust. J. Agric. Res. 56, 715.

Zachow C, Tilcher R, Berg G (2008) Sugar beet-associated bacterial and fungal communities show a high indigenous antagonistic potential against plant pathogens. Microb. Ecol. 55, 119-129.

Zahir ZA, Ghani U, Naveed M, Nadeem SM, Asghar HN (2009) Comparative effectiveness of *Pseudomonas* and *Serratia* sp. containing ACC-deaminase for improving growth and yield of wheat (*Triticum aestivum* L.) under salt-stressed conditions. Arch. Microbiol. 191, 415-424.

Zancarini A, Mougel C, Voisin AS, Prudent M, Salon C, Munier-Jolain N (2012) Soil nitrogen availability and plant genotype modify the nutrition strategies of *M. truncatula* and the associated rhizosphere microbial communities. PLoS ONE 7, e47096.

Zhang JL, Shi H (2013) Physiological and molecular mechanisms of plant salt tolerance. Photosynth. Res. 115, 1-22.

Zhang Y, Lu Z, Liu S, Yang Y, He Z, Ren Z (2013) Geochip-based analysis of microbial communities in alpine meadow soils in the Qinghai-Tibetan plateau. BMC Microbiol. 13, 72.

Zhou T, Chen D, Li C, Sun Q, Li L, Liu F, Shen Q, Shen B (2012) Isolation and characterization of *Pseudomonas brassicacearum* J12 as an antagonist against *Ralstonia solanacearum* and identification of its antimicrobial components. Microbiol. Res. 167, 388-394.

Zhu JK (2001) Plant salt tolerance. Trends Plant Sci. 6, 66-71.

Zhu JK (2002) Salt and drought stress signal transduction in plants. Annu. Rev. Plant. Biol. 53, 247-273.

3

Bioconversion of Municipal Solid Waste and its Use in Soil Fertility

Poonam Verma and Jamaluddin

Abstract

With increasing population and subsequent human requirements; the generation of municipal solid waste (MSW) has increased all over the world. In India, the safe disposal of the municipal solid waste still remains a problem for the authorities concerned. Depending upon the nature and type of the MSW, there are physical and chemical methods for the disposal of the waste, but these methods have their own limitations. Those limitations may be overcome by biological methods. Higher number of microorganisms are present in MSW which may either be pathogenic or nonpathogenic. These microorganisms may play a vital role in bioconversion of MSW into compost. The compost thus produced is useful in increasing soil fertility and reducing the concentration of the waste.

Keywords: Decomposition, Microorganisms, Municipal solid waste

1. Introduction

Municipal solid waste (MSW) is a term usually applied to a heterogeneous collection of wastes produced in urban areas. The nature of which varies from region to region (Gautam 2010b; Moreno *et al.* 2013) means it is an unwanted or undesired material. Also described as the waste that is produced from residential and industrial (non-process wastes), commercial and institutional sources with the exception of hazardous and universal wastes, construction and demolition wastes and liquid wastes (water, wastewater, industrial processes). A general estimation of MSW generation from the metropolitans of India is listed by Sahu (2007). In India, there is no segregation of waste and it is dumped in a mixed form in an unscientific manner on open wasteland or low-lying areas even near creeks, forests, rivers, ponds and other ecological sensitive regions. This practice is commonly known as 'Open dumping' which is disastrous for the ecology of the area.

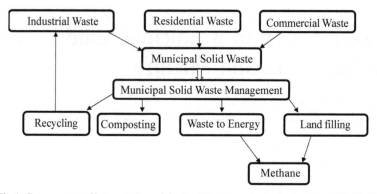

Fig 1: Components of integrated municipal solid waste management system (Warith 2003).

Municipal solid waste management (MSWM) is one of the major environmental problems of Indian megacities. It involves activities associated with generation, storage, collection, transfer and transport, processing and disposal of solid wastes (Mahar *et al*. 2007). But, in most cities, only four activities, i.e., waste generation, collection, transportation, and disposal are practiced without considering its environmental consequences. The management of MSW requires proper infrastructure, maintenance and upgradation for all activities are done. This becomes increasingly expensive and complex due to the continuous and unplanned growth of urban centers (Bundela *et al*. 2010). A considerable amount of money goes into managing such huge volumes of solid waste. Asian countries alone spent about US$25 billion on solid waste management per year in the early 1990s; the figure is expected to rise to around US$50 billion by 2025 (Hoornweg and Thomas 1999). MSW is also classified in to two classes

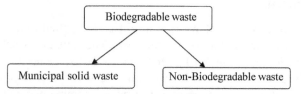

Biodegradable waste can be classified into three broad categories: putrescible, fermentable, and non-fermentable. Putrescible wastes tend to decompose rapidly and unless carefully controlled, decompose with the production of objectionable odours and visual unpleasantness. Fermentable wastes tend to decompose rapidly, but without the unpleasant accompaniments of putrefaction. Non-fermentable wastes tend to resist decomposition and, therefore, break down very slowly. A major source of putrescible waste is food preparation and consumption. As such, its nature varies with lifestyle, standard of living and seasonality of foods. Fermentable wastes are typified by crop and market debris (Warith 2003). About 80-90% waste is organic matter in MSW (Carboo and Fobil 2005).

The organic matter found in solid waste mainly includes cellulose (C), hemicellulose (H), and Lignin (L) (Barlaz *et al.* 2002). In 1996, NEERI, Nagpur carried out a study on the characterization of Indian MSW, demonstrating that it contains large organic fraction (30–40%), ash and fine earth (30–40%), paper (3–6%) along with plastic, glass and metal (each less than 1%) (Shirbhate and Malode, 2012). However, heavy metals (HMs) such as Cd, Cu, Pb and Zn are found in all MSW compost (Table 1) (Verma *et al.* 2012; Gautam *et al.* 2010; Gillet 1992; Roghanian *et al.* 2012).

Table 1: Classification of MSW on the basis of waste component

Types of waste	Components
Putrescibles	Food waste (vegetables, meats scrap, egg shell, dairy product) and Yard waste (leaves and grasses pruning and trimming, branch and stems, crop residues, manure and other).
Paper	News paper, Office paper, Ledgers, Magazines, Books, Cardboard.
Plastics	Plastic containers, Bags, HDPE, PETE, Other composites.
LWTR	Leather, Wood, Textiles, Rubber, Threads, Yarns, Fabrics, Clothes, Other.
Ferrous metals	Tins, Steel cans, Iron rods, and other ferrous metals.
Non Ferrous	Aluminum foil, Aluminum cans, other.
Glass	Clear and Colored bottles, Glasses.
Inert	Stones, Rocks, Soils, Ash, Construction and Demolition waste
Miscellaneous	Nappies, Sanitary napkins and Others.

Source: (Gidarakos *et al.* 2010)

Nonhazardous solid waste treatment	• Open dumps, land fills, Sanitary land fills, Incineration, Plasma Arc Furnaces, Stabilization and Solidification, Underground injection, Heat drying, Wet oxidation and Composting
Hazardous solid waste treatment	•Physical treatment:- Gravity separation, Gravity floatation, Dissolved air floatation, air stripping, Steam stripping, Solvent extraction, Sorption on activated carbon •Chemical treatment:- Neutralization, Chemical precipitation, Solidification and stabilization, Oxidation, Ozonation, Thermal treatment. •Biological treatment

MSW is also divided in to two classes: nonhazardous and hazardous

Treatment of nonhazardous and hazardous solid waste: The major methods of waste management are: recycling-the recovery of materials from products after they have been used by consumers, composting- an aerobic, biological process of degradation of biodegradable organic matter, sewage treatment-process of treating raw sewage to produce non-toxic liquid effluent, incineration-a process of combustion designed to recover energy and reduce the volume of waste going to disposal, landfill- the deposition of waste in a specially designated area, which in modern sites consists of a pre-constructed 'cell' lined with an impermeable layer (man-made or natural) and with controls to minimize emissions (Rushton 2003; Awasthi *et al.* 2012).

In incineration ash may contain heavy metal and organic compound which are highly harmful for nature because in many countries, the major portion of the total solid waste is biodegradable organic matter and other inert material, results in high waste density and high moisture content. Wastes with a high water or inert content will have low calorific value and thus may not be suitable for incineration (Gary *et al.* 1971; Marugg *et al.* 1993; Boni and Musmeci 1998). This makes the recycling of organic waste as soil amendments a useful alternative to incineration, landfill or rubbish dumps (Hassan *et al.* 2002). In most countries, sanitary landfilling is nowadays the most common way to eliminate MSW. In spite of many advantages, generation of heavily polluted leachates, presenting significant variations in both volumetric flow and chemical composition, constitutes a major drawback (Renou *et al.* 2008).

The utilization of biowaste relieves stresses on the environment; but it should not be ignored that a variety of heavy metals and pathogens can be found in raw biowastes (Hassan *et al.* 2002). The amounts of heavy elements should not reach thresholds, which may damage either soil fertility (toxic effect against microorganisms, inhibition of mineralization and humification, perturbation of bio-geochemical cycles) or the food chain (heavy metals uptake by plants, crop species cultivated and ingested by the human or animals) (Gillet 1992; Roghanian *et al.* 2012).

2. Microorganisms present in municipal solid waste

Different microbial communities predominate during various composting stage. Microorganisms present in waste are either beneficial or pathogenic. Pathogenic microorganisms present in municipal and sewage sludge are listed by Nell *et al.* 1983; De Bertoldi *et al.* 1988; Strauch 1991. In waste material, the following fungi can be present *viz. Aspergillus niger, A. flavus, Aspergillus* sp., *A. ustus, Mucor* sp., *Alternaria alternata, Cladosporium* sp. (Ashraf *et al.*

2007) *Aspergillus glaeucus, Sarcinella* sp., (Roy *et al.* 2005; Verma *et al.* 2012), *Trichoderma, Alternaria, Penicillium, Ulocladium* and *Aspergillus* (Rebollido *et al.* 2008), *Chaetomium thermophilum, Trichoderma viride,* white-rot fungi, *Candida rugopelliculosa* (Gautam *et al.* 2010a).

Rebollido *et al.* (2008) isolated following groups of microorganisms in organic waste compost; Gram-negative aerobes of genera *Pseudomonas, Azotobacter, Azospirillum,* Gram-positive aerobes of genera *Micrococcus,* Gram-positive spore forming bacteria of genus *Bacillus,* Gram-positive bacteria of genera *Streptomyces* and *Actinomyces. Salmonella, Shigella, Escherichia coli, Enterobacter, Yersinia, Streptococci* and *Klebsiella* (Hassen 2001) and *Bacillus licheniformis, B. casei* and *Lactobacillus buchneri* (Gautam *et al.* 2010; Gautam *et al.* 2010a) can be present in compost. In thermophilic phase of composting Gram-positive cocci *Micrococcus, Planococcus* and *Staphylococcus* were isolated (Hassan *et al.* 2002).

The fungi, *Aspergillus fumigatus* is predominant because of its cellulolytic and thermotolerant properties (Hassan *et al.* 2002). *Aspergillus* sp. was most dominant genus represented by 5 isolates (Roy *et al.* 2005; Ashraf *et al.* 2007; Verma *et al.* 2012). It is important to mention that a large variety of mesophilic, thermotolerent, thermophilic aerobic microorganisms, including bacteria, actinomycetes, yeast and various other fungi have been extensively reported in composting (Amner *et al.* 1988; Faure and Deschanps 1991; Finstein and Morris 1975; Strom 1985).

3. Composting

The practice of producing organic fertilizer through the biological decomposition of organic wastes is known as composting (Tang *et al.* 2004; Khalil *et al.* 2001). It is dependent on many biotic and abiotic factors. During composting, the microorganisms use the organic matter as a food source. The process produces heat, carbon dioxide, water vapour and humus as a result of growth and activities of microorganisms (Pathak *et al.* 2012). In biological processes, aerobic microorganisms biodegrade or mineralize solid waste completely into C, H compounds and mineral salts (Gautam *et al.* 2011). For these reasons composting is a better option because it is a process in which microbial communities can multiply and mature. (Verma and Jamaluddin 2012a; Gautam *et al.* 2010; Hassen *et al.* 1998; Shyamala and Belagali 2012). Composting can decrease or eliminate the toxicity of MSW (Mondini *et al.* 2006).

3.1 Role of microorganisms in composting

There are several naturally occurring microorganisms that are able to convert organic waste into valuable resources such as plant nutrients, and reduce the C:N ratio to support soil productivity (Pan *et al.* 2012). These microorganisms are also important to maintain nutrient flows from one system to another and to minimize ecological imbalance (Novinsak *et al.* 2008; Umsakul *et al.* 2010). Bacteria and fungi are frequent microbes in soil, manure and decaying plant tissues that are able to degrade domestic wastes and their distribution patterns are correlated with the substrate organic matter (Alexander 1961). But fungi play important role because they use many carbon sources mainly lignocellulosic polymers and can survive in extreme conditions (Miller 1996). For utilization of carbon source fungi secrete many enzymes that are responsible for breakdown of the large molecules like cellulose, pectin, keratin, collagen, lignin, chitin, casein and starch to small and useful compounds (Brindha *et al.* 2011; Goldbeck *et al.* 2012). Fungi have two types of extracellular enzymatic systems; the hydrolytic system, which produces hydrolases that are responsible for polysaccharide degradation and a unique oxidative and extracellular ligninolytic system, which degrades lignin and opens phenyl rings (Sanchez 2009). Organic residues from wood, grass, agricultural and forestry wastes and municipal solid wastes are particularly abundant in nature and have a potential for bioconversion. *Thermomonospora curvata* isolated from MSW secreting cellulase was responsible for this (Stutzenberger 1971).

Many species of fungi have been reported to produce the enzyme cellulase and help the decomposition process, such as *Strobilurus ohshimae* (Homma *et al.* 2007), *Phanerochaete chrysosporium* (Kersten and Cullen 2007), *Trametes versicolor, Pleurotus ostreatus* (Quintero *et al.* 2006), *P. ostreatus, P.pulmonarius* (Marnyye *et al.* 2002), *Aspergillus niger* (Park *et al.* 2002), *Bjerkandera adusta* (Kimura *et al.* 1991), *Fusarium merismoides* (Fernandez-Martin *et al.* 2007) *Streptomyces, Penicillium* sp., *Clonostachys rosea, Fusarium oxysporum* (Mikan and Castellanos 2004), *Pycnoporus cinnabarinus* (Alves *et al.* 2004), *Xylaria hypoxylon* (Liers *et al.* 2006), *X. polymorpha* (Xing-Na *et al.* 2005), *Fomitopsis palustris* (Yoon *et al.* 2007). *Phanerochaete chrysosporium, Botrytis cinerea, Stropharia coronilla, Sclerotium rolfsii, Aspergillus niger, Achlya bisexuals, Orpinomyces* sp., *Rhizopus chinensis, Penicillium brefeldianum* show highest specific activity for cellulose and *Aspergillus niger,Trichoderma longibrachiatum, Aspergillus nidulans, Aspergillus niger, Sclerotium rolfsii, Phanerochaete chrysosporium, Schizophyllum* show activity for hemicellulose (Howard *et al.* 2003).

3.2 Maturation of organic waste

The degree of biodegradability of urban solid waste in the environment depends largely on the type, nature of macromolecules and bond type present in the waste. It is also important to differentiate between organic material of biological origin and synthetic materials (Carboo and Fobil 2005). Maturation of organic waste is considered by many physical and chemical parameters like pH, moisture content, ash, electric conductivity, weight loss, organic carbon and nitrogen concentration, C/N ratio, phosphorus, sodium, potassium, calcium and magnesium concentration. It also includes heavy metal concentration and number of pathogenic microorganisms. The changes in chemical parameters C:N ratio is the main important parameter that determines the quality of compost and degree of compost maturity. In this process C:N ratio decreases in the final product as nitrogen remains in the system while some of carbon is released as carbon dioxide (Gautam *et al*. 2010). Further nitrogen fixing microbes indirectly help in decreasing C: N ratio by making more nitrogen available from added organic matter (Gautam *et al*. 2010).

3.3 Elimination of pathogenic microorganisms

Several authors have quantified the pathogens of concern during composting (Nell *et al*. 1983; De Bertoldi 1988; Strauch 1991). In contrast, MSW composting has received little attention. Pathogens found in MSW, includes viruses, bacteria, protozoa or helminths. As they are heat-sensitive, the heat increase occurring during the composting process should eliminate them. Several parameters (e.g. humidity and oxygen) have an influence on the heat increase.

3.4 Elimination of heavy metals

Heavy metals commonly present in MSW are B, Cu, Cr, Co, Cd, Fe, Mn, Hg, Ni, Al, Pb and Zn (Gautam *et al*. 2010; Verma *et al*. 2012; Turner *et al*. 1994; Shyamala and Belagali 2012; Shirbhate and Malode 2012; Carboo and Fobil 2005; Flyhammar 1997). Application of MSW compost rich in heavy metals to agricultural soils may cause heavy metal accumulation to toxic levels (King *et al*. 1990). The concentration of heavy metals in MSW compost showed values significantly lower than standard values prescribed by Central Pollution Control Board. This clearly brought out that composting decreases the concentration of heavy metals. Reason for decreased concentration of heavy metals during composting is because microbial cells can accumulate heavy metals, radionuclides and organometalloid compounds by physicochemical and biological processes.

Different parts of cell through which microorganisms can uptake heavy metals are:

1. *Cell membrane/ periplasmic space*: Adsorption, Ion exchange, Redox reaction, Transformation, Precipitation, Diffusion and transport (influx and efflux).

2. *Cell wall*: Adsorption, Ion exchange and covalent bonding, Entrapment of particles, Precipitation.

3. *Cell associated materials*: Ion exchange, Particulate entrapment, Precipitation, Non-specific bonding.

4. *Extracellular reaction*: Precipitation with excreted products (e.g. oxalate, Siderophores.

5. *Intracellular*: Redox reaction, Transformation, Non-specific bonding, Sequestration.

On the basis of toxicity heavy metals can be considered at different levels.

Fe, Mo, Mn	Low toxicity
Zn, Ni, Cu, V, Co, W, Cr	Average toxicity
As, Ag, Sb, Cd, Hg, Pb, U	High toxicity

(*Source:* Thakur, 2006).

In fungi chitin present in the cell wall has been found to be an effective metal and radionuclide biosorbent (Tsezos 1983). Chitosan and other chitin derivatives also have a significant biosorptive capability (Wales and Sagar 1990). Biosorption is mainly studied in uranium and many fungi such as *Saccharamyces cerevisiase* (Strandberg *et al.* 1981), *Trichoderma harzianum* (Gadd and Fomina 2011), white rot fungus *Lentinus sajor-caju* (Bayramoglu *et al.* 2006), immobilized and dried powdered fungal mycelia of *Trameter versicolour* and *Phanerochaete chrysosporium* (Genc *et al.* 2003), *Penicillium digitatum* (Galun *et al.* 1987), *Cladosporium oxysporum, Aspergillus flavus, Curvularia clavata* (Mishra *et al.* 2009), *Aspergillus ochraceus* and *Penicillium funiculosum* were able to make up large amounts of uranium from rocks in their mycelium (Gadd and Fomina 2011). U uptake by *Mucor mehei* at pH 3 was considerable, but was increased two times at pH 4 and three times at pH 5 (Guibal *et al.* 1992). Fomina *et al.* (2007) showed that saprotrophic ericoid and ectomycorrhizal fungi, *Hymenoscyphus ericae* and *Rhizopogon* were also responsible for absorption.

Arbuscular mycorrhizal fungal (AMF) vesicles accumulate Mn, Cu, Ni, U and Fe. It was also observed that AMF can transfer Cd, Zn, Ni and radionuclides such as Cs, Se or U (Gadd and Fomina 2011). Rufyikiri *et al*. (2003) showed that U concentration was 5-10 times higher in AMF hyphae than in non-mycorrhizal roots. *Glomus mosseae* and *G. intraradices* are also known to accumulate heavy metals (Chen *et al*. 2006).

Lichens directly grow on uranium minerals and concentrate uranium within the thallus. *Trapelia innoluta, Peltigera membranacea* (Griffiths and Greenwood 1972), *Cladonia rangiferina, Cladonia mitis* (Gadd and Fomina 2011) and epiphytic lichens *Hypogymnia physodes* (Golubev *et al*. 2005), *Parmoterma tinctorum* (Ohnuki *et al*. 2004) are known to accumulate uranium. High concentration of mucopolysaccharides and P recorded inside the melanized apothecia may protect the reproductive tissue from toxic effects of metals (Purvis *et al*. 2004).

Conclusion

- Recent global problems of food shortage have been caused by rising cost of chemical fertilizers. Composting at low cost has been re-evaluated as an important alternative fertilizer production method. High-quality compost is produced by interaction of many organisms that have suitable properties for the composting process.

- Composting brings down the concentration of toxic substances particularly heavy metals and other xenobiotics.

- The use of fungi in low cost bioremediation projects might be attractive.

References

Alguacil MD, Kohler J, Caravaca F, Roldan A (2009) Differential effects of *Pseudomonas mendocina* and *Glomus intraradices* on lettuce plants physiological response and aquaporin PIP2 gene expression under elevated atmospheric CO_2 and drought. Microbial Ecol. 58, 942-951.

Alexander M (1961) Introduction to Soil Microbiology. Wiley, New York.

Alves MCRA, Record E, Lomascolo A, Scholtmeijer K, Asther M, Wessels GHJ (2004) Highly efficient production of laccase by the basidiomycete *Pycnoporus cinnabarinus*. Appl. Environ. Microbiol. 70, 6379–6384.

Amner W, McCarthy AJ, Edwards C (1988) Quantitative assessment of factor affecting the recovery of indigenous and release thermophilic bacteria from compost. Appl. Environ. Microbiol. 54, 3107-3112.

Ashraf R, Shahid F, Ali TA (2007) Association of fungi, bacteria and actinomycetes with different composts. Pakistan J. Bot. 39(6), 2141-2151.

Awasthi MK, Bundela PS, Pandey AK, Jamaluddin (2012) Monitoring of microbial population and their activities during composting of organic municipal solid wastes at central India. Int. J. Plant Animal Environ. Sci. 2(2), 26-36.

Barlaz MA, Rooker AP, Kjeldsen P, Gabr MA, Borden RC (2002) A critical evaluation of factors required to terminate the post-closure monitoring period at solid waste landfills. Environ. Sci. Tech. 36, 3457–3464.

Bayramoglu G, Celik G, Yakup Arica M (2006) Studies on accumulation of uranium by fungus *Lentinus sajor-caju*. J. Hazard. Mater. 136, 345-353.

Boni MR, Musmeci L (1998) Organic fraction of municipal solid waste (OFMSW): extent of biodegradation. Waste Manage. Res. 16 (2), 103-107.

Brindha JR, Mohan ST, Immanual G, Jeeva GS, Lekshmi NCJP (2011) Studies on amylase and cellulose enzyme activity of the fungal organism causing spoilage in tomato. Eur. J. Exp Biol. 1(3), 90-96.

Bundela PS,Gautam SP, Pandey AK, Awasthi MK, Sarsaiya S (2010) Municipal solid waste management in Indian cities: a review. Int. J. Environ. Sci. 1(4), 591-606.

Carboo D, Fobil JN (2005) Physico chemical analysis of municipal solid waste (MSW) in the Accra metropolis. West Afr. J. Appl. Ecol. 7, 31-39.

Castaldi P, Santona L, Melis P (2005) Evolution of heavy metals mobility during municipal solid waste composting. International Symposium on Environmental Pollution and its Impact on Life in the Mediterranean Region (MESAEP), pp 1133-1140.

Chen BD, Zhu YG, Smith FA (2006) Effects of arbuscular mycorrhizal inoculation on uranium and arsenic accumulation by Chinese brake fern (*Pteris vittata* L.) from a uranium mining impacted soil. Chemosphere 62, 1464-1473.

De Bertoldi M, Zucconi F, Civilini M (1988) Temperature, pathogen control and product quality. Biocycle 29(2), 43–47.

Erses S, Onay TT (2003) In situ heavy metal attenuation in landfills under landfills to accelerate completion. Waste Manage. 28, 1039-1048.

Erses S, Onay TT, Yenigun O (2008) Comparison of aerobic and anaerobic degradation of municipal solid waste in bioreactor landfills. Bioresour. Technol. 99, 5418–5426.

Faure D, Deschamps AM (1991) The effect of bacterial inoculation on the initiation of composting grape pulps. Bioresour. Technol. 37, 235-238.

Fernandez-Martin R, Domenech C, Cerda-Olmedo E, Avalos J (2007) Ent-Kaurene and squalene synthesis in *Fusarium fujikuroi* cell-free extracts. Phytochemistry 54, 723–728.

Finstein MS, Morris ML (1975) Microbiology of municipal solid waste composting. Adv. Appl. Microbiol. 19, 113-153.

Flyhammar P (1997) Estimation of heavy metal transformations in municipal solid waste. Sci. Total Environ. 198, 123-133.

Fomina M, Charnock JM, Hillier S, Alvarez R, Gadd GM (2007) Fungal transformations of uranium oxides. Environ Microbiol. 9, 1696-1710.

Gadd GM, Fomina M (2011) Uranium and fungi. Geomicrobiol J. 28(5-6), 471-482.

Galun M, Siegel SM, Cannon MI, Barbara Z, Siegel BZ, Galun E (1987) Ultrastructure localization of uranium biosorption in *Penicillium digitatum* by stem X-ray microanalysis. Environ. Pollut. 43, 209-218.

Gautam SP, Bundela PS, Pandey AK, Awasthi MK, Sarsaiya S (2010) Physicochemical analysis of municipal solid waste compost. J. Appl. Sci. Res. 6(8), 1034-1039.

Gautam SP, Bundela PS, Pandey AK, Awasthi MK, Sarsaiya S (2010a) Effect of different carbon sources on production of cellulases by *Aspergillus niger*. J. Appl. Sci. Environ. Sanitation 5(3), 277-281.

Gautam SP, Bundela PS, Pandey AK, Awasthi MK, Sarsaiya S (2010b) Composting of municipal solid waste of Jabalpur city. Global J. Environ. Res. 4(1), 43-46.

Gautam SP, Bundela PS, Pandey AK, Awasthi MK, Sarsaiya S (2011) Isolation, identification and cultural optimization of indigenous fungal isolates as a potential bioconversion agent of municipal solid waste. Ann. Environ. Sci. 5, 23-34.

Genc O, Yalcinkaya Y, Buyuktuncel E, Denizli A, Arica MY, Bektas S (2003) Uranium recovery by immobilized and dried powdered biomass: characterization and comparison. Int. J. Mineral Process. 68, 93-107.

Gidarakos E, Havas G, Ntzamalis P (2010) Municipal solid waste composition determination supporting integrated solid waste management system in Crete. Waste Manage. 26, 668-79.

Gillett JW (1992) Issues in risks assessment of compost from municipal solid waste: occupational health and safety, public health and environmental concerns. Biomass Bioenergy 3, 145–162.

Goldbeck R, Andrade CCP, Pereira GAG, Filho FM (2012) Screening and identification of cellulose producing yeast like microorganism from Brazilian biomes. Afr. J. Biotechnol. 11(53), 11595-11603.

Golubev AV, Golubeva VN, Krylov NG, Kuznetsova VF, Mavrin SV, Aleinikov AY, Hoppes WG, Surano KA (2005) On monitoring anthropogenic airborne uranium concentration and $^{235}U/^{238}U$ isotopic ratio by Lichen bio indicator technique. J. Environ. Radioactivity 84, 333-342.

Gray KR, Sherman K, Biddlestone AJ (1971) A review of composting. Part 1 Microbiology and Bio-chemistry. Process Biochem. 6(6), 32-36.

Griffiths HB and Greenwood AD (1972) The concentric bodies of lichenized fungi. Arch. Microbiol. 87, 285-302.

Guibal E, Roulph C, Le Cloirec P (1992) Uranium biosorption by a filamentous fungus *Mucor miehei* pH effect on mechanisms and performances of uptake. Water Res., 26, 1139-1145.

Hassen A, Belguith K, Jedid N, Cherif A, Cherif M, Boudabus A (2001) Microbial characterization during composting of municipal solid waste. Bioresour. Technol. 80, 217-225.

Hassen A, Belguith K, Jedidi N, Cherif M, Boudabous A (2002) Microbial characterization during composting of municipal solid waste. Proceedings of International Symposium on Environmental Pollution Control and Waste Management 7-10 January 2002, Tunis (EPCOWM'2002), pp 357-368.

Hassen A, Jedidi N, Cherif M, Mhiri A, Boudabous A, Cleemput OV (1998) Mineralization of nitrogen in a clay loamy soil amended with organic waste residues enriched with Zn, Cu and Cd. Bioresour. Technol. 64, 39-45.

Homma H, Shinoyama H, Nobuta Y, Terashima Y, Amachi S, Fujii T (2007) Lignin-degrading activity of edible mushroom *Strobilurus ohshimae* that forms fruiting bodies on buried soil (Cryptomeria japonica) twigs. J. Wood Sci. 53, 80–84.

Hoornweg D, Thomas L (1999) What a waste: Solid waste management in Asia. Washington, DC, USA: World Bank Urban waste management working paper series no.1.

Howard RL, Abotsi E, Jansen van Rensburg EL, Howard S (2003) Lignocellulose biotechnology: issues of bioconversion and enzyme production. Afr. J. Biotechnol. 2, 602–19.

Kersten P, Cullen D (2007) Extracellular oxidative systems of the lignin-degrading Basidiomycete *Phanerochaete chrysosporium*. Forest Genet. Biol. 44, 77–87.

Khalil AI, Beheary MS, Salem EM (2001) Monitoring of microbial populations and their cellulolytic activities during the composting of municipal solid wastes. World J. Microbiol. Biotechnol. 17, 155–161.

Kim H (2005) Comparative studies of aerobic and anaerobic landfills using simulated landfill lysimeters. Ph.D. Thesis, University of Florida, USA.

Kimura Y, Asada Y, Oka T, Kuwahara M (1991) Molecular analysis of a *Bjerkandera adusta* lignin peroxidase gene. Appl. Microbiol. Biotechnol. 35, 510–514.

King LD, Burns JC, Westerman PW (1990) Longterm swine lagoon effluent applications on 'Coastal' bermudagrass: Effect on nutrient accumulation in soil. J. Environ. Res.19, 756-760.

Liers C, Ullrich R, Steffen KT, Hatakka A, Hofrichter M (2006) Mineralization of 14C-labelled synthetic lignin and extracellular enzyme activities of the wood-colonizing ascomycetes *Xylaria hypoxylon* and *Xylaria polymorpha*. Appl. Microbiol. Biotechnol. 69, 573–579.

Mahar A, Malik RN, Qadir A, Ahmed T, Khan T, Khan MA (2007) Review and Analysis of Current Solid Waste Management Situation in Urban Areas of Pakistan, In : Proceedings of the International Conference on Sustainable Solid Waste Management, 5-7 September, Chennai, India. pp 34-41.

Marnyye A, Velasquez C, Mata G, Michel SJ (2002) Waste-reducing cultivation of *Pleurotus ostreatus* and *Pleurotus pulmonarius* on coffee pulp: changes in the production of some lignocellulolytic enzymes. World J. Microbiol. Biotechnol. 18, 201–207.

Marugg C, Gerbus M, Hansen RC, Keener HM, Hoitink HAJ (1993) A kinetic model of the yard waste composting process. Compost Sci. Util. 10, 38-51.

Metcalf E, Eddy H (1991) Waste water engineering: treatment, disposal and methanogenic conditions. J. Hazard. Mater. 99, 159-175.

Mikan VJF, Castellanos SDE (2004) Screening for isolation and characterisation of microorganisms and enzymes with useful potential for degradation of cellulose and hemicellulose. Rev Colomb. Biotechnol. 6, 58–67.

Miller FC (1996) Composting of municipal slide waste and its components. In: Microbiology of solid waste. (Ed) Palmisano AC, Barlaz MA, CRS Press, pp 115-154.

Mishra A, Pradhan N, Kar RN, Sukla LB, Mishra BK (2009) Microbial recovery of uranium using native fungal strains. Hydrometallurgy 95, 175-177.

Mondini C, Sanchez-Monedero MA, Sinicco T, Letia L (2006) Evolution of extract organic carbon and microbial biomass as stability parameters in Lingo-cellulosic waste composts. J. Environ. Qual. 35, 2313-2320.

Moreno AD, Rodriguez MG, Velasco AR, Enriquez JCM, Lara RG, Gutierrez AM, Hernandez NAD (2013) Mexico city's municipal solid waste characteristics and composition analysis. Revista Internacional de Contaminacion Ambiental 29(1), 39-46.

Murphy RJ, Jones DE, Stesse RI (1995) Relationship of microbial mass and activity in biodegradation of solid waste. Waste Manage. Res. 13, 485-497.

Nell JH, Steer AG, Van Rensburg PAJ (1983) Hygienic quality of sewage sludge compost. Water Sci. Technol. 15, 181–194.

Novinsak A, Surette C, Allain C, Filion M (2008) Application of molecular technologies to monitor the microbial content of biosolids and composted biosolids. Water Sci. Technol. 57, 471–477.

Ohnuki T, Aoyagi H, Kitatsuji Y, Samadfam M, Kimura Y and Purvis OW (2004) Plutonium (VI) accumulation and reduction by lichen biomass: correlation with U (VI). J Environ. Radioactivity, 77, 339-353.

Pan I, Dam B, Sen SK (2012) Composting of common organic wastes using microbial inoculants. 3 Biotech. 2, 127–134.

Park YS, Kang SW, Lee JS, Hong SI, Kim SW (2002) Xylanase production in solid state fermentation by *Aspergillus niger* mutant using statistical experimental design. Appl Microbiol. Biotechnol. 58, 762–6.

Pathak AK, Singh MM, Kumara V, Arya S, Trivedi AK (2012) Assessment of physico- chemical properties and microbial community during composting of municipal solid waste (Viz. Kitchen waste) at Jhansi City, U.P. (India). Recent Res. Sci. Technol. 4(4), 10-14.

Purvis OW, Bailey EH, McLean J, Kasama T, Williamson BJ (2004) Uranium biosorption by the lichen *Trapelia involuta* at a uranium mine. Geomicrobiol. J. 21, 159-167.

Quintero DJC, Gumersindo FEJOOC, Lemar RJM (2006) Production of ligninolytic enzymes from basidiomycete fungi on lignocellulosic materials. Rev. Facult. Quim. Farmaceut. 13, 61–7.

Rebollido R, Martinez J, Aguilera Y, Melchor K, Koerner I, Stegmann R (2008) Microbial populations during composting process of organic fraction of municipal solid waste. Appl. Ecol. Environ. Res. 6(3), 61-67.

Renou S, Givaudan JG, Poulain S, Dirassouyan F, Moulin P (2008) Landfill leachate treatment: review and opportunity. J. Hazard. Mater. 150, 468–493.

Rich C, Gronow J, Voulvoulis N (2008) The potential for aeration of MSW Reuse. 3rd Edn, McGraw Hill Publishing Co., New York, USA.

Roghanian S, Hosseini HM, Savaghebi G , Halajian L, Jamei M, Etesami H (2012) Effects of composted municipal waste and its leachate on some soil chemical properties and corn plant responses. Int. J. Agric. Res. Rev. 2(6), 801-814.

Roy T, Pandey AK (2005) Proteolytic potential of some fungi isolated from slaughter house at Jabalpur (M.P.). J. Basic Appl. Mycol. 4, 161-164.

Rufyikiri G, Thiry Y, Declerck S (2003) Contribution of hyphae and roots to uranium uptake and translocation by arbuscular mycorrhizal carrot roots under root organ culture condition. New Phytol.156, 275-281.

Rushton L (2003) Health hazards and waste management. British Medical Bulletin 68, 183-197.

Sahu AK (2007) Present scenario of municipal solid waste (MSW) dumping grounds in India. In: Proc. Intl. Con. Sustainable Solid Waste Management, 5-7 September 2007, Chennai, India, pp 327-333.

Sanchez C (2009) Lignocellulosic residues: biodegradation and bioconversion by fungi. Biotechnol. Adv. 27, 185–194.

Saviozzi A, Levi- Minzi R and Raffaldi R (1988) Maturity evolution of organic waste. Biocycle 29(3), 54-56.

Shirbhate N, Malode SN (2012) Municipal solid waste management: a survey and physicochemical analysis of contaminated soil from sukali compost and landfill depot, batkuli road, Amravati. Global J. Biosci. Biotechnol. 2, 215-219.

Shyamala DC, Belagali SL (2012) Studies on variations in physico-chemical and biological characteristics at different maturity stages of municipal solid waste compost. Int. J. Environ. Sci. 2(4), 1984-1997.

Soumare M, Demeyer A, Tack FMG, Verloo MG (2002) Chemical characteristics of Malian and Belgian solid waste composts. Bioresour. Technol. 81, 97-101.

Strandberg GW, Shumate SE, Parrott JR (1981) Microbial cells as biosorbents for heavy metals-accumulation of uranium by and *Pseudomonas aeruginosa*. Appl. Environ. Microbiol. 41, 237-245.

Strauch D (1991) Survival of pathogenic micro-organism and parasites in excreta, manure and sewage sludge. Rev. Sci. Tech. OIE 10, 813–846.

Strom PF (1985) Effect of temperature on bacterial diversity in thermophillic solid waste composting. Appl. Environ. Technol. 50, 906-913.

Stutzenberger FJ (1971) Cellulase production by *Thermomonospora curvata* isolated from municipal solid waste. Appl. Microbiol. 22(2), 147-152.

Taiwo LB, Oso BA (2004) Influence of composting techniques on microbial succession, temperature and pH in a composting municipal solid waste. Afr. J. Biotechnol. 3(4), 239-243.

Tang JC, Inoue Y, Yasuta T, Yoshida S, Katayama A (2003) Chemical and microbial properties of various compost products. J. Soil Sci. Plant Nutr. 49, 273–280.

Thakur IS (2006) Bioabsorption of metals. In: Environmental Biotechnology (Ed) Makhijani K, I.K. International Pvt. Ltd. pp 472.

Tsezos M (1983) The role of chitin in uranium adsorption by *Rhizopus arrhizus*. Biotechnol. Bioeng. 25, 2025-2040.

Turner MS, Clarck GA, Standey CD, Smaystrla AG (1994) Physical characterization of sandy soil amended with municipal solid waste compost soil. Crop. Sci. Soc. Florida Proc. 53, 24-26.

Umsakul K, Dissara Y, Srimuang N (2010) Chemical physical and microbiological changes during composting of the water hyacinth. Pakistan J. Bio. Sci. 13, 985–992.

Verma P and Jamaluddin (2012a) A study on correlation between different physicals parameter after composting, which change after composting. Biomed. Pharmacol. J. 5(2), 1-6.

Verma P and Jamaluddin (2012b) A study on correlation between different chemical parameter after composting. Golden Res. Thoughts 2(3), 1-6.

Verma P, Chandrakar V, Jaiswal M and Jamaluddin (2012) Bioconversion of municipal solid waste and analysis of many physicochemical parameters. J. Trop. Fores. 28(3), 19-24.

Wales DS, Sagar BF (1990) Recovery of metal ions by microfungal filters. J. Chem. Technol. Biotechnol. 49, 345-356.

Warith MA (2003) Solid waste management: new trends in landfill design Emirates. J. Eng. Res. 8(1), 61-70.

Xing-NaW, Ren-Xiang T and Ji-Kai L (2005) Xylactam, a new nitrogen-containing compound from the fruiting bodies of ascomycete *Xylaria euglossa*. J. Antibiotics, 58, 268–270.

Yoon JJ, Cha CJ, KimYS, Son DW and KimYK (2007) The brown-rot basidiomycete *Fomitopsis palustris* has the endo-glucanases capable of degrading microcrystalline cellulose. J. Microbiol. Biotechnol. 5, 800–805.

4

Microorganisms for Abiotic Stress Management in Crop Plants: Recent Developments in India

Minakshi Grover, Venkateswarlu B., Desai S., Yadav S.K.
Maheswari M. and Srinivasa Rao Ch

Abstract

Impact of climate change and related stresses on plant productivity are being witnessed worldwide with developing countries being more vulnerable due to more dependency on agriculture. Microorganisms can play important role as economic and farmer's friendly strategy to combat abiotic stress in agriculture. A wide range of microorganisms associate with plants, and many of them confer beneficial effects to their hosts through different mechanisms like production of growth promoting substances, nutrient acquisition, protection against plant pathogens etc. In the last few decades, role of microorganisms in imparting tolerance to plants against different abiotic stress has been reported in many crop plants. Under stresses conditions, plant-microbe interactions become more common towards coping with the stress. Selection and application of efficient abiotic stress tolerant microorganisms can be a fruitful strategy to alleviate negative effects of abiotic stresses in plants.

Keywords: Bioinoculants, Climate change, Drought, Heat, Stress tolerance

1. Introduction

Plants including agricultural crops are exposed to a plethora of biotic and abiotic stresses during their growth cycle. With the change in global climate pattern and anthropogenic activities incidences of environmental stresses are on increasing trend causing heavy losses to agricultural production worldwide, thus, posing threat to food security. Agriculture sector in India provides food and livelihood activities to much of the population and is the major contributor to the country's economy. Rainfed agriculture in India (nearly 55% of the net sown area, supports nearly 40% of India's population and meets approximately 44% food demand) is considered more vulnerable to climate change due to uncertainty of rainfall, increasing frequency of droughts, extreme events such as heat waves, floods, hail storms etc. (Srinivasarao *et al.* 2015). Various abiotic stresses are expected to impact on agricultural productivity and crop patterns. Under such conditions, it becomes very challenging to feed ever-increasing population. Globally, different strategies are being used to develop measures to combat environmental stresses in agricultural crops. Plant breeding has been conventionally used to develop plants with desirable traits. Genetic engineering is the advanced approach to develop transgenic plants expressing one or more genes from other organism/s for abiotic stress tolerance. Genetic improvement of crop plants for better adaptability to different stress is expensive and time consuming. Further it is unlikely to genetically improve all the crop plants for different types of stresses. Therefore, alternate strategies that are quick, cheap and environmental friendly are needed to address the issue. Microorganisms can play a very important role in this direction. A wide range of microorganisms associate with plants, and many of them confer beneficial effects to their hosts through different mechanisms like production of growth promoting substances, nutrient acquisition, protection against plant pathogens etc. These micro-organisms include those inhabit rhizosphere, rhizoplane, phyllosphere, phylloplane, endophytes and symbiotic that operate through a variety of mechanisms (Grover *et al.* 2011). As plants share intimate relations with microorganisms, climate change and related stresses that influence plants are also bound to impact plant-microbe interactions (Grover *et al.* 2015).

Recently, role of microorganisms in imparting tolerance to plants against different abiotic stresses has been reported in many crop plants. Under stress, plants or microorganisms need a stronger network to survive in a harsher environment, so their interactions become more common towards coping with the stress conditions. The term induced systemic tolerance has been used for microbial mediated abiotic stress tolerance in plants (Yang *et al.* 2009). Reports have accumulated on plant growth promotion by microorganisms under different abiotic stress conditions like drought, high and low temperature, salinity, flooding, nutrient deficiency (Kohler *et al.* 2009; Grover *et al.* 2011; Selvakumar *et al.* 2012). Bacteria belonging to different genera including *Rhizobium, Bacillus, Pseudomonas, Pantoea, Paenibacillus, Burkholderia, Achromobacter, Azospirillum, Microbacterium, Methylobacterium, Variovorax, Enterobacter* etc. have been reported to provide tolerance to host plants under different abiotic stress environments. Selection and application of efficient abiotic stress tolerant microorganisms can be a fruitful strategy to alleviate negative effects of abiotic stresses in plants.

2. Mechanisms behind microbial induced abiotic stress tolerance in plants

A variety of mechanisms have been proposed behind microbial elicited stress tolerance in plants (Grover et al. 2011). Microorganisms are known to produce phytohormones, solubilize and mobilize nutrients, compete with plant pathogens for nutrients and space, and antagonize plant pathogens. These properties directly or indirectly help plants grow better under normal as well as stress conditions. Production of stimulatory phytohormones like indole acetic acid, gibberellins and some unknown determinants by PGPR, result in increased root length, root surface area and number of root tips, that enhance uptake of nutrients resulting in improved plant health under abiotic stress conditions (Egamberdieva and Kucharova 2009). Production of cytokinin and antioxidants by microorganisms result in abscisic acid (ABA) accumulation and degradation of reactive oxygen species. Many aspects of plant life are regulated by ethylene levels and under stress conditions the plant hormone ethylene endogenously regulates plant homoeostasis resulting in reduced root and shoot growth. However, ACC deaminase producing bacteria can reduce the effect of stress ethylene by sequestering and degrading plant ACC (which is an immediate precursor for ethylene production) to get nitrogen and energy. Thus inoculation with ACC deaminase producing bacteria can help in ameliorating plant stress and promoting plant growth (Glick et al. 2007). The role of ACC deaminase producing PGPR in stress agriculture has been reviewed by Saleem et al. (2007). Inoculation with ACC deaminase containing bacteria induce longer roots which might increase water use efficiency of the plant due to enhanced uptake of water from deep soil (Zahir et al. 2008). Certain microorganisms produce exopolysaccharides which can bind soil particles to form microaggregates and macroaggregates. Plant roots and fungal hyphae fit in the pores between microaggregates and thus stabilize macroaggregates. Application of EPS producing microorganisms can help improve the soil structure thus improving water and nutrient retention (Sandhya et al. 2009). EPS can also bind to cations including Na^+ thus making it unavailable to plants under saline conditions. Accumulation of osmoprotectants has been correlated with abiotic stress tolerance in plants. Proline as a compatible solute helps in maintaining osmotic turgor, stabilizes macromolecules, acts as a sink of carbon and nitrogen for use after relief of stress, help in free radical detoxification etc. (Mohammadkhani and Heidari 2008). Transgenic plants with proline over-accumulation are more tolerant to abiotic stresses (Verdoy *et al*. 2006). Similarly *Rhizobium* mediated trehalose accumulation has been related with abiotic stress tolerance in legumes (Figueiredo *et al*. 2008). Volatiles organic compounds (VOCs) like 2R, 3R-butanediol, salicylic acid (SA), and jasmonic acid emitted by microorganisms are reported to be involved in induced systemic tolerance (IST) (Cho *et al*. 2008). Further role of RNA chaperones have been reported in abiotic stress tolerance. The expression of bacterial cold shock proteins (CSPs) (Csp A and Csp B) improved tolerance of transgenic rice, maize and arabidopsis plants to a number of abiotic stresses including cold, heat and water deficit resulting in improved yields under field conditions (Castiglioni *et al*. 2008). The plant-microbe interactions under stress conditions need to be studied at molecular levels for better understanding and applications.

3. Microbes for management of abiotic stress: recent developments in India

The field of microbial management of abiotic stress is relatively new. Timmusk and Wagner (1999) were among the firsts to report on this subject. They observed that inoculation of *Paenibacillus polymyxa* confers drought tolerance in *Arabidopsis thaliana* through the induction of drought responsive gene. In the last two decades, many reports have been published emphasizing role of bacteria, fungi and virus in conferring tolerance against different types of abiotic stresses in many crop plants. In India, Indian Council of Agricultural Research (ICAR) launched a network project on Application of Microorganisms in Agriculture and Allied Sectors (AMAAS) in August 2006. Under AMAAS network project, research projects on role of microorganisms in management of drought, heat, cold and salinity stress in crop plants were started at different centers under the theme "Microbial management of abiotic stress". Significant progress was made under this theme during the 11th plan, as many abiotic stress tolerant microorganisms were isolated and identified and their role in alleviating effects of abiotic stress/es on different crop plants was studied. Work on microbial management of abiotic stresses continued in 12th five year plan under AMAAS project and under ICAR-National initiative on Climate Resilient Agriculture (NICRA). In the last decade many reports have been published from India on role of microorganisms in the management of abiotic stress in crop plants (Table 1). The subject has been extensively reviewed by Indian workers in the recent years (Venkateswarlu *et al*. 2008; Grover *et al*. 2011; Selvakumar *et al*. 2012; Arora *et al*. 2012; Tewari and Arora 2013). Among the first reports on abiotic stress management in plants using microorganisms are by Saravanakumar and Samiyappan (2007), who reported that inoculation with ACC deaminase producing *Pseudomonas fluorescens* could protect the groundnut plants from salinity stress. Recently, Bharti *et al*. (2014) reported role of three salt-tolerant PGPR, *Bacillus pumilus*, *Halomonas desiderata* and *Exiguobacterium oxidotolerans* in augmenting salt tolerance in medicinal plant *Mentha arvensis* (menthol) through improved foliar nutrient uptake and enhanced antioxidant machinery.

Under AMAAS project, substantial work was carried out on isolation and characterization of heat and drought tolerant *Pseudomonas* and *Bacillus* spp. and their role in alleviating heat and drought stress effects in rainfed crops. Srivastava *et al*. (2008) isolated a thermotolerant *Pseudomonas putida* strain NBR10987 from drought stressed rhizosphere of chickpea. The strain showed over expression of stress sigma factor (RpoS) and enhanced biofilm formation at high temperatures which might have contributed towards the thermotolerance. Sandhya *et al*. 2009 reported the role of exopolysacchride producing *Pseudomonas putida* strain GAP-45 in alleviating drought stress effects in sunflower seedlings. Simultaneously, Ali *et al*. (2009) reported the role of a thermotolerant plant growth-promoting *Pseudomonas aeruginosa* strain AKM-P6 in alleviating heat stress effects in sorghum through induction of high molecular weight proteins. The inoculation resulted in reduced membrane injury, and improved the levels of cellular metabolites like proline, chlorophyll, sugars, amino acids, and proteins. The results indicated that inoculation with AKM-P6 could enhance tolerance of sorghum seedlings to elevated temperatures by inducing physiological and biochemical changes in the plant (Ali *et al*. 2009). Subsequently, Ali *et al*. (2011) reported positive effect of plant growth promoting thermotolerant *Pseudomonas putida* strain AKMP7 inoculation on the growth of wheat plants under heat stress. Sandhya *et al*. (2010, 2011) reported role of *Bacillus* and *Pseudomonas* spp. in alleviating effects of drought stress in maize seedlings. Bacterial inoculation improved plant biomass, relative water content, leaf water potential, root adhering soil/root tissue ratio, aggregate stability under stress conditions. Inoculated plants showed higher levels of osmoprotectants and reduced activity of antioxidant enzymes ascorbate peroxidase, catalase, glutathione peroxidase compared to uninoculated plants, indicating that probably the inoculated plants felt less stress as compared to uninoculated plants. However, Saravanakumar *et al*. (2011) reported a contrasting observation on the ability of *P. fluorescens* Pf1 to increase the activity of catalase and peroxidase in water stressed green gram plants when compared to untreated plants. The bacterized plants were found to tolerate stress better than the uninoculated controls. Yandigeri *et al*. (2012) reported the positive effect of drought tolerant endophytic actinobacteria *Streptomyces coelicolor* DE07, *S. olivaceus* DE10 and *S. geysiriensis* DE27 on growth of wheat under drought stress conditions. Production of phytohormones, plant growth promotion traits combined with water stress tolerance potential in these endophytic actinobacteria contributed towards enhanced plant growth promotion of wheat in the stressed soil. Grover *et al.* 2014 evaluated four *Bacillus* spp. strains for growth promotion of sorghum seedlings under moisture stress conditions. Successful root surface colonization by inoculated bacteria was observed under stress conditions. Plants inoculated with *Bacillus* spp. strains showed higher biomass, chlorophyll and relative water contents in leaves and soil moisture content. Tiwari *et al.* 2015 demonstrated the role of *Pseudomonas putida* MTCC5279 in ameliorating drought stress on cv. BG-362 (*desi*) and cv. BG-1003 (*kabuli*) chickpea cultivars under *in vitro* and greenhouse conditions. Bacterial inoculation modulated the effect of drought on various growth parameters, water status, membrane integrity, osmolyte accumulation, reactive oxygen species (ROS) scavenging ability and stress-responsive gene expressions. Quantitative real-time (qRT)-PCR analysis showed differential expression of genes involved in transcription activation (*DREB1A* and *NAC1*), stress response (*LEA* and *DHN*), ROS scavenging (*CAT, APX, GST*), ethylene biosynthesis (*ACO* and *ACS*), salicylic acid (*PR1*) and jasmonate (*MYC2*) signalling in both chickpea cultivars exposed to drought stress and recovery in the presence or absence of *P. putida* MTCC5279. The observations imply that *P. putida* MTCC5279 confers drought tolerance in chickpea by altering various physical, physiological and biochemical parameters, as well as by modulating differential expression of at least 11 stress-responsive genes (Tiwari *et al.* 2015).

Microorganisms can impart tolerance to crop plants against cold stress as demonstrated by work of Selvakumar *et al.* (2008a, b, 2009, 2010, 2011). They isolated cold tolerant plant growth promoting rhizobacterial species *viz.*, *Pantoea dispersa, Serratia marcescens, Pseudomonas fragi, Exiguobacterium acetylicum* and *Pseudomonas lurida* from Indian Himalayan regions and demonstrated their plant growth promoting potential under cold stress conditions. The observed positive effects on plant growth under cold stress conditions were attributed mainly due to auxin production and phosphate solubilization by the cold tolerant bacterial species. Mishra *et al.* (2009a) inoculated wheat seedlings with cold tolerant pseudomonads, and observed changes in parameters critical to the plant's ability to tolerate cold stress conditions. Inoculation with cold tolerant pseudomonads significantly enhanced total chlorophyll, anthocyanin, free proline, total phenolics, and starch contents; while a decrease in Na+/K+ ratio and electrolyte leakage was observed in bacterized seedlings. In another study, they observed that inoculation with cold tolerant bacterium *Pseudomonas* spp. strain PPERs23 significantly improved vegetative growth, total chlorophyll, total phenolics, and amino acid contents of wheat seedlings under cold stress. Inoculated plants showed higher levels of iron, protein concentration, anthocynine,

Table 1: List of reports from India on role of microorganisms in conferring abiotic stress tolerance to plants

Microorganism	Crop	Stress	Mechanism observed	Reference
Pseudomonas fluorescens	Groundnut	Salinity	Enhanced ACC deaminase activity	Saravanakumar and Samiyappan (2007)
Pseudomonas putida P45	Sunflower	Drought	Improved soil aggregation due to EPS production	Sandhya *et al.* (2009)
Pseudomonas sp. AMK-P6	Sorghum	Heat	Induction of heat shock proteins and improved plant biochemical status	Ali *et al.* (2009)
P. fluorescens Pf1	Green gram	Drought	Increase the activity of catalase and peroxidase	Saravanakumar *et al.* (2011)
Pseudomonas spp.	Maize	Drought	Increased relative water content, Higher levels of osmoprotectants	Sandhya *et al.* (2010)
Bacillus spp.	Maize	Drought	Increased plant biomass, relative water content, RAS/RT ratio	Sandhya *et al.* (2011)
Pseudomonas putida AKMP7	Wheat	Heat	Reduced membrane injury, increased levels of osmoprotectants	Ali *et al.* (2011)
Bacillus cereus	Mungbean, chickpea, rice	Salinity	Enhanced antioxidants levels	Chakraborty *et al.* (2011)
Pseudomonas spp.	Wheat	Cold	Improved level of chlorophyll, anthocyanin, free proline, total phenolics, starch content, physiologically available iron, proteins, and amino acids, relative water content. Reduced membrane injury and Na^+/K^+ ratio	Mishra *et al.* (2011)
Streptomyces spp.	Wheat	Drought	Plant growth promotion	Yandigeri *et al.* (2012)
Bacillus spp.	Maize	Drought	Improved plant growth and chlorophyll and relative water content in leaves, soil moisture content	Grover *et al.* (2014)

Microorganism	Crop	Stress	Mechanism observed	Reference
Bacillus pumilus, Halomonas desiderata and *Exiguobacterium oxidotolerans*	Menthol	Salinity	Improved foliar nutrient uptake and enhanced antioxidant machinery	Bharti *et al.* (2014)
Citricoccus zhacaiensis	Onion	Drought	Improved the percent germination, seedling vigour and germination rate of onion seeds (cv. Arka Kalyan) at osmotic potentials up to -0.8 MPa	Selvakumar *et al.* (2015)
Pseudomonas putida	Chickpea	Drought	Improved growth parameters, water status, membrane integrity, osmolyte accumulation, ROS scavenging ability and influenced stress-responsive gene expression	Tiwari *et al.* (2015)

proline, relative water content; and decreased the Na+/K+ ratio and electrolyte leakage, thereby enhancing the cold tolerance of wheat plants (Mishra *et al.* 2009b, 2011).

Most of the reports on microbial management of abiotic stress present controlled conditions studies at seedling stage. More studies are required on performance of the microorganisms under field conditions.

4. Conclusion and future prospects

Climate change and abiotic stresses are posing serious threat to agricultural productivity worldwide with developing countries like India being more vulnerable. It is evident from different reports that microorganisms have the potential to help crop plants in coping with abiotic stress. However , the underlying mechanisms for microbially induced abiotic stress tolerance in plants need to be understood well for better exploitation of plant-microbe interaction for agricultural benefits. Further, selection of efficient microorganisms and development of suitable formulations which can perform under a variety of stress conditions, in different crops can prove to be an economic and farmers' friendly strategy to cope with climate change induced abiotic stresses.

Acknowledgements

The authors of this manuscript are thankful to Indian Council of Agricultural Research for providing support under AMAAS (Application of Microorganisms in Agriculture and Allied Sectors) project.

References

Ali SZ, Sandhya V, Grover M, Rao LV, Kishore VN and Venkateswarlu B (2009) *Pseudomonas* spp. Strain AKM-P6 enhances tolerance of sorghum seedlings to elevated temperature. Biol. Fertil. Soil. 46, 45-55.

Ali SZ, Sandhya V, Grover M, Rao LV and Venkateswarlu B (2011) Effect of inoculation with a thermotolerant plant growth promoting *Pseudomonas putida* strain AKMP7 on growth of wheat (*Triticum* spp.) under heat stress. J Plant Interact. pp 1-8 DOI: 10.1080/17429145.2010.545147

Arora NK, Tewari S, Singh S, Lal N and Maheshwari DK (2012) PGPR for protection of plant health under saline conditions. In: Bacteria in Agrobiology: Stress Management (Ed) Maheshwari DK, Springer-Verlag, pp 239-258.

Bharti N, Barnawal D, Awasthi A, Yadav A and Kalra A (2014) Plant growth promoting rhizobacteria alleviate salinity induced negative effects on growth, oil content and physiological status in *Mentha arvensis*. Acta Physiol. Plant. 36(1), 45-60.

Castiglioni P, Warner D, Benson RJ, Anstrom DC, Harrison J, Stoecker M, Abad M, Kumar G, Salvador S, D'Ordine R, Navarro S, Back S, Fernandes M, Targolli J, Dasgupta S, Bonin C, Luethy MH and Heard JE (2008) Bacterial RNA chaprones confer abiotis stress tolerance to plants and improved grain yield in maize under water-limited conditions. Plant Physiol. 147, 446-455.

Chakraborty U, Royl S, Chakraborty AP, Dey P and Chakraborty B (2011) Plant growth promotion and amelioration of salinity stress in crop plants by a salt-tolerant bacterium. Rec. Res. Sci. Tech. 3(11), 61-70.

Cho S, Kang BR, Han SH, Anderson AJ, Park JY, Lee YH, Cho BH, Yang KY, Ryu CM, Kim YC (2008) 2R, 3r-butanediol, a bacterial volatile produced by *Pseudomonas chlororaphis* O6, is invoplved in induction of systemic tolerance to drought in *Arabdopsis thaliana*. Mol. Plant Microbe. Interact. 21(8), 1067-1075.

Egamberdieva D and Kucharova Z (2009). Selection for root colonizing bacteria stimulating wheat growth in saline soils. Biol. Fertil. Soil. 45(6), 563-571.

Figueiredo, VB, Burity HA, Martinez CR and Chanway CP (2008) Alleviation of drought stress in the common bean (*Phaseolus vulgaris* L.) by co-inoculation with *Paenibacillus polymyxa* and *Rhizobium tropici*. Appl. Soil Ecol. 40, 182-188.

Glick BR, Cheng Z, Czarny J and Duan J (2007) Promotion of plant growth by bacterial ACC deaminase. Crit Rev Plant Sci. 26, 227-242.

Grover M, Ali Sk Z, Sandhya V, Rasul A and Venkateswarlu B (2011) Role of microorganisms in adaptation of agriculture crops to abiotic stress. World J. Microbiol. Biotechnol. 27, 1231-1240.

Grover M, Madhubala R, Ali SkZ, Yadav SK, Venkateswarlu B (2014) Influence of *Bacillus* spp. strains on seedling growth and physiological parameters of sorghum under moisture stress conditions. J. Basic Microbiol. 54, 951-961.

Grover M, Maheswari M, Desai S, Gopinath KA, Venkateswarlu B (2015). Elevated CO_2: Plant associated microorganisms and carbon sequestration. App. Soil. Ecol. 95, 73-85.

Kohler J, Caravaca F, del Mar Alguacil M and Roldan A (2009) Elevated CO_2 increases the effect of an arbuscular mycorrhizal fungus and a plant-growth-promoting rhizobacterium on structural stability of a semiarid agricultural soil under drought conditions. Soil Biol. Biochem. 41, 1710-1716.

Mishra PK, Bisht SC, Pooja R, Joshi P, Suyal P, Bisht JK and Srivastva AK (2009a) Enhancement of chilling tolerance and productivity of inoculated wheat with cold tolerant plant growth promoting *Pseudomonas* spp. PPERs23. In: Abstracts of 4th USSTC, 10-12 Nov 2009

Mishra PK, Bisht SC, Ruwari P, Selvakumar G and Bisht JK (2009b) Enhancement of chilling tolerance of inoculated wheat seedlings with cold tolerant plant growth promoting *Pseudomonads* from NW Himalayas. In: Abstract of first Asian PGPR conference. ANGRAU, Hyderabad, 22-24 June 2009.

Mishra PK, Bisht SC, Pooja R, Selvakumar G, Joshi P, Bisht JK, Bhatt JC and Gupta HS (2011) Alleviation of cold stress in inoculated wheat (*Triticum aestivum* L.) seedlings with psychrotolerant pseudomonads from NW Himalayas. Arch. Microbiol. 193(7), 497-513.

Mohammadkhani N and Heidari R (2008). Water stress induced by polyethylene glycol 6000 and sodium chloride in two maize cultivars. Pak. J. Biol. Sci. 11(1), 92-97.

Nanjo T, Kobayashi TM, Yoshida Y, Kakubari Y, Yamaguchi-Shinozaki K and Shinozaki K (1999) Antisense suppression of proline degradation improves tolerance to freezing and salinity in *Arabidopsis thaliana*. FEBS Lett. 461, 205-210.

Saleem M, Arshad M, Hussain S and Bhatti AS (2007) Perspective of plant growth promoting rhizobacteria (PGPR) containing ACC deaminase in stress agriculture. J. Indian Microbiol. Biotechnol. 34, 635-648.

Sandhya V, Ali SZ, Grover M, Reddy G and Venkateswarlu B (2009) Alleviation of drought stress effects in sunflower seedlings by the exopolysaccharides producing *Pseudomonas putida* strain GAP-P45. Biol. Fertil. Soil. 46, 17-26.

Sandhya V, Ali SZ, Grover M, Reddy G and Venkateswaralu B (2010) Effect of plant growth promoting *Pseudomonas* spp. on compatible solutes antioxidant status and plant growth of maize under drought stress. Plant Growth Regul. 62, 21-30 DOI:10.1007/s 10725-010-9479-4

Sandhya V, Ali SZ, Grover M, Reddy G and Venkateswaralu B (2011) Drought-tolerant plant growth promoting *Bacillus* spp.: effect on growth, osmolytes, and antioxidant status of maize under drought stress. J. Plant Interact. 6, 1-14 DOI: 10.1080/17429145.2010.535178

Saravanakumar D and Samiyappan R (2007) ACC deaminase from *Pseudomonas fluorescens* mediated saline resistance in groundnut (*Arachis hypogea*) plants. J. Appl. Microbiol. 102,1283-1292.

Saravanakumar D, Kavino M, Raguchander T, Subbian P and Samiyappan R (2011) Plant growth promoting bacteria enhance water stress resistance in green gram plants. Acta Physiol. Plant. 33, 203-209.

Selvakumar G, Kundu S, Joshi P, Nazim S, Gupta AD, Mishra PK and Gupta HS (2008a) Characterization of a cold-tolerant plant growth-promoting bacterium *Pantoea dispersa* 1A isolated from a sub-alpine soil in the North Western Indian Himalayas. World J. Microbiol. Biotechnol. 24, 955-960.

Selvakumar G, Mohan M, Kundu S, Gupta AD, Joshi P, Nazim S and Gupta HS (2008b) Cold tolerance and plant growth promotion potential of *Serratia marcescens* strain SRM (MTCC 8708) isolated from flowers of summer squash (*Cucurbita pepo*). Lett. Appl. Microbiol. 46, 171-175.

Selvakumar G, Joshi P, Nazim S, Mishra PK, Bisht JK and Gupta HS (2009) Phosphate solubilization and growth promotion by *Pseudomonas fragi* CS11RH1 (MTCC 8984) a psychrotolerant bacterium isolated from a high altitude Himalayan rhizosphere. Biologia. 64(2), 239-245.

Selvakumar G, Kundu S, Joshi P, Nazim S, Gupta AD and Gupta HS (2010b) Growth promotion of wheat seedlings by *Exiguobacterium acetylicum* 1P (MTCC 8707) a cold tolerant bacterial strain from the Uttarakhand Himalayas. Ind. J. Microbiol. 50, 50-56.

Selvakumar G, Joshi P, Suyal P, Mishra PK, Joshi GK, Bisht JK, Bhatt JC and Gupta HS (2011) *Pseudomonas lurida* M2RH3 (MTCC 9245), a psychrotolerant bacterium from the Uttarakhand Himalayas, solubilizes phosphate and promotes wheat seedling growth. World J. Microbiol. Biotechnol. 27, 1129-1135.

Selvakumar G, Panneerselvam P and Ganeshamurthy AN (2012) Bacterial mediated alleviation of abiotic stress in crops. In: Bacteria in Agrobiology: Stress Management (Ed) Maheshwari DK, Springer-Verlag, pp 205-224.

Selvakumar G, Bhatt RM, Upreti KK, Bindu GH, Shweta K (2015) *Citricoccus zhacaiensis* B-4 (MTCC 12119) a novel osmotolerant plant growth promoting actinobacterium enhances onion (*Allium cepa* L.) seed germination under osmotic stress conditions. World J. Microbiol. Biotechnol. 31, 833-839.

Srinivasarao Ch, Lal R, Prasad JVNS, Gopinath KA, Singh R, Jakkula VS, Sahrawat KL, Venkateswarlu B, Sikka AK, Virmani SM (2015) Potential and challenges of rainfed farming in India. In: Advances in Agronomy (Ed) Sparks DL, Elsevier 133, 113-181.

Srivastava S, Yadav A, Seem K, Mishra S, Chaudhary V and Srivastava CS (2008) Effect of high temperature on *Pseudomonas putida* NBRI0987 biofilm formation and expression of stress sigma factor RpoS. Curr. Microbiol. 56(4), 453-457.

Tewari S and Arora NK (2013) Plant growth promoting rhizobacteria for ameliorating abiotic stresses triggered due to climatic variability. Climate Change Environ. Sustain. 1(2), 95-103.

Timmusk S and Wagner EGH (1999) The plant growth-promoting rhizobacterium *Paenibacillus polymyxa* induces changes in *Arabidopsis thaliana* gene expression: a possible connection between biotic and abiotic stress responses. Mol. Plant Microbe Interact. 12, 951-959.

Tiwari S, Lata C, Chauhan PS, Nautiyal CS (2015) *Pseudomonas putida* attunes morphophysiological, biochemical and molecular responses in *Cicer arietinum* L. during drought stress and recovery, Plant Physiol. Biochem. DOI: 10.1016/j.plaphy.2015.11.001

Venkateswarlu B, Desai S and Prasad YG (2008) Agriculturally important microorganisms for stressed ecosystems: Challenges in technology development and application. In: Agriculturally Important Microorganisms, Vol 1 (Ed) Kachatourians GG, Arora DK, Rajendran TP and Srivastava AK, Academic World, Bhopal, pp 225-246.

Verdoy D, Coba De La Pena T, Redondo FJ, Lucas MM and Pueyo JJ (2006) Transgenic *Medicago truncatula* plants that accumulate proline display nitrogen-fixing activity with enhanced tolerance to osmotic stress. Plant Cell Environ. 29, 1913-1923.

Yandigeri MS, Meena KK, Singh D, Malviya N, Singh DP, Solanki MK, Yadav AK and Arora DK (2012) Drought-tolerant endophytic actinobacteria promote growth of wheat (*Triticum aestivum*) under water stress conditions. Plant Growth Regul. 68, 411-420.

Yang J, Kloepper J and Ryu CM (2009) Rhizosphere bacteria help plants tolerate abiotic stress. Trend. Plant Sci. 14, 1-4.

Zahir ZA, Munir A, Asghar HN, Arshad M and Shaharoona B (2008) Effectiveness of rhizobacteria containing ACC-deaminase for growth promotion of peas (*Pisum sativum*) under drought conditions. J. Microbiol. Biotech. 18(5), 958-963.

5

Endurance to Stress: An Insight into Innate Stress Management Mechanisms in Plants

Krishna Sundari Sattiraju* and Srishti Kotiyal

Abstract

In its natural environment, a plant faces various types of stressors. In fact, stress in moderation is a welcome event in plants. Only when it is in excess, would it compromise the productivity of plants. Beyond a threshold, stress (biotic and abiotic) has a detrimental effect on plant growth causing extensive yield losses in crop plants worldwide. Plants being sessile, have evolved a complex set of mechanisms to perceive the changes in the environment. Plants respond to minimize the potential damage due to stress while simultaneously preserving resources vital for their growth and development. The chapter presents an overview of various types of biotic and abiotic factors that act as plant stressors and the mechanisms by which plants respond to these stresses. The chapter concludes with a perspective on how the plant's innate stress response machinery can be exploited for achieving higher tolerance to stress in crop plants.

Keywords: Abiotic stress, Biotic stress, Hypersensitive response, Plant hormones, PR proteins, Stress signalling mechanisms, Systemic acquired resistance

1. Introduction

World population is expected to reach the 8 billion mark by the year 2020. This ever-increasing population has led to the prognostication that food scarcity will prove to be a major challenge in meeting humanity's existential needs. Boosting agricultural production takes precedence in meeting this predicted food crisis. Based on a 2007 FAO report, only 3.5% of the global land area is unaffected by any environmental constraint. It was suggested that environmental conditions can limit up to 70% of the crop production. The situation is expected to exacerbate due to the continued reduction in arable land, reducing water resources, increasing global warming trends and climate change (Cramer *et al.* 2011). In this chapter, we start by understanding the different types of environmental stressors and the major mechanisms by which plants respond to biotic and abiotic challenges. The roles of various hormones and molecules in stress-response pathways are also discussed. Gene expression and regulation under stress as well as the post-transcriptional role of miRNAs in plant stress responses were further deliberated. In the native habitats, plants are subjected to a variety of stresses simultaneously which necessitates a crosstalk in plant's responses to biotic and abiotic stresses. Simultaneous stresses induce plant responses which are not in direct correlation with their responses to individual stresses, thus, a brief view of such responses is also given as the penultimate topic of the chapter. The chapter concludes with a note on the significance of understanding the physiological, transcriptional and translational stress-responses of plants as these pathways converge into a uni-functionary mechanism. Insights into these mechanisms are crucial in order to design effective strategies for stress-management in plants.

2. Types of plant stresses

Environmental excesses can be of abiotic or biotic nature, both of which are equally harmful for plant growth. Biotic environmental stress involves unfavourable interactions with other organisms like pathogen infection, mechanical damage, herbivory, obligate symbiosis or parasitism (Schulze *et al.* 2005). Abiotic stress results from extreme environmental conditions that decrease plant growth and yield below optimum levels (Cramer *et al.* 2011). Reduced water availability, extreme temperatures (heating or freezing), lack or limited availability of soil nutrients, excess of toxic ions, excess of light and soil compaction, all constitute abiotic stress for plants (Duque *et al.* 2013).

2.1 Biotic stress

Biotic stress in plants is caused by organisms such as bacteria, fungi, viruses, nematodes, insects, mites and animals (Flynn 2003). Worldwide, biotic stress is responsible for causing greater than 42% loss in the potential crop yield (15% attributed to insects, 13% to weeds, and 13% to other pathogens) (Balconi *et al.* 2012). When it comes to pollination and pollen dispersal, animals can also act as biotic stress factors (Web resource 1). Plant pathogens are categorized based on their mode of nutrition: necrotrophic pathogens (which kill host tissue while they colonize and thrive on dead or dying cells), biotrophic pathogens (which maintain host viability as they derive nutrients from living cells) and hemibiotrophs (which use both above mentioned methods to derive nutrition through biotropy in early phases followed by nectrophy during later stages of disease). The biotroph-plant host relationship is highly specialized and a structurally and biochemically complex one. Obligate biotrophs penetrate cell wall of host, colonize intercellular spaces and suppress host defences without disrupting the plasma membrane. Before causing any necrosis, they multiply in host tissues for some time (Alfano *et al.* 1996). On the other hand, infection structures in necrotrophs are less specialized; instead they overpower the host by secreting diverse pathogenicity and virulence factors throughout infection (Laluk *et al.* 2010), rapidly killing parenchymatous tissues during active pathogenesis (Alfano *et al.* 1996). The various types of biotic stressors as well as their morphological effects are shown in Figure 1.

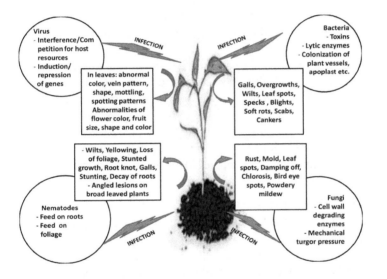

Fig. 1: Different types of biotic stressors and their morphological effects.

a)   **Bacteria:** Higher plants possess various nutrient sources for the diverse bacterial species in their environment. The miniscule size of bacteria allows them to pass through stomata and other natural openings, into the apoplast where they can access the nutrient resources (Alfano *et al.* 1996). Plant pathogenic bacteria can cause a wide range of symptoms in plants, including galls and overgrowths, wilts, leaf spots, specks and blights, soft rots, scabs and cankers (Sarah *et al.* 2008). Gram negative bacteria specialize in colonizing the apoplast and produce rots, spots, wilts, cankers, blights, etc., infecting, majority of crop plants (Alfano *et al.* 1996). Phytopathogenic bacteria cause diseases by various mechanisms; including: production of toxins, injecting special proteins that kill host cell, production of lytic enzymes targeting structural components of plant cells, while others colonize xylem vessels leading to wilting and death. Apoplastic colonizers spend their parasitic life in the intercellular spaces of various plant organs or in the xylem and cause necrosis (Alfano *et al.* 1996).

b)   **Fungi:** All soil-borne fungi are mostly necrotrophic. As environmental conditions in the soil are generally not conducive for fungal growth, pathogens survive in soil as resistant propagules or survive in plant roots and crop residues. Fungal diseases can be difficult to diagnose in plants as mostly symptoms occur under-ground, and even the above-ground symptoms are not unique enough and can be confused with those caused by other abiotic stresses. Some general signs and symptoms of fungal diseases include leaf and stem rust, white mold, birds-eye spot, powdery mildew, damping-off of seedlings, spots on leaves and chlorosis (Isleib 2012). Under favourable conditions, or as a seed or root approaches, the fungus germinates due to the stimulation *via* root or seed exudates and grows chemo-tactically towards the plant. Some fast-growing pathogens attack seeds and embryos even before they emerge. Fungi penetrate intact cell walls by secreting cell wall degrading enzymes and/or by using mechanical turgor pressure. They colonize the root cortex, kill the roots and form spores within the root tissue. Majority of soil-borne fungi attack young, juvenile roots as compared to secondary woody roots (Raaijmakers *et al.* 2009).

c)   **Viruses:** Plant viruses are biotrophic pathogens. Though viral infections are not lethal to plants in majority of situations, they can severely impact the host in terms of quantity, quality and longevity. In some cases, viral infections do not even cause apparent pathological effects. However, in many instances they cause pathological phenotypes, which can be specific to the host or the invading virus, including abnormalities of colour, vein patterns, mottling, spotting patterns, and shape of leaves. Abnormalities are also seen in flower colour, fruit size, shape and colour; although with some viral diseases, the symptoms are masked. These disease associated phenotypes are caused due to interference/competition for the host resources, disturbing the host physiology and causing disease. This interference and competition can affect a number of genes by inducing or repressing them. Plant to plant transmission of viruses can be from the parent plant to an offspring, either through vegetative propagation or through grafting, budding, transmission by seeds and mechanical spread by insects/man (Web resource 2).

d)   **Nematodes:** Nematodes are soil inhabiting free-living macrobionts which consume bacteria, fungi, and other nematodes. Some can also be parasitic on plants (Raaijmakers *et al.* 2009). Nematodes which feed on roots cause above-ground symptoms resembling those caused by other kinds of root injuries, such as loss of lustre in foliage, yellowing, wilting and eventual loss of foliage. This leads to stunted and weak growth and susceptibility of plants to other stressors increases. Nematodes feeding on foliage cause angled lesions on broad leaved plants. Root symptoms of nematode infestation might include galls, stunting, and root decay (Web resource 3). Nematodes can be migratory ectoparasitic (feeding on the outside of the root), migratory endoparasitic (penetrating and moving in the interior of the root) and sedentary endoparasites (setting up a feeding site in the interior of the root and remaining there for reproduction) (Raaijmakers *et al.* 2009). Some examples of economically significant plant diseases, along with their causal organisms and conditions which favour that disease, are given in Table 1.

2.2 Abiotic stress

Abiotic stress conditions can significantly limit productivity and yield quality in many crop plants (Fraire-Velaìzquez *et al.* 2013). Abiotic stressors prominently discussed in this chapter include pH, salinity, drought, temperature (high and low) and light. Plants exhibit some common physiological responses regardless of the specific abiotic stressor whereas some other stressors induce very unique response. Not only the type of stressor but also the time of incidence of stress, its duration and magnitude also have a significant role in determining host's reaction to stress. Adverse effects of some of these stressors on host plants are briefly discussed in the following paragraphs.

a)   **pH:** Soil pH governs many biochemical processes affecting nutrient bioavailability and toxicity. Based on nutrient solubility concerns, soils that are either too acidic (pH near 4.0) or too basic (pH near 9.0) are usually unsuitable for plant growth. Soils in near neutral range (pH 5.5 to 7.0) are suitable. At pH lower than 5.0, many metal ions including iron (Fe), copper (Cu), manganese (Mn) and zinc (Zn) have very high solubility in soil solution and thus can be toxic (particularly for vegetable crops). Moreover, at pH this low, they can also form precipitates with phosphate (P), make it unavailable and cause phosphate limiting conditions. At pH greater than 7.0, calcium (Ca) and magnesium (Mg) have high solubility and can assist in fixing P. However at alkaline pH, bioavailability of Fe, Mn and Zn cannot satisfy requirements of most vegetable crops. Elucidating this with

Table 1: Economically significant plant diseases and conditions favouring disease.

Disease	Pathogens	General host	Facilitating conditions	Reference
Bacterial diseases				
Shoot and flower blights, cankers, diebacks; specks; bleeding cankers	*Pseudomonas syringae pathovars*	Monocots, herbaceous dicots, and woody dicots	Frost damage or injury, cool, wet weather and rainstorms	Kennelly *et al.* (2007); Mansfield *et al.* (2012); Web resource 4
Granville wilt, wilt in tomato, Moko disease, brown rot	*Ralstonia solanacearum*	Almost 200 plant species in 33 different plant families, many dicots and few monocots	Average rainfall > 100 cm/year (39 in/year), average growing season > 6 months, average winter temperatures >10°C, average summer temperatures >21°C and average yearly temperature <23°C	Web resource 5; Web resource 6
Crown gall	*Agrobacterium tumifaciens*	Many woody and herbaceous plants, including fruit, vegetables and ornamental plants	Hot, moist conditions	Bell *et al.* (2010); Web resource 7
Bacterial leaf blight	*Xanthomonas oryzae* pv. *oryzae*	Most importantly rice	Severe wind and temperature of 25-30°C, heavy rain, heavy dew, flooding, deep irrigation water	Web resource 8
Fire blight disease	*Erwinia amylovora*	Apple, pear, several rosaceous ornamentals	High temperatures and presence of free water	Johnson, (2000); Wale *et al.* (2008)
Fungal diseases				
Rice blast	*Magnaporthe oryzae*	Wide variety of monocotyledonous hosts, especially rice	Moderate temperatures (24°C) and periods (>/= 12 hours) of high moisture	TeBeest *et al.* (2007); Park *et al.* (2009); Dean *et al.* (2012)

Disease	Pathogens	General host	Facilitating conditions	Reference
Grey mould	*Botrytis cinerea*	More than 200 plant species especially dicots, including important protein, oil, fibre and horticultural crops	High humidity and temperatures between 17°C and 25°C	Williamson *et al* (2007); Web resource 9
Wheat rust	*Puccinia* spp.	Cereal crops, especially wheat	Humid conditions and warmer temperatures (15-35°C)	Singh *et al.* (2002); Web resource 10
Fusarium head blight	*Fusarium graminearum*	Cereal crops, especially wheat	High humidity and temperatures between 24°C and 29°C	Wise *et al.* (2015)
Fusarium wilt	*Fusarium oxysporum*	More than 100 different hosts, provoking severe losses in crops such asmelon, tomato, cotton and banana, among others	Temperatures around 25°C	Dean *et al.* (2012); Wise *et al.* (2015)
Viral diseases				
Rice tungro disease	Rice tungro baciliform virus (RTBV) and Rice tungro spherical virus (RTSV)	Rice	Synchronization of viral sources, vector, age and susceptibility (more in vegetative stage) of host plants.	Rybick, (2015); Web resource 11
Barley yellow dwarf	Barley yellow dwarf disease *Luteovirus* complex	Barley, oats, wheat, maize, rice and other grasses	High light intensity and relatively cool temperatures (15-18°C), leaf discoloration attracts aphid vectors resulting in virus spread	D'Arcy *et al.* (2000); Rybick, (2015)
Maize streak disease	Maize streak *Mastrevirus*	Maize	Sensitive to increase in temperature	Damsteegt, (1984); Rybick, (2015)
Tomato spotted wilt	Tomato spotted wilt virus (*Tospovirus*)	Over 1000 species in over 85 families, including many	Summers following abnormallydry springs	Sherwood *et al.* (2003); Bost, (2013)

Disease	Pathogens	General host	Facilitating conditions	Reference
African cassava mosaic disease	African cassava mosaic disease virus (*Begomovirus* complex)	vegetables, peanut and tobacco. Cassava	High temperature, high light intensity and wind speed increase activity of vector; moderate rainfall	Fauquet *et al.* (1990); Fargette *et al.* (1994)
Nematode diseases				
Root knot	Root-knot nematodes (*Meloidogyne* spp.)	Many important fruit, vegetable and ornamental crops	Optimal soil temperature (21-26°C), coarser soils	Web resource 12; Mai *et al.* (1981); Noling (1999); Jones *et al.* (2013)
Cyst nematode disease	Cyst nematodes (*Heterodera glycines* and *Globodera pallida* and *G rostochiensis*)	Soybean, potato and cereals	Warm temperatures, medium to heavy clay soil, well drained and aerated sands, silts and peat soils, moisture content 50-75% of water capacity	Mai *et al.* (1981); Jones *et al.* (2013)
Root lesion nematode disease	Root lesion nematodes (*Pratylenchus* spp.)	Cereals, sugarcane, coffee, banana, maize, legumes, potato, many vegetables and fruit trees	Species-specific temperature requirements, coarse textured soil, soil moisture level where soil particles/ aggregates are surrounded by a film of water with free intercellular spaces	Mai *et al.* (1981); Jones *et al.* (2013)
Burrowing nematode disease, toppling disease, blackhead disease, spreading decline, yellows, slow wilt	Burrowing nematode *Radophilus similis*	Over 350 plant hosts in tropical and subtropical regions including banana, citrus, black pepper, aroids, ginger, tea, coconut palms	Tropical and subtropical temperatures	Brooks, (2008)
Ufra, potato rot, peanut pod nematode disease	*Ditylenchus dipsaci*	About 450 plant species, including peas, celery, strawberries, beetroot, pumpkin, ornamental bulbs, oats, rye and some weeds	Temperature in the range 15-20°C.	Jones *et al.* (2013)

an example, most crop species can utilize ferrous iron but not ferric iron. If the pH is relatively high, the primary form available is ferric iron which may cause iron deficiency in crop plants. Soil pH can also affect nutrient bioavailability by changing soil particle properties. At adversely low soil pH, the positive charges on soil particle surfaces would bind nutrients like P, causing P deficiency in vegetable plants. Some soil-borne diseases are also associated with soil pH. For instance, the clubroot disease of crucifers caused by *Plasmodiophora brassicae* becomes epidemic at pH lower than 5.7 but is drastically reduced in the pH range 5.7 to 6.2 and is almost eliminated at soil pH greater than 7.3 (Liu *et al.* 2015).

b) **Temperature:** Temperature is one of the primary factors affecting the rate of plant development. Rate of plant growth and development depends on the temperature surrounding the plant and every species has a specific optimal temperature range which is represented by a minima, maxima, and optima. The process of pollination is very sensitive to temperature extremes (across species) and thus any exigencies in temperature during this stage, can hugely impact crop production. Warm temperatures increase the rate of phenological development and make a major impact during the reproductive stage. A report showed that in wheat (*Triticum aestivum* L.), frost could cause sterility and abortion of formed grains while excessive heat could cause reduction in number of grains and duration of grain-filling period. Moreover, temperature effects are exacerbated by both water scarcity and excess soil water (Hatfield *et al.* 2015). Cold stress can be of two types: non-freezing chilling stress (above freezing point and below 15°C) and freezing stress which is below freezing point (Shi 2014). While freezing temperatures are harmful to most plants, temperatures above freezing but below 10°C may kill or damage tropical or subtropical plants (Decoteau 1998). Chilling stress was reported to inhibit the activity of enzymes involved in photosynthesis, respiration, and biochemical processes such as ROS (reactive oxygen species) scavenging, causing oxidative damage that may lead to accumulation of toxic compounds and inhibition of metabolic reactions. It also leads to formation of intracellular ice crystals inducing cellular dehydration and osmotic stress, which can result in membrane damage and finally, tissue death (Shi 2014). Susceptibility to cold damage may differ amongst inter or intra-species and varies with stage of plant development too. Plants show increased sensitivity to cold temperatures during flowering period (Decoteau 1998). Soil temperature also has a wide ranging impact on microbial growth/development, organic matter decomposition, seed germination, root development, absorption of water and nutrients. Generally, these processes would occur faster with an increase in temperature. Soil temperature also affects size, quality, and shape of storage organs in plants. Soil temperature is in turn influenced by soil types. For instance, clay soil releases heat to the surface faster than dry sandy soils and these heat losses can be more rapid at lower air temperature. On the other hand, light coloured soils absorb lesser solar energy (than dark-coloured soils) and release lesser heat to the atmosphere owing to low water-holding capacity of the soil (Decoteau 1998).

c) **Drought:** Erratic rainfall, compounded by present climatic changes, often leads to drought stress in terrestrial plants. Plant responses to water stress can either result in an injurious change or serve as a tolerance index (Kar 2011). Water deficit reduces water uptake into the expanding cells, and enzymatically alters the rheological properties of the cell wall; negatively affecting plant growth. Water deficit also alters cell walls non-enzymatically, say by the interaction of pectate and calcium. Long term water stress not only leads to decline in photosynthesis as decreased stomatal aperture limits CO_2 uptake (Cornic 2000); it also increases photo-inhibition due to difficulties in dissipating excess light energy (Cramer *et al.* 2011). Besides, drought stress upsets both the content and the activity of photosynthetic carbon reduction cycle enzymes including RuBisCO (ribulose-1,5-bisphosphate carboxylase/oxygenase). Furthermore, drought stress induces generation of ROS which is kept under tight regulation through increased antioxidative systems (Reddy *et al.* 2004).

d) **Light:** Light plays a critical role in plant growth and development. Quality, quantity and direction of light can affect plants by affecting photosynthetic rate and assimilate accumulation, or play a regulatory role controlling growth and development. Fluctuation in light intensity can cause stress in plants affecting phytochrome action, photoperiodism, breaking of dormancy, flowering and many such aspects. Several authors validated the importance of light intensity and showed that reduced light intensity reduced growth in trees (e.g. root size and weight) and considered it to be one of the important factors leading to mortality of young plants. Extreme solar radiation was also shown as harmful as it could cause water shortage by evaporation, causing dehydration (Ologundudu *et al.* 2013). Excessive exposure to photosynthetically active radiation causes photoinhibition; decreasing photosynthesis and/or photosynthetic capacity (Duque *et al.* 2013).

e) **Salinity:** Most plants do not grow in saline soils. Salt stress can be caused due to accumulation of several ions/salts like sodium, chlorine and boron (Shrivastava *et al.* 2015). High salt concentrations negatively affect seed germination and plant growth, eventually killing growing plants. Though sensitivity or tolerance to salts varies with plant species, shoot growth is repressed more than root growth by soil salinity. High salt content decreases osmotic potential of soil water and consequently the availability of soil water to plants causing salt-induced water deficit. In salt-stressed plants, many nutrient interactions occur that can have a considerable impact on plant growth (Bhatt et al. 2008). For example, high salt levels in soil can upset the nutrient balance in the plant or interfere with uptake of certain nutrients as many salts are plant nutrients too. Also, plant P uptake would be significantly reduced under soil salinity as phosphate ions precipitate with Ca^{2+} ions (Shrivastava et al. 2015). Protein content can also be affected by salt stress (Qados 2011). K^+ acts as cofactor in several enzymes and cannot be substituted by Na^+. Under highly saline conditions, K^+ is replaced by Na^+ in biochemical reactions, and Na^+ and Cl^- induce conformational changes in proteins causing ion toxicity. Ion toxicity and osmotic stress thus create metabolic imbalance, which in turn leads to oxidative stress. Additionally, salinity has a higher impact on plant development during the reproductive phase and adversely affects reproductive development by inhibition of microsporogenesis, elongation in stamen filament, increased programmed cell death in certain tissues, ovule abortion and senescence of fertilized embryos. Post-translational inhibition during salt stress results in diminished activity of cyclin-dependent kinases. Reduced expression, activity of cyclins and cyclin-dependent kinases arrests cell cycle transiently which results in fewer cells in the meristem, thereby limiting growth. Salinity impacts photosynthesis by causing reduction in leaf area, chlorophyll content and stomatal conductance, and partly through reduced efficiency of photosystem II (Shrivastava et al. 2015).

3. Plant response to biotic stress

At cellular level, response of plants to pathogen infection brings down convergence of many factors including specific pathogen response proteins, plant hormones like SA (salicylic acid), JA (jasmonic acid), ABA (abscissic acid), ET (ethylene) etc. and production of specific metabolites. Interaction between a pathogen and a plant can fit the model of 'gene for gene' system wherein avirulence gene product (Avr) secreted by the pathogen interacts with the resistance gene product (Res), thereby stimulating a cascade of events in the plant cell which further activate plant defense responses. In the absence of complementary resistance and avirulence genes, plants still exhibit these responses but at a lower frequency and magnitude (Web resource 1). Immune response by plants at a basal level involves NB-LRR (Nucleotide-Binding Leucine Rich Repeats) which detect PAMPs (pathogen-associated molecular patterns) directly or indirectly, initiating a signalling cascade that activates plant defences inducing PTI (PAMP-triggered immunity) (Boatwright 2013). P/MAMPs are evolutionarily conserved pathogen components, which resemble the innate immune system patterns of mammals and insects, directed mainly against epitopes characteristic for fungi or bacteria. They serve as non-self recognition mechanisms (Laluk *et al.* 2010). In PTI, signal transduction from receptors to downstream components occurs via the MAP (mitogen-activated protein) kinase pathway and several known PAMPs activate MAP kinases. Flagellin derived peptide flg22 rapidly and strongly activates MPK3, MPK4 and MPK6. MPK4 and MPK6 get activated by hairpin proteins, encoded by HR (hypersensitive response) and pathogenicity genes present in several plant pathogenic bacteria. Following this activation, PR (pathogenesis-related) genes get induced which encode for anti-microbial proteins (Pitzschke *et al.* 2009). Pathogens respond to this with effector molecules to suppress PTI and plants further counter it with R (resistance) proteins that recognize pathogen effectors and promote ETI (effector-triggered immunity) (Salas-Marina *et al.* 2015). ETI is often characterized by an early defence response, which regulates PCD (programmed cell death) in order to restrict pathogen growth, known as HR (Boatwright 2013). ETI is widespread in the plant kingdom as a form of resistance effective against (hemi) biotrophic pathogens but plays a limited role in resistance against necrotrophs (Laluk *et al.* 2010). Even though distinct plant receptors detect different elicitors, the downstream responses merge to elicit a common regulatory control and induce PTI (Laluk *et al.* 2010). A schematic of plant response mechanisms to biotic stress is given in Figure 2.

HR is a form of programmed plant cell death, facilitating cell-wall strengthening, and the expression of various defence-related R genes that mediate recognition of pathogen effectors. The R genes activate a cascade of defence signalling responses and PR gene expression to generate SAR (systemic acquired resistance) which primes plants against a broad spectrum of pathogens. These defences are activated only after the detection of a

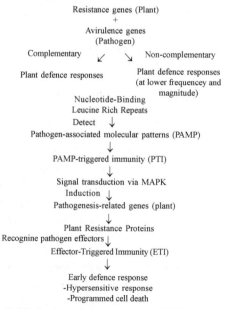

Fig. 2: Mechanism of plant response to biotic stress

prospective invasion and kept under stringent genetic control (Balconi *et al.* 2012). HR is executed in a stepwise manner. The first stage, ion leakage, is marked by calcium and hydrogen influx alongside hydroxide and potassium efflux. The second stage, oxidative burst, involves production of ROS which cause lipid peroxidation and other oxidative damages. The final stage involves ROS mediated breakdown of cellular components, cross-linking between cell walls and progression to cell death (Boatwright 2013). Various hormones are either up-regulated or down-regulated in response to pathogen detection to confer immunity to plants. A brief summary of some of these hormones is given (Table 2) in the following paragraphs.

Table 2: Different plant hormones which are up-regulated/down-regulated in response to abiotic stress

Hormone	Metabolic Effect	Stressor
ABA (abscisic acid)	Increase in endogenous concentration to regulate cellular osmotic pressure	Drought stress
JA (jasmonic acid)	Accumulation induces plant stress genes	Drought, low temperature, salinity
IAA (indole-acetic acid)	Increases root and shoot growth of plants	Salinity, heavy metal stress
ET (ethylene)	Promotes stress tolerance	Salt stress
GA (gibberellic acid)	Accumulation improves photosynthetic efficiency, nutrient usage, decrease in stromatal resistance, increased growth & yield response, saline conditions	General abiotic stress
CK (cytokinin)	Accumalation increases photosynthetic rate, improves grain quality, antioxidative mechanisms	Salinity, drought, high temperature
BR (brassinosteroid)	Regulates various physiological processes	Heavy metal stress, salinity, drought & heat stress
SA (salicylic acid)	Stimulates RuBisCO, increases IAA levels and decreases ABA, protects from oxidative damage	Drought, salinity, heavy metal tolerance, temperature extremes
Triazoles	Inhibits gibberellin synthesis, reduces ET levels, increases CKs, increases photosynthetic rate and antioxidant enzymes	Drought, heat, chilling, ozone, SO_2, salinity

Sources: Rajalekshmi *et al.* (2009); Duque *et al.* (2013); Fahad *et al.* (2015)

SA (salicylic acid) is a phenolic compound produced by plants which functions as a plant hormone. It plays a critical role in plant growth regulation, development, and interaction with other organisms. Majority of genes responding positive to SA treatments are linked to stress signalling pathways which ultimately lead to cell death. These genes include genes encoding: chaperone proteins, heat shock proteins, antioxidants and those involved in biosynthesis of secondary metabolites (Fahad et al. 2015). Plants can up-regulate expression of SA both locally and systemically in response to biotrophic or hemibiotrophic pathogens (Boatwright 2013; Dorantes-Acosta et al. 2012). SA acts as an endogenous signal and its accumulation locally leads to LAR (Local Acquired Resistance) while systemic accumulation leads to SAR (Boatwright 2013; Dorantes-Acosta et al. 2012). When plants activate their defence mechanisms under challenge of pathogen infection, especially if the pathogen induces HR, SAR development takes place (Compant et al. 2005). Increase in production and secretion of plant defence proteins (eg: PR proteins) is common to both LAR and SAR with NPR1 (non-inducible pathogenesis-related 1) acting as a key regulator of PR1 expression and SAR establishment. NPR1 interacts with TGA (TGACGTCA cis-element-binding protein) transcription factors, ultimately leading to the activation of SA-dependent responses (Song et al. 2013). Multiple members of the NPR family of proteins facilitate SA binding and pathogen-specific defence responses are in their joint regulation (Boatwright 2013). NPR3 and NPR4 are SA receptors which bind SA with different affinities and regulate the degradation of NPR1 in a SA-dependent manner (Dorantes-Acosta et al. 2012). Mutants that show inability to accumulate SA or are insensitive to SA have increased susceptibility to biotrophs (Spoel et al. 2007).

The R protein-mediated HR and SA-mediated basal resistance are generally ineffective against necrotrophic pathogens as infection often results in the rapid accumulation of JA (jasmonic acid), which activates a set of PR genes distinct from those induced by SA. These genes encode proteins exhibiting antifungal activities (Spoel et al. 2007). JA and JA-related compounds, including MeJA (methyl-jasmonate) and JA-Ile (jasmonate-isoleucine conjugate), play integral roles in endogenous regulation of plant resistance to mechanical wounding and herbivory by modulating gene expression in plants. A combination of JA and ET (ethylene) signalling also induces resistance against necrotrophic pathogens (Dorantes-Acosta et al. 2012) and an accumulation of JA and ET in response to necrotrophic infections leads to Induced Systemic Resistance (Salas-Marina et al. 2015). The secondary indole metabolite, camalexin, is also involved in defence against necrotrophs and it has been shown that camalexin-deficient mutant exhibit increased susceptibility to necrotrophs. However, SA can antagonize JA signalling and vice versa. For instance, SA promoted disease development caused by a necrotrophic pathogen (Botrytis cinerea) in tomato via NPR1 (using transcription factor TGA1), and suppressed modulated expression of proteinase inhibitors I and II which are JA-dependent defence genes (Rahman et al. 2012). Also, activation of SA- or JA-dependent signalling pathways is not necessarily exclusive in response to biotrophs or necrotrophs. For eg: the biotrophic bacterial leaf-pathogen *Pseudomonas syringae* pv. *tomato* (*Pst*) DC3000 can simultaneously induce both SA and JA synthesis. In response to such a pathogen, plants cross-talk to adjust the response in favour of the more effective pathway (i.e. the SA pathway) and this cross-talk is mediated by NPR1 as it is required for SA-mediated activation of expression of *PR* genes, as well as for suppression of synthesis of JA and JA-responsive gene expression (Spoel *et al.* 2007). Apart from the action of a wide range of hormone and secondary metabolites, resistance to necrotrophs is also mediated by plant cell wall composition. Mutants with defected primary and secondary wall cellulose synthase (CESA) expression or function have shown increased resistance to necrotrophic fungi (Dobón *et al.* 2015).

High ABA (abscisic acid) concentration is known to affect plant disease resistance playing both a positive and a negative role in stimulating plant response to pathogen infection. ABA exerts positive responses predominantly in pre-invasive defence against pathogens. It increases penetration resistance through rapid closure of stomata and exerts positive response in early post-invasive defence. It has been reported that exogenous application of ABA contributes to the resistance of *Arabidopsis* plants against fungal pathogens like *Pythium irregulare* and *Alternaria brassicicola*, and of barley against *Blumeria graminis* f. sp. *hordei*. This enhanced resistance may be due to reduction in spread of pathogen achieved by ABA-mediated callose biosynthesis or inhibition of its degradation. On the other hand, higher levels of ABA in plants can repress expression of defence gene by regulating SA, JA or ethylene mediated signalling. Exogenous application of ABA on *Arabidopsis* plants has also been reported to increase virulence of *P. syringae* pv. tomato. Similarly, ABA application suppressed transcription of defence genes which increases susceptibility of *Arabidopsis* to *Fusarium oxysporum* (causal agent of wilt) and *Erwinia chrysanthemi* (causal agent of bacterial wilt) infection. Studies have also used mutants defective in ABA biosynthesis and perception to demonstrate the effects of ABA in disease resistance (Ramegowda *et al.* 2015).

Apart from the above-listed plant hormones, several other molecules are also produced in response to biotic stress which helps in regulating plant stress response pathways. ROS, RNS (reactive nitrogen species) and AOX (alternative oxidase) come under this category of bioactive molecules which have a significant role in bringing about active defence responses in plants. ROS is an inclusive term for oxygen radicals like O^{2-} (superoxide) and OH^- (hydroxyl) radicals, and non-radicals such as H_2O_2 (hydrogen peroxide), singlet oxygen etc. which are rapidly produced in plants as a defence response to pathogen attack. ROS is involved in regulation of various biological processes such as growth, development, response to biotic stress, programmed cell death, and is also involved in regulating environmental stresses (Río 2015). Though there are a number of potential sources of ROS, such as protoplastic sources centred upon mitochondrial, chloroplastic or peroxisomal generating systems, they have largely been studied only in relation to abiotic stresses. At the cell-surface level, superoxide anion is believed to originate at plasma membrane in several species in response to biotic stresses from a NADPH (nicotinamide adenine dinucleotide phosphate) oxidase or from NADH-dependent superoxide synthase. Superoxide dismutase subsequently dismutates this superoxide to H_2O_2. Potential sources of apoplastic origin of H_2O_2 include peroxidases, amine oxidases and oxalate oxidases. Prospective mechanisms for it include a system analogous to mammalian NADPH oxidase system based upon evidences like defence response inhibition by DPI (diphenylene iodonium), which is a specific inhibitor of the mammalian oxidative burst (Bolwell *et al.* 2002).

The term RNS (reactive nitrogen species) includes both radicals like NO (nitric oxide) and NO_2 (nitric dioxide) and non-radicals like HNO_2 (nitrous acid) and N_2O_4 (dinitrogen tetroxide) etc. Amongst RNS produced in plants, NO along with ROS is a key mediator in the early defence response to pathogen attacks in plants (Río 2015). Less is known about the other NO-derived molecules during the plant-pathogen interaction. RSNO (*S*-nitrosothiols) were reported to show a regulating influence on sunflower hypocotyls dependent on the plant's susceptibility to the infection by downy mildew. It is proposed that RNS may be involved in influencing post-translational protein modification (tyrosine nitration of proteins) during plant-pathogen interaction (Chaki *et al.* 2009).

SA, ROS (particularly H_2O_2 and O_2-) and RNS (particularly NO and peroxynitrite), which are important signalling molecules in initiation and coordination of plant defence responses against bacterial pathogens, were found to increase AOX (alternative oxidase) levels in several plant species. AOX expression acts as a potential key regulator of a mitochondrial O_2- based signalling pathway which has a subsequent impact on plant responses to bacterial infection. However, establishing a relationship between changes in AOX amount and activity, type of bacterial interaction (compatible or incompatible) and downstream responses, has been difficult. A study using transgenic tobacco plants with silenced Aox1a gene expression was useful in evaluating the role of AOX. In AOX-silenced plants, more leaf damage was caused by the piercing-sucking insect and in the bacterial infection, a HR-like cell death response was said to occur more rapidly. In both these cases higher content of SA was observed in plants that were lacking AOX as compared to wild type plants. This elevated levels of SA may have promoted cell death by expediting mitochondrial dysfunction. This report suggested that AOX levels may directly or indirectly influence SA levels which in turn would act as an important determinant of numerous biotic stress responses (Vanlerberghe 2013).

Infection by fungal pathogens elicits characteristic defence responses, prominent of which is induction of chitinases (which accumulate at penetration site) that hydrolyze chitin (constituent of fungal cell walls). Fragments generated from chitin degradation are then detected by host cells leading to PTI (*PAMP-triggered immunity*) activation. The LysM Receptor Kinase1 (LysM RLK1) (also known as CERK1) is involved in early recognition events and is required for chitin-elicited immune responses. *It* exhibits a pathogen-dependent function in resistance (Laluk *et al.* 2010). It was reported that post-chitin treatment, expression of chitooligosaccharide-responsive genes was completely blocked in *LysM RLK1* mutant plants but it still showed induced expression upon pathogen infection. This indicated that other fungal elicitors induce immune responses which overlap with the genes required for chitin responses (Wan *et al.* 2008). This could account for weak susceptibility of the mutants to necrotrophic fungi. *LysM RLK1* mutants also showed altered chitin-induced MAP kinase signalling and altered *WRKY* (transcription factors) gene expression involved in PTI response pathway. Therefore it is possible that LysM RLK1 is an upstream regulator of chitin-triggered immunity which has overlapping components with other PTI responses.

Nectrotrophic fungi produce pectolytic enzymes that can hydrolyze the homogalacturonan of plant cell wall pectin. Plants counteract their action by producing PGIPs (polygalacturonase inhibiting proteins) which contribute to basal resistance by blocking cell wall degradation and enhancing accumulation of long chain oligogalacturonides (OGs). OGs are fragments released due to partial degradation of primary cell wall which are capable of activating innate immune responses. As they are derived from host tissue rather than pathogen components, they are considered typical damage-associated molecular patterns (DAMPs). Depending on the degree of chain polymerization, OGs induce oxidative bursts, cell wall lignification, accumulation of phytoalexin, protease inhibitor expression, and changes in ion fluxes as well as SA, JA, and ET biosynthesis. Necrotrophic pathogens commonly activate DAMP-triggered immune responses as they actively release OGs during the course of infection (Laluk *et al.* 2010).

In response to viral infections, plants possess both active and passive means of defence. Passive defences are when the plant fails to produce one or more host factors required for virus reproduction and spread whereas active defences involve detection and destruction of virus-infected cells by specific plant resistance genes. Resistance genes are specific for particular viruses. The plant defence system functions *via* RNA silencing wherein viral RNAs are detected and degraded (Gergerich *et al.* 2006).Viral infection is limited to the site of entry *via* induction of host immune response. Symptoms of viral infection like chlorotic lesions or spots, ring spots and necrotic lesions are basically manifestations of host defence responses activated in infected cells. Generally, viruses do not encode P/MAMPs or effector proteins and the R protein mediated antiviral immune response is not classified as an ETI response. Similar to non-viral infections, viral infections also can trigger HR response. However, in susceptible (or compatible) virus infections, HR response and necrotic lesions do not occur but another form of necrosis, termed systemic necrosis, is observed. Similar yet distinct from HR-associated necrosis, systemic necrosis occurs at a much later stage of infection and is mainly seen in upper non-inoculated tissues. The ubiquitin proteasome system (UPS) is a key factor influencing virus-host interactions throughout. It is prominently involved in regulation of various cellular activities like cell cycle, transcription, and signal transduction. In turn, viruses also utilize numerous strategies to modulate UPS processes. However, an ambiguity exists on whether UPS processes are employed by plants against virus infections or viruses use UPS to promote virulence. Viral infections also activate SAR but more research is required to understand molecular pathways leading to SAR in virus-host interactions (Mandadi *et al.* 2013).

Over and above the systems explained so far, there exist certain other mechanisms of defence against biotic stress. One such mechanism involves dynamics of protein folding. At molecular level, many defence responses against pathogenic infections lead to accumulation of unfolded or misfolded proteins. Nascent protein synthesis and folding demands exert stress on the ER (endoplasmic reticulum) and induce stress-relieving responses, jointly referred as UPR (unfolded protein response). In order to recover homeostatic functions, it is necessary for cells to activate UPR signalling pathways. Under non-stress conditions, ER chaperones bind to the ER luminal domains suppressing ER membrane spanning UPR pathways. During stress, unfolded proteins accumulate in ER and these ER chaperones dissociate to facilitate protein folding which activates transmembrane signalling proteins. Though much research is needed to decipher role of UPR in the regulation of other plant cellular responses, it has nevertheless been linked to the SA pathway in *Arabidopsis* and rice (Boatwright 2013).

4. Plant response to abiotic stress

Plant responses to abiotic stresses are dynamic and complex; they are both elastic (reversible) and plastic (irreversible). Responses of plants to stress are dependent on the tissue or organ affected by the stress, and the level and duration of stress (acute vs chronic), both of which have a significant effect on the complexity of the response (Cramer *et al.* 2011). The MEKK1-MKK2-MPK4/MPK6 module is the most complete MAP kinase cascade functioning in abiotic stresses in *Arabidopsis*. Environmental stresses like low temperature, high salinity and mechanical stresses lead to accumulation of MEKK1 mRNA. Protein-protein interactions were shown between- MEKK1: MKK1/MKK2, MKK1/MKK2: MPK4 and MEKK1: MPK4. MPK4 and MPK6 show rapid and transient activation by low temperature, low humidity, hyper-osmolarity, touch and wounding. During salt and cold stress, MKK2-MPK4/MPK6 functions downstream of MEKK1. Moreover, stresses like salt, drought and wounding activate MKK1 which can phosphorylate MPK4, suggesting its involvement in abiotic stress signalling. Oxidative stress activates MPK1 and MPK2, MPK3 and MPK6, MPK4 and MPK7. H_2O_2-dependent activation of ANP1 (a type of *Arabidopsis* cDNA encoding putative protein kinase (Nishihama *et al.* 1997)) was shown in protoplasts (Danquah *et al.* 2014).

Various plant hormones are up-regulated or down-regulated in response to various environmental stressors. Endogenous levels of ABA increase under drought stress as ABA biosynthesis genes are induced which reprogram gene expression to regulate water relations by adjusting cellular osmotic pressure, stomatal closure, reduction of leaf canopy, deeper root growth and variation in architecture of root system. Stomatal closure is one of the earliest plant responses to water deficit condition, and is regulated mainly in an ABA-dependent manner (Duque *et al.* 2013). During drought stress, ABA accumulated in the vascular tissue is transported to guard cells, *via* passive diffusion in response to pH changes or specific transporters (Yuriko *et al.* 2014), where it acts on guard cells and induces closure of stomata *via* efflux of potassium and anions from the guard cells. ABA also regulates membrane ion channels *via* increased cytosolic Ca^{2+} obtained from intracellular stores and a Ca^{2+} influx from the extracellular space. It has been reported that drought stress up-regulates NCED3 (Neoxanthin epoxy-carotenoid cleavage dioxygenase 3) expression (an enzyme involved in biosynthesis of ABA) in *Arabidopsis*, maize, tomato, bean and avocado. In response to drought stress NCED3 is accumulated in the vascular parenchyma cells of the leaves and not detected in the absence of stress conditions. ABA can diffuse passively across biological membranes in its protonated state and its ability to travel long distances allows it to serve as a critical stress messenger. ABA is meant to be transported to the apoplastic space, which has a pH around 5.0-6.0, before reaching the target cell but during stress response, strong alkalization of apoplastic pH slows diffusive transport of ABA from apoplastic space to target cells. Thus, ABA transporters facilitate movement of non-protonated ABA. After receiving ABA from ABC transporters, the specific receptor PYR/PYL/RCAR-ABA (pyrabactin-resistance1/pyrabactin resistance like/regulatory component of aba receptor) complex identifies ABA, forming ternary complexes. These complexes further inhibit ABA insensitive 1 (ABI1), ABA insensitive 2 (ABI2), hypersensitive to aba1 (HAB1), the negative regulators of ABA signalling, thereby facilitating activation of down-stream PP2C targets [SnRK2 (the sucrose non-fermenting 1- related subfamily 2 protein kinase)] that play key roles in the regulation of ABA signalling. Lastly, the enzyme OST1 (open stomatal) shows dominant kinase activity during response to drought stress when ABA signal is transmitted to guard cells. Reports have shown that OST1 mutants display a wilted phenotype under water deficit conditions while mutants for the other two ABA-activated kinases, SnRK2.2 and 2.3, do not exhibit a drought-sensitive phenotype. The triple mutant snrk2.2/d snrk2.3/I snrk2.6/e displays an extremely sensitive phenotype under water deficit conditions. ABA efficacy is regulated not only by the duration of drought stress or the previous stress history of a given plant, but also by other phytohormones like jasmonates, cytokinins (CKs) and ET (Duque et al. 2013).

ABA-dependent stress signalling activates AERB (leucine zipper transcription factors), which bind to ABA-responsive element (ABRE: major cis-element for ABA-responsive gene expression) (Nakashima et al. 2013) to induce the stress responsive gene RD29B which encodes a late embryogenesis abundant (LEA)-like protein. SNF1-related protein kinases 2 are key regulators of ABA signalling including the AREB/ABF regulon. ABA receptors and group A 2C-type protein phosphatases govern ABA signalling pathway (Nakashima et al. 2013). Transcription factors like DREB2A and DREB2B transactivate the DRE cis-element of osmotic stress genes and maintain the osmotic equilibrium of the cell. The drought-inducible expression of DREB1D is regulated by ABA-dependent pathways, indicating that it functions in slow response to drought that depends upon the accumulation of ABA (Tuteja 2007). Moreover, salinity stress also increases ABA concentration as it is associated with leaf/soil water potential (due to water deficit). Nevertheless increase in ABA levels may also occur in association with slowly increasing salinity stresses in nature or field situations (Bari and Jones 2009, Danquah et al. 2014, Fahad et al. 2015).

Apart from its role during biotic stresses, JA (jasmonic acid) also mediates plant defence responses to environmental stressors like drought, low temperature, and salinity. Water scarcity induces expression of several genes which also respond to JA such as the expression of genes encoding the soybean vegetative storage protein (VSP) acid phosphatases. A study showed that VSPá and VSPâ accumulate in soybean vacuoles and, to a smaller amount, in cell walls. It was seen that accumulation of JA increased VSP messenger RNA (mRNA) levels in wounded tissue. In fact, the VSP acid phosphatases B promoter contains a DNA domain that mediates responses to JA and another DNA domain which mediates responses to sugars (positive), auxin (negative), and phosphate (negative). Exogenous application of JA to plants can induce expression of stress related genes. Upon exogenous application, JA or methyl jasmonate (MeJA) gets converted to a biologically active form (+)-7-iso-Jasmonoyl-L-isoleucine (JA-Ile) which binds to the receptor SCFCOI complex. This leads to degradation of the repressor protein – JAZ (Jasmonate ZIM-domain), and allows MYC2 (myc domain transcription factor 2) activation of distinct JA response genes. Up-regulation of MYC2 can take place both by ABA and drought. These interactions suggest an important regulatory role of JA in an ABA-dependent response to drought. Moreover, jasmonates also trigger stomatal closure in drought stress response. Low endogenous ABA content impair MeJA (methyl-jasmonate)-stimulated Ca^{2+} elevation and MeJA stimulates the expression of NCED3. Presence of endogenous ABA is required for MeJA signalling in guard cells. JA also up-regulates pyrabactine like 4 (AtPYL4), pyr abactine like 5 (AtPYL5) and pyrabactine like 6 (AtPYL6), members of the PYR/PYL/RCAR ABA receptor family providing yet another example of JA and ABA crosstalk. Elevated JA levels were seen in soybean and *Pinus pinaster* when under drought stress and in tomato and *Iris hexagona* under saline stress. It has been reported that both drought and increased salinity elevated JA levels in rice plant inducting stress-related PR and JA biosynthetic genes. Exogenous application of JA ameliorates salt stressed rice seedlings and can significantly decrease sodium concentration. Post salt treatment, exogenous application of JA might affect the endogenous hormone balance, such as ABA, thereby providing clues for understanding the protection mechanisms against salt stress (Bari and Jones 2009, Fahad *et al.* 2015).

IAA (indole-3-acetic acid) plays a major role in regulating plant growth including cell elongation, vascular tissue development, and apical dominance. IAA levels show similar variation as that of ABA under stress condition. Increase in IAA levels correlate with reduced growth, indicating that the reduction in plant growth under stress conditions could be due to altered hormonal balance. Under salinity or heavy metal stresses, IAA modulates root and shoot growth of plants. It was seen that salinity stress reduced IAA levels in maize plants; however under salt stress application of SA results in elevated IAA levels suggesting that hormonal balance is critical, thus influencing plant growth and development. By stimulating transcription of a large number of genes called primary auxin response genes, auxins play a significant role in defence responses mediating crosstalk between abiotic and biotic stress responses. Biosynthetic pathway of IAA still has some open ended questions at the genetic level and remains to be elucidated (Bari and Jones 2009, Fahad *et al.* 2015).

GA (gibberellic acid) is an important plant growth regulator which increases the stress tolerance of many crop plants and is accumulated under abiotic stress. At a certain concentration, it is beneficial for plant physiology and metabolism as it regulates the metabolic processes as a function of sugar signalling and anti-oxidative enzymes. By influencing photosynthetic enzymes, GAs improve the photosynthetic efficiency of plants, light interception, leaf area index, and efficiency of nutrient usage. GAs are involved in stress-induced growth modulation, and signalling and would either suppress or promote growth depending on the particular abiotic stress. Under stress condition, GA increases the photosynthetic rate leading to increased source potential and redistribution of photosynthates, thereby increasing sink strength. GA also alleviates the adverse effects of environmental stresses on plant water relations (Bari and Jones 2009, Fahad *et al.* 2015).

Two more hormones implicated with stress response in plants are SA and brassinosteroids (BRs). Apart from its involvement in biotic stresses, SA is also involved in responses to several abiotic stresses like drought, salinity, extremely low and high temperature and heavy metal tolerance. SA increases photosynthetic capacity *via* stimulatory effects on RuBisCO activity and pigment content. It also increases IAA levels and decreases ABA content in maize plants under saline stress. The exact regulatory role and mechanism of action of SA remains an active area of research. Brassinosteroids (BRs) are steroidal plant hormones regulating plant growth and development by controlling a wide range of physiological changes including seed germination, vascular formation, cell, reproductive growth, flowers and fruit production. They are involved in various types of biotic and abiotic stress responses including those induced by salinity, drought, and heat stress. BRs also play a vital role in regulation of ion uptake and its application reduces accumulations of heavy metals and radioactive elements (Bari and Jones 2009, Fahad *et al.* 2015).

CKs (cytokinins) are involved in regulation of cell proliferation and differentiation. Abiotic stresses like drought reduce the biosynthesis and transportation of CKs from roots to shoots. Apart from maintaining a better photosynthetic rate, increase in CK levels enhances activity of cell-cycle genes which consequently increases cell number resulting in improved grain filling. While increased CK levels improve grain quality and photosynthesis rate, high levels of ABA increase root extension rate, osmoprotectant activity, and solute biosynthesis. Therefore obtaining high amounts of ABA and CKs is desirable, particularly during weaker drought conditions. CKs also help the plants resist against salinity and high temperature. Under increased concentration of NaCl, concentrations of zeatin (Z), zeatin riboside (ZR), isopentenyl adenine (iP), and isopentenyl adenine (iPA) are decreased in shoots and roots of barley cultivars. Kinetin acts as free radical scavenger, or is involved in anti-oxidative mechanism which is related to the safety of purine breakdown. CKs regulate the expression of various stress-induced genes; however, the exact mechanisms are not well understood (Duque *et al.* 2013; Fahad *et al.* 2015).

ET (ethylene) is a gaseous plant hormone which inhibits root growth/development and is involved in stress-induced leaf senescence contributing to reduced rate of photosynthesis. Influence of ethylene on stomatal conductance is further modulated by ABA. However, contradictory reports of the role of ethylene in stomatal action exist. At times ethylene was shown to induce stomatal closure, while in certain other reports it has also been shown to antagonize ABA action in the stomata. This indicates that the ratio of ABA to ethylene can be more important than the concentration of either hormone (Duque *et al.* 2013). Ethylene signalling functions in multiple stress responses and is shown to promote salt tolerance in *Arabidopsis*. Plant response to salt stress also depends on the balance and/or interaction between ET receptor and ET. Under prevalent receptor signalling, plant is susceptible to salt stress and shows large rosette and late flowering whereas in prevalent ethylene signalling, the plant is tolerant to salt stress but exhibits small rosette and early flowering. Plants need to strike a balance between these two extremes in order to regulate salt stress responses (Bari and Jones 2009, Fahad *et al.* 2015).

Triazole compounds protect plants from stress-related damages caused by drought, heat, chilling, ozone, and sulphur dioxide. They augment the plant dry mass of radish and pigeon pea under saline stress (Fahad *et al.* 2015). Triazoles alter certain plant hormones by inhibition of gibberellin synthesis, reduction of ethylene evolution and increase in cytokinin levels (Rajalekshmi *et al.* 2009). They enhance photosynthesis rate by increasing chlorophyll content and more tightly packed chloroplasts within a smaller leaf area. Under salt stress, triazole raises the net photosynthetic rate and intercellular CO_2 concentration in radish and drought-prone wheat plants. Triazole was also shown to protect bean plants from heat stress and stopped electrolyte leakage. Under stressful conditions, triazole stimulates nitrate reductase and protease activity and enhances activity of antioxidant enzymes like peroxidase, superoxide dismutase, and polyphenol oxidase (Fahad *et al.* 2015). Triazole compounds have both fungi-toxic and plant growth-regulating properties. Additionally, they can also protect plants against various environmental stresses (Rajalekshmi *et al.* 2009).

5. Metabolic adjustments during stress

Under stress, early plant response mechanisms would be focused towards re-achievement of homeostasis and continuation of growth, equilibrium recovery of energetic, osmotic, redox imbalances imposed by the stressor. This is achieved by various metabolic and physiological adjustments including changes in amino acid, carbohydrate, and amine metabolic pathways. Metabolic activities respond faster than transcriptional activities during stress conditions and help cells in restoring chemical and energetic balances which are disturbed by stress (Cramer *et al.* 2011). Transcriptional activation of several genes takes place under stress conditions. Among these, many code for stress-induced proteins that improve water movement through membranes, detoxification enzymes and enzymes required for osmolyte biosynthesis. Accumulation of low-molecular weight osmolytes (in response to water scarcity) protects macromolecular structure from stress-induced damage. However, increased intracellular osmolyte concentrations would affect protein structure and dynamics. Some stress proteins can facilitate protein folding and thus offer protection from misfolding and aggregation. Their targets are partially unfolded or misfolded proteins with patches of hydrophobic residues exposed to the surface. In order to maintain protein performance under abiotic stress, the prime factors involved in catalytic mechanisms of plants are hydration status and temperature. Hydration is essential for catalytic function of enzymes for diffusion of substrate and product. Temperature is a fundamental environmental stress, as flexibility and functionality of enzymes are temperature dependent. Various osmolytes protect enzymes against heat-induced loss of activity by a variety of means such as: stabilizing protein global folds, supporting retention of secondary structure of elements and facilitating refolding of thermally unfolded proteins (Ortbauer 2013).

5.1 Energy sensor protein kinase

The SnRKs (sucrose non-fermenting-1 (SNF1)-related protein kinases) gene family in plants is sub-divided into three sub-families: SnRK1, SnRK2, and SnRK3. SnRKs play a key role in metabolic and stress signalling and this ability can be manipulated to improve crop performance in unfavourable environments (Coella *et al.* 2010). SnRK1 is a protein kinase that regulates expression of genes related to energy-depleting conditions and is also activated by various abiotic stresses such as drought, salt, flooding, or nutrient scarcity. It modifies the expression of over a thousand stress-responsive genes to facilitate re-establishment of homeostasis by repressing energy-expensive processes, thereby promoting stress tolerance. Upon curbing of energy-expensive processes, energy resources can be utilized to activate protective mechanisms (Fraire-Velazquez *et al.* 2013). The SnRK2 subfamily includes plant-specific kinases which are involved in abiotic stress response as well as ABA-dependent plant development. The SnRK2 members are classified into three groups based on their relation to ABA. ABA doesn't activate group 1 kinases while it does not or very weakly activates group 2 kinases. Group 3 kinases are strongly activated by ABA and are considered integral in regulation of plant response to ABA. Though group 2 kinases share certain cellular functions with group 3 but their role in ABA-related responses has not been determined clearly. They are believed to positively up-regulate plant responses to drought stress and might play a complementary role to the ABA-dependent kinases in order to regulate plant response against environmental stressors (Kulik *et al.* 2011). The SnRK3 family members are also known as CIPKs (CBL-interacting protein kinases) as they interact with CBL (calcineurin B-like) calcium-binding proteins and CIPKs are reportedly involved in plant responses to several environmental stresses like drought, low temperature, ABA, salinity, and pH variation. One of the most widely known pathway in which there is CIPK involvement is the salt overly sensitive (SOS) pathway which regulates salt sensitivity in *A. thaliana* by controlling the expression of many tolerance determinant genes specific for salt stress (Kulik *et al.* 2011).

5.2 Osmolyte accumulation

Production and accumulation of compatible solutes is a common defence mechanism in plants under stress. Osmoprotectants have multiple protective functions during environmental stress such as scavenging of ROS, restoring redox metabolism, maintaining cellular turgor by restoration of osmotic balance and associated stabilization of proteins and cellular structures. Low molecular weight organic osmoprotectants have diverse chemical natures and include: amino acids (asparagine, proline, serine), amines (polyamines and glycinebetaine), GABA (γ-amino-N-butyric acid), carbohydrates (fructose, sucrose, trehalose, raffinose), polyols (myo-inositol, D-pinitol) and anti-oxidants [GSH (glutathione) and ascorbate]. These solutes have high level of solubility in the cellular environment and their enzyme activities are not inhibited even at higher concentrations (Fraire-Velaìzquez *et al.* 2013). Proline accumulation correlates with stress tolerance and it serves as an osmoprotectant, cryoprotectant, signalling molecule, stabilizer of protein structure, and as ROS scavenger in response to dehydration causing stresses such as salinity, freezing, heavy metals, and drought. Oxidation of proline provides energy to sustain the metabolic energy requirements of plant reproduction after stress has ended. Balance between proline biosynthesis and catabolism is achieved by core enzymes such as P5CS, P5C, P5CR, ProDH, and ornithine aminotransferase (OAT) and its regulation occurs at the transcriptional level of genes encoding key enzymes. During dehydration/osmotic stress, genes for P5CS and P5C are transcriptionally up-regulated to increase proline synthesis from glutamate and genes for P5CR and ProDH are down-regulated to inhibit proline catabolism. Moreover, post-translational regulation of core enzymes is very much influenced by level of proline and other environmental signals. Overexpression of biosynthetic proline enzymes increases the levels of the compatible solute thereby improving plant tolerance against abiotic stress (Fraire-Velaìzquez *et al.* 2013).

6. Changes in plant bioenergetic status

Changes in the bioenergetic status of plant may include either rise in energy demand or a depleted energy production owing to metabolic adjustments designed to overcome stress. to overcome stress. Plants use various methods for energy dissipation and avoidance or minimisation of photo-inhibition, when the production of ATP and NADPH exceeds the capacity for utilization in CO_2 fixation. This includes alternative electron sinks dependent on O_2 like the oxygenase reaction catalysed by RuBisCO for initiation of photorespiration. Under drought, photosynthesis and photorespiration is reduced due to lower CO_2 and O_2 availability in the chloroplast though the photo-respiratory pathway is less decreased than photosynthesis. As photo-respiration is the major cause of lower bioenergetic balance in photosynthetic tissues of C3 plants, increase in plant growth by overcoming of photosynthesis limitations imposed by Rubisco is an important target of research and plant improvement. In case of photo-inhibition, non-photochemical quenching of chlorophyll fluorescence modulates energy dissipation and protects against photo-damage. Alternative electron cycling like photorespiration also defends against photo-damage. However, when photo-oxidation is not preventable, the photosynthetic apparatus gets damaged. Mainly the reaction centre D1/D2 heterodimer in PSII is affected, especially D1 while D2 is affected to a lesser extent. As D1 has a high turnover rate, repair of damaged components is activated but if the rate of repair is not proportional to the rate of damage, photosynthesis is decreased and photo-inhibition happens. There also might be the involvement of nuclear-encoded early-light inducible proteins (ELIPs) in the protection mechanism mentioned above (Duque *et al.* 2013).

7. Gene expression and regulation under stress

Regulation of plant genes can either be at the transcriptional, post-transcriptional or post-translational level. Chromatin modification and remodelling were frequently reported in plant abiotic stress responses. The process of priming (sensitization of stress responsiveness) bolsters plant's defensive capacity and brings it into an alarmed state of defence. It is correlated with chromatin modification of promoter region of WRKY transcription factors. The contribution of epigenetic mechanisms to environmental cues and different types of abiotic stresses has been documented. Recent reports have shown that different environmental stresses lead to alteration in methylation of DNA and modification of nucleosomal histones. Involvement of dehydration-responsive element-binding

(DREB) or C-repeat binding factor (CBF), MYB, basic-leucine zipper (bZIP), and zinc-finger family members in regulation of plant defence and stress responses has been shown and most of these transcription factors regulate target gene expression by binding to the cognate cis-element in promoters of stress-related genes. WRKY transcription factors are one of the best-characterized classes of plant transcription factors and several WRKY proteins are involved in plant drought and salinity stress responses (Duque *et al.* 2013). NAC (N-acetylcysteine) proteins are plant-specific transcription factors involved in plant development and in stress responses. The cDNA which encodes NAC protein is responsive to dehydration 26 (RD26) gene in *Arabidopsis*. It is reported that OsNAC6 expression is induced by cold, drought, high salinity, and ABA. High sequence similarity between OsNAC6 and the *Arabidopsis* stress-responsive NAC proteins ANAC019, ANAC055, and ANAC072 (RD26) has also been shown. Abiotic stress-responsive NAC-type transcription factors, particularly the SNAC group genes, have important roles in controlling tolerance against environmental stresses such as drought (Duque *et al.* 2013).

Post-transcriptional regulation is the second level of gene expression modulation. Alternative splicing regulates gene expression in plants challenged by temperature extremes. Stabilized1 (STA1) is a gene coding for a splicing factor and it plays a significant role in overcoming cold stress in *A. thaliana*. Alternative splicing has been reported upon water scarcity as well. Stress can also regulate microRNA (miRNA) levels in plants. Several plant miRNAs play major roles in conferring resistance plant resistance to abiotic stresses. In plants, they repress gene expression by directing mRNA degradation or translational arrest. MiRNAs guided binding of Argonaute (AGO) to target promotes cleavage of mRNAs with almost perfect base complementarity and/or by inhibition of translation of lower complementarity miRNAs (Duque *et al.* 2013). Table 3 provides a list of miRNAs with known involvement in abiotic stress responses.

At the post-translational level, phosphorylation, sumoylation and ubiquitination of proteins play vital role in modulating plant responses to abiotic stress. Under water scarcity/osmotic stress phosphorylation of specific residues activates various signal transduction cascades formed by MAPKs and SnRKs. The transcriptional, post-transcriptional and post-translational mechanisms combine to ensure appropriate downstream gene expression patterns. They ultimately regulate the transcriptome and proteome of plants subjected to stress in order to exhibit an adaptive response. Elucidation of these three regulatory pathways and their interplay is required to fully understand the molecular mechanisms by which plants adapt to their environment and for plant improvement for stress tolerance (Duque *et al.* 2013).

Table 3: miRNAs involved in abiotic stress responses.

S. No.	Stressor	MiRNA	Host	
			Up-regulated	Down-regulated
1	Drought responsive	MiR395	Rice	*Prunus persica* and *Populus tomentosa*
		MiR396	-	Rice, *Prunus persica* and *Medicago truncatula*
		MiR408	*A. thaliana* and *Medicago truncatula*	Rice, *Populus trichocarpa, Populus tomentosa* and *Prunus persica*
		MiR398	*Medicago truncatula* and *Triticum dicoccoides*	*Prunus persica*
		MiR399	*Medicago truncatula*	-
		MiR156	*Arabidopsis, Prunus persica,* barley, *Panicum virgatum* and *Triticum dicoccoides*	Rice and maize
		MiR157	Cotton	-
		MiR159	*Arabidopsis thaliana*	Rice and *Prunus persica*
		MiR160	*Populus tomentosa* and *Prunus persica*	*Populus trichocarpa*
		MiR162	*Populus tomentosa*	-
		MiR164	-	*Medicago truncatula, Populus trichocarpa, Brachypodium distachyon*
		MiR165	-	*Prunus persica*
		MiR166	*Glycine max*	*Triticum dicoccoides*
		MiR167	*A. thaliana, Populus tomentosa*	*Prunus persica* and maize
		MiR168	*Arabidopsis*	Rice and *Z. mays*

S. No.	Stressor	MiRNA	Host	
			Up-regulated	Down-regulated
		MiR168a and b	-	Arabidopsis
		MiR169a and c	-	A. thaliana
		MiR169g and MiR169n/o	Rice	-
	Prunus persica	MiR169	Tomato, Glycine max, Populus euphratica	Medicago truncatula, Populus tomentosa,
		MiR170	-	Arabidopsis and rice
		MiR171	A. thaliana, Prunus persica	Triticum dicoccoides, Medicago truncatula,
	Populus tomentosa	MiR172	A. thaliana and rice	Populus tomentosa
		MiR319	Populus tomentosa, A. thaliana and rice	-
		MiR390	-	Populus tomentosa
		MiR393	A. thaliana, Phaseolus vulgaris, O. sativa	Prunus persica
		MiR394	Populus tomentosa and Populus trichocarpa	Glycine max
		MiR397	A. thaliana	Rice, Populus tomentosa and Prunus persica
		MiR403	Populus tomentosa	-
		MiR474	Rice, maize, M. truncatula, Triticum dicoccoides	-
		MiR528	-	Z. mays
		MiR827	Z. mays	-
		MiR1432	Triticum dicoccoides	-
		MiR2118	Medicago truncatula	-
		MiR530a,	-	Populus trichocarpa

S. No.	Stressor	MiRNA	Host	
			Up-regulated	Down-regulated
		miR1445, miR1446a-e and MiR1447		
		MiR1450	-	*Populus trichocarpa*
2	Salt stress	MiR395	*P. tremula* and *Z. mays*	-
		MiR396	*A. thaliana*	*Z. mays*
		MiR398	*P. tremula*	*A. thaliana*
		MiR399	*P. tremula*	-
		MiR156	-	*Z. mays*
		MiR157	-	Cotton
		MiR159	*A. thaliana*	-
		MiR169a and c	-	*A. thaliana*
		MiR169g and n/o	Rice	-
		MiR169	-	*P. tremula*
		MiR167	*A. thaliana, Populus tomentosa*	*Prunus persica* and maize
		MiR168	*A. thaliana, Z. mays, Populus tomentosa*	-
		MiR171	*A. thaliana*	-
		MiR319	*A. thaliana*	-
		MiR393	*A. thaliana, Phaseolus vulgaris*	-
		MiR397	*A. thaliana*	-
		MiR530a,	-	*Populus trichocarpa*

S. No.	Stressor	MiRNA	Host	
			Up-regulated	Down-regulated
		miR1445, miR1446a-e and miR1447		
		MiR1450	*Populus trichocarpa*	-
3	Oxidative stress	MiR398	-	*Arabidopsis*
4	UV-B	MiR398	*Populus tremula*	-
		MiR395	-	*Populus tremula*
5	Copper starvation	MiR408	-	-
		MiR398	-	-
		MiR857	*Arabidopsis* and other plants	-
6	Sulphate starvation conditions	MiR395	-	-
7	Phosphate starvation conditions	MiR399	-	-
8	Iron deficiency	MiR854 family	*Arabidopsis*	-

Sources: Burkhead *et al.* (2009); Covarrubias *et al.* (2010); Lima *et al.* (2012); Ding *et al.* (2013); Duque *et al.* (2013); Ferdous *et al.* (2015); Xie *et al.* (2015).

8. Response of plants to simultaneous biotic and abiotic stresses

Simultaneous occurrence of biotic and abiotic stresses can either exacerbate (i.e., susceptibility) or ameliorate (i.e., tolerance) the effect of stress on plants. Depending on the particular stress and pathogen, plants exhibit a complex and differential response leading to resistance or susceptibility of plants. Taking the case of simultaneous drought stress and pathogen interaction, generally both these stresses do not occur together instantaneously as drought stress develops gradually during the course of which pathogens can infect plants or drought stress can occur on pathogen infected plants due to which plants have to face combined pathogen and drought stress. Plant exposure to mild drought stress activates their basal defence response which defends them against pathogen infection. On the other hand, severe drought can facilitate pathogen infection by causing leakage of cellular nutrients into apoplast. Combined tolerance might be due to the inherent capacity of plants to customize the existing mechanisms while susceptibility could be due to the failure to tailor tolerance mechanisms and due to worsening of damage caused by one stress. This scenario gets even more complex when multiple stresses are involved. ABA, the primary regulator of drought stress response also alters pathogen response of plants. Drought-induced pathogen resistance may be a result of antimicrobial and PR-proteins activated by drought which protect plants during early stages of infection, while susceptibility might be caused due to elevated levels of ABA in drought stressed plants which can meddle with pathogen-induced plant defence signalling reducing the expression of defence-related genes. On the other hand, subjecting pathogen infected plants to drought stress can lead to tolerance to drought stress through pathogen-induced SA-dependent ROS signalling. It can also result in susceptibility of the plants to drought stress due to SA or JA-mediated reduction in responsiveness of plants to ABA. A study employed transcriptome analysis of *Arabidopsis* subjected to individual drought and nematode infections and their combinations. Results suggested that plants subjected to combined stress prioritize that abiotic stress which is more potentially damaging stress (Suzuki *et al.* 2014; Ramegowda and Senthil-Kumar 2015).

9. Conclusion and future prospects for developing stress tolerant plant varieties

Methods of hybridization and breeding followed by specific procedures to select plants expressing quantitative trait loci (QTLs) for stress tolerance have been extensively studied by several authors. Conventional plant resistant variety development methods involving a step-wise selection procedure can be employed where many genotypes can be evaluated under specific stress conditions. The best performing genotypes selected can further be propagated through breeding (for many generations) till a stable genotype is evolved. However, as many of the abiotic stresses may have common physiological expression, selection of stress tolerant traits through breeding is a complex process (Bänziger *et al.* 2000; Balconi *et al.* 2012).

Modern molecular tools have enlarged the biologist's tool kit to exploit the inherent ability of plants to counter stress. Using differential hybridization and expression analysis techniques, specific stress responsive gene were identified, isolated and cloned in many host plants such as *Arabidopsis, Nicotiana tabacum, Oryza sativa, Cassia tora* etc., amongst others (Cramer *et al.* 2011; Danquah *et al.* 2014; Ortbauer 2013). Some studies involved either up-regulation or down-regulation of stress responsive genes to obtain improved plant varieties (Roy *et al.* 2011). While describing the mechanisms of stress tolerance, several individual genes and gene products were named in the previous sections of the chapter and their role in offering protection against various types of stresses explained. Each of those genes can be probable candidates for developing transgenic plants resistant to stress. For instance, biosynthesis of hormones (IAA, JA, SA, ABA), metabolites (proline, betaine, polyamines, mannitol, glycine betaine) and specific PR proteins can be a sure approach to engineer stress tolerance in plants (Nuccio *et al.* 1999; Sakamoto and Murata 2000; Fahad *et al.* 2015). Scientists have been successful in creating transgenic plants through gene manipulation by targeting specific genes (within the host) involved in stress metabolism. Many studies have also attempted heterologous gene expression where candidate genes with known stress tolerance abilities were transferred across the plant genera and species. Examples for the former include: studies targeting carotenoid metabolism, targeting NF-YB1 homolog (a transcription factor conferring significant drought tolerance), the SNAC1 gene (transcription factor which up-regulates several stress-tolerance genes) etc. (Davison *et al.* 2002; Cramer *et al.* 2011). Some successful examples for heterologous gene expression are: salt tolerant rice with vacuolar ATPase gene from halophytic grass (Baisakh *et al.* 2012), salt tolerant transgenic *Brassica juncea* and tomato expressing NHX1 gene from *Arabidopsis thaliana* and other hosts (Roy *et al.* 2011). However, concerns were raised by multiple authors regarding production/yield compromises in transgenic plants. This is explainable as introduction of a transgene would mean reallocation of plants metabolic resources which may have an impact on plants yield /productivity. So careful engineering of thought is needed before venturing upon manipulating host genome for stress tolerance. Recent advances at the interception of molecular biology systems biology and informatics are opening new avenues deciphering the genome, transcriptome and metabolome of higher organisms including plants (Cramer *et al.* 2011; Samjova *et al.* 2012). Future plant biotechnology can greatly benefit from such knowledge inflows facilitating redesigning genetic networks of stress metabolism.

Acknowledgement

Authors thankfully acknowledge the Vice Chancellor, Jaypee Institute of Information Technology (JIIT), Noida, for all necessary support.

References

Alfano JR, Collmer A (1996) Bacterial Pathogens in Plants: Life up against the Wall. 1996 Plant Cell. 8, 1683-1698.

Balconi C, Stevanato P, Motto M, Biancardi E (2012) Breeding for biotic stress resistance/tolerance in plants. In: Crop Production for Agricultural Improvement (Ed) Ashraf M, Ozturk M, Ahmad MSA and Aksoy A, Springer, pp 57-114.

Baisakh N, Mangu V, Rao R, Rajasekaran K, Subudhi P, Janda J, Galbraith D, Vanier C, Pereira A (2012) Enhanced salt stress tolerance of rice plants expressing a vacuolar H+-ATPase subunit c1 (SaVHAc1) gene from the halophyte grass *Spartina alterniflora* Loisel. Plant Biotechnol. J. 10, 453-464.

Bänziger M, Edmeades GO, Beck D, Bellon M (2000) Breeding for drought and nitrogen stress tolerance in maize: From theory to practice. Mexico, D.F.: CIMMYT. http://repository.cimmyt.org/xmlui/bitstream/handle/10883/765/68579.pdf

Bari R Jones JDG (2009) Role of plant hormones in plant defence responses. Plant Mol. Biol. 69(4), 473-488.

Bell AA, Howell CR, Stipanovic RD (2010) Cotton host-microbe interactions. In: Physiology of Cotton (Ed) Stewart JM, Oosterhuis DM, Heitholt JJ and Mauney JR, Springer, Netherlands, pp. 187-205.

Bhatt MJ, Patel AD, Bhatti PM, Pandey AN (2008) Effect of soil salinity on growth, water status and nutrient accumulation in seedlings of *Ziziphus mauritiana* (*Rhamnaceae*). J. Fruit Ornam. Plant Res. 16, 383-401.

Boatwright JL (2013) Regulatory mechanisms of pathogen-mediated cellular stress signaling in *Arabidopsis thaliana*. Dissertation, The University of Alabama at Birmingham.

Bolwell GP, Bindschedler LV, Blee KA, Butt VS, Davies DR, Gardner SL, Gerrish C, Minibayeva F (2002) The apoplastic oxidative burst in response to biotic stress in plants: a three-component system. J. Exp. Bot. 53, 1367-1376.

Bost S (2013) Tomato Wilt Problems. UT Extension. https://ag.tennessee.edu/EPP/Extension%20Publications/Tomato%20Wilt%20Problems.pdf

Brooks FE (2008) Burrowing Nematode. *The Plant Health Instructor.* DOI: 10.1094/PHI-I-2008-1020-01.

Burkhead JL, Reynolds KA, Abdel-Ghany SE, Cohu CM, Pilon M (2009) Copper homeostasis. New Phytol. 182, 799-816.

Chaki M, Fernández-Ocaña AM, Valderrama R, Carreras A, Esteban FJ, Luque F, Gómez-Rodríguez MV, Begara-Morales JC, Corpas FJ, Barroso J (2009) Involvement of reactive nitrogen and oxygen species (RNS and ROS) in sunflower-mildew Interaction. Plant Cell Physiol. 50, 665-679.

Coella P, Hey SJ, Halford NG (2010) The sucrose non-fermenting-1-related (SnRK) family of protein kinases: potential for manipulation to improve stress tolerance and increase yield. J. Exp. Bot. 62, 883-893.

Compant S, Duffy B, Nowak J, Clément C, Barka EA (2005) Use of plant growth-promoting bacteria for biocontrol of plant diseases: Principles, mechanisms of action, and future prospects. Appl. Environ. Microbiol. 71(9)4951-4959.

Cornic G (2000) Drought stress inhibits photosynthesis by decreasing stomatal aperture – not by affecting ATP synthesis. Trends Plant Sci. 5, 187-188.

Covarrubias AA, Reyes JI (2010) Post-transcriptional gene regulation of salinity and drought responses by plant microRNAs. Plant Cell Environ. 33, 481-489.

Cramer GR, Urano K, Delrot S, Pezzotti M, Shinozaki K (2011) Effects of abiotic stress on plants: a systems biology perspective. BMC Plant Biol. 11, 163-177.

Damsteegt VD (1984) Maize streak virus: effect of temperature on vector and virus. Phytopathology 74, 1317-1319.

Danquah A, de Zelicourt A, Colcombet J, Hirt H (2014) The role of ABA and MAPK signaling pathways in plant abiotic stress responses. Biotechnol Adv. 32, 40–52.

D'Arcy CJ, Domier LL (2000) Barley yellow dwarf. *The Plant Health Instructor*. DOI: 10.1094/PHI-I-2000-1103-01.
http://www.apsnet.org/edcenter/intropp/lessons/viruses/Pages/BarleyYelDwarf.aspx

Davison PA, Hunter CN, Horton P (2002) Over expression of β-carotene hydroxylase enhances stress tolerance in *Arabidopsis*. Nature 418, 203–206.

Dean R, Van Kan JA, Pretorius ZA, Hammond-Kosack KE, Di Pietro A, Spanu PD, Rudd JJ, Dickman M, Kahmann R, Ellis J, Foster GD (2012) The Top 10 fungal pathogens in molecular plant pathology. Mol. Plant Pathol. 13, 414-30.

Decoteau D (1998) Plant Physiology: Environmental Factors and Photosynthesis. Greenhouse Glazing & Solar Radiation Transmission Workshop.
http://aesop.rutgers.edu/~horteng/workshop/lecture2.pdf

Ding Y, Tao Y, Zhu C (2013) Emerging roles of microRNAs in the mediation of drought stress response in plants. J. Exp. Bot. 64, 3077-3086.

Dobón A, Canet JV, García-Andrade J, Angulo C, Neumetzler L, Persson S, Vera P (2015) Novel disease susceptibility factors for fungal necrotrophic pathogens in *Arabidopsis*. PLoS Pathog. 11, e1004800.

Dorantes-Acosta AE, Sánchez-Hernández CV, Arteaga-Vázquez MA (2012) Biotic stress in plants: Life lessons from your parents and grandparents. Front. Genet. 3, 256.

Duque AS, de Almeida AM, da Silva AB, da Silva JM, Farinha AP, Santos D, Fevereiro P, de Sousa Araùijo S (2013) Abiotic stress responses in plants: Unravelling the complexity of genes and networks to survive in abiotic stress - Plant responses and applications in agriculture (Ed) Vahdati K, InTech, DOI: 10.5772/52779.

Fahad S, Hussain S, Bano A, Saud S, Hassan S, Shan D, Khan FA, Khan F, Chen Y, Wu C, Tabassum MA, Chun MX, Afzal M, Jan A, Jan MT, Huang J (2015) Potential role of phytohormones and plant growth-promoting rhizobacteria in abiotic stresses: consequences for changing environment. Environ Sci. Pollut. Res. Int. 22, 4907-21.

Fargette D, Jeger M, Fauquet C, Fishpool LD (1994). Analysis of Temporal Disease Progress of African Cassava Mosaic Virus. Phytopathology 84, 91-98.

Fauquet C, Fargette D (1990) African Cassava Mosaic Virus: Etiology, Epidemiology, and Control. Plant Dis. 74, 404-411.

Ferdous J, Hussain SS, Shi B (2015) Role of microRNAs in plant drought tolerance. Plant Biotechnol. J. 13, 293-305.

Flynn P (2003) Biotic vs. Abiotic - Distinguishing disease problems from environmental stresses. ISU Entomology. http://www.ipm.iastate.edu/ipm/hortnews/2003/9-12-2003/stresses.html

Fraire-Velaìzquez S, Balderas-Hernaìndez VE (2013) Abiotic stress in plants and metabolic responses. In: Abiotic stress - Plant responses and applications in agriculture (Ed) Vahdati K, In Tech, DOI: 10.5772/54859.

Gergerich RC, Dolja VV (2006). Introduction to plant viruses, the invisible foe. The Plant Health Instructor. DOI: 10.1094/PHI-I-2006-0414-01.

Hatfield JL, Prueger JH (2015) Temperature extremes: Effect on plant growth and development. Weather and Climate Extremes. 10, 4-10.

http://www.apsnet.org/edcenter/intropp/lessons/fungi/ascomycetes/Pages/RiceBlast.aspx

Isleib J (2012) Signs and symptoms of plant disease: Is it fungal, viral or bacterial? Michigan State University Extension.

http://msue.anr.msu.edu/news/signs_and_symptoms_of_plant_disease_is_it_ fungal_ viral_ or_ bacterial

Noling JW (1999) Nematode management in tomatoes, peppers, and eggplant. University of Florida IFAS extension. https://edis.ifas.ufl.edu/ng032

Johnson KB (2000) Fire blight of apple and pear. *The Plant Health Instructor.* DOI: 10.1094/PHI-I-2000-072601. http://www.apsnet.org/edcenter/intropp/lessons/prokaryotes/Pages/FireBlight.aspx

Jones JT, Haegeman A, Danchin EG, Gaur HS, Helder J, Jones MG, Kikuchi T, Manzanilla-López R, Palomares-Rius JE, Wesemael WM, Perry RN (2013) Top 10 plant-parasitic nematodes in molecular plant pathology. Mol. Plant Pathol. 14, 946-61.

Kar RK (2011) Plant responses to water stress: Role of reactive oxygen species. Plant Signalling Behav. 6, 1741-1745.

Kennelly MM, Cazorla FM, de Vicente A, Ramos C, Sundin GW (2007) *Pseudomonas syringae* Disease of fruit trees: Progress toward understanding and control. Plant Dis. 91, 4-17.

Kulik A, Wawer I, Krzywiñska E, Bucholc M, Dobrowolska G (2011) SnRK2 protein kinases—key regulators of plant responses to abiotic stresses. OMICS J. Integr. Biol. 15, 859-72.

Laluk K, Mengiste T (2010) Necrotroph Attacks on Plants: Wanton Destruction or Covert Extortion? The *Arabidopsis* Book. Am. Soc. Plant Biol. 8:e0136.

Lima JC, Loss-Morais G, Margis R (2012) MicroRNAs play critical roles during plant development and in response to abiotic stresses. Genet. Mol. Biol. 35, 1069-1077.

Liu G, Hanlon E (2015) Soil pH Range for Optimum Commercial Vegetable Production. IFAS extension, University of Florida. http://edis.ifas.ufl.edu/pdffiles/HS/HS120700.pdf

Mai WF, Brodie BB, Harrison MB Jatala P (1981) Nematodes. In: Compendium of Potato Diseases (Ed) Hooker WJ. Int. Potato Center Vol. 8.

Mandadi KK, Scholthof KBG (2013) Plant Immune Responses Against Viruses: How Does a Virus Cause Disease?. *Plant Cell.* 25, 1489-1505.

Mansfield J, Genin S, Magori S, Citovsky V, Sriariyanum M, Ronald P, Dow M, Verdier V, Beer SV, Machado MA, Toth I, Salmond G, Foster GD (2012) Top 10 plant pathogenic bacteria in molecular plant pathology. Mol. Plant Pathol. 13, 614-29.

Nakashima K, Yamaguchi-Shinozaki K (2013) ABA signalling in stress-response and seed development. Plant Cell Rep. 32, 959-70.

Nishihama R, Banno H, Kawahara E, Irie K, Machida Y (1997) Possible involvement of differential splicing in regulation of the activity of *Arabidopsis* ANP1 that is related to mitogen-activated protein kinase kinase kinases (MAPKKKs). Plant J. 12, 39-48.

Nuccio ML, Rhodes D, McNeil SD, Hanson AD (1999) Metabolic engineering of plants for osmotic stress resistance. Curr. Opin. Plant Biol. 2, 128-134.

Ologundudu AF, Adelusi AA, Adekoya KP (2013) Effect of light stress on germination and growth parameters of *Corchorus olitorius, Celosia argentea, Amaranthus cruentus, Abelmoschus esculentus* and *Delonix regia.* Not. Sci. Biol. 5, 468-475.

Ortbauer M (2013) Abiotic stress adaptation: Protein folding stability and dynamics. In: Abiotic stress - Plant Responses and Applications in Agriculture (Ed) Kourosh V, InTech, DOI: 10.5772/53129.

Park JY, Jin JM, Lee YW, Kang S, Lee YH (2009) Rice blast fungus (*Magnaporthe oryzae*) Infects

Arabidopsis via a mechanism distinct from that required for the infection of rice. Plant Physiol. 149, 474-486.

Pitzschke A, Schikora A, Hirt H (2009) MAPK cascade signalling networks in plant defence. Curr. Opin. Plant Biol. 12, 1-6.

Qados AMSA (2011) Effect of salt stress on plant growth and metabolism of bean plant *Vicia faba* (L.). J. Saudi Soc. Agric. Sci. 10, 1-52.

Raaijmakers JM, Paulitz TC, Steinberg C, Alabouvette C, Moënne-Loccoz Y (2009) The rhizosphere: a playground and battlefield for soil borne pathogens and beneficial microorganisms. Plant Soil 321, 341-361.

Rahman TA, Oirdi ME, Gonzalez-Lamothe R, Bouarab K (2012) Necrotrophic pathogens use the salicylic acid signalling pathway to promote disease development in tomato. Mol Plant Microbe Interact. 25, 1584-93.

Rajalekshmi KM, Jaleel CA, Azooz MM, Panneerselvam R (2009) Effect of triazole growth regulators on growth and pigment contents in *Plectranthus aromaticus* and *Plectranthus vettiveroids*. Adv. Biol. Res. 3, 117-122.

Ramegowda V, Senthil-Kumar M (2015) The interactive effects of simultaneous biotic and abiotic stresses on plants: Mechanistic understanding from drought and pathogen combination. J. Plant Physiol. 176, 47-54.

Reddy AR, Chaitanya KV, Vivekanandan M (2004) Drought-induced responses of photosynthesis and antioxidant metabolism in higher plants. J. Plant Physiol. 161, 1189-1202.

Río DLA (2015) ROS and RNS in plant physiology: an overview. J Exp Bot. 66, 2827-37.

Roy B, Noren SK, Mandal Asit B, Basu AK (2011) Genetic engineering for abiotic stress tolerance in agricultural crops. Biotechnol. 10, 1-22.

Rybick EP (2015) A Top Ten list for economically important plant viruses. Arch. Virol. 160, 17-20.

Salas-Marina MA, Isordia-Jasso MI, Islas-Osuna MA, Delgado-Sánchez P, Jiménez-Bremont JF, Rodríguez-Kessler M, Rosales-Saavedra MT, Herrera-Estrella A and Casas-Flores S (2015) The Epl1 and Sm1 proteins from *Trichoderma atroviride* and *Trichoderma virens* differentially modulate systemic disease resistance against different life style pathogens in *Solanum lycopersicum*. Front. Plant Sci. 6, 77.

Sakamoto A, Murata N (2000) Gentic engineering of glycinebetanine synthesis in plants: Current status and implications for enhancement of stress tolerance. J. Exp. Bot. 51, 81-88.

Šamajová O, Plíhal O, Al-Yousif M, Hirt H, Šamaj J (2012) Improvement of stress tolerance in plants by genetic manipulation of mitogen-activated protein kinases. Biotechnol. Adv. DOI:10.1016/j.biotechadv.2011.12.002.

Sarah Ellis D, Boehm MJ, Coplin D (2008) Bacterial Diseases of Plants. Department of Plant Pathology. Agriculture and Natural Resources. PP401.06.

Schulze ED, Beck E, Muller-Hohenstein K (2005) Environment as Stress Factor: Stress Physiology of Plants. In Plant ecology, Springer, Berlin, pp 7-11.

Sherwood JL, German TL, Moyer JW, Ullman DE (2003) Tomato spotted wilt. *The Plant Health Instructor*. DOI:10.1094/PHI-I-2003-0613-02.

Shi Y, Yang S (2014) ABA Regulation of the cold stress response in plants. In: Abscisic acid: Metabolism, Transport and Signalling (Ed) Zhang DP, Springer, pp 337-363.

Shrivastava P, Kumar R (2015) Soil salinity: A serious environmental issue and plant growth promoting bacteria as one of the tools for its alleviation. Saudi J. Biol. Sci. 22, 123-131.

Singh RP, Huerta-Espino J, Roelfs AP (2002) The wheat rusts. FAO Corporate document repository. http://www.fao.org/docrep/006/y4011e/y4011e0g.htm

Song GC, Ryu C (2013) Two volatile organic compounds trigger plant self-defense against a bacterial pathogen and a sucking insect in cucumber under open field conditions. Int J. Mol. Sci. 14, 9803-9819.

Spoel SH, Johnson JS, Dong X (2007) Regulation of trade-offs between plant defences against pathogens with different lifestyles. Proc. Nat. Acad. Sci. USA 104, 18842-18847.

Suzuki N, Rivero RM, Shulaev V, Blumwald E, Mittle R (2014) Abiotic and biotic stress combinations. New Phytol. 203, 32–43.

TeBeest DO, Guerber C, Ditmore M (2007) Rice blast. *The Plant Health Instructor.* DOI: 10.1094/ PHI-I-2007-0313-07.

Tuteja N (2007) Abscisic acid and abiotic stress signaling. Plant Signalling Behav. 2, 135-138.

Vanlerberghe GC (2013) Alternative oxidase: A mitochondrial respiratory pathway to Maintain Metabolic and Signalling Homeostasis during Abiotic and Biotic Stress in Plants. Int. J. Mol. Sci. 14, 6805-6847.

Wale SJ, Plat HWt, Cattlin N (2008) Bacterial diseases. In: Diseases, Pests and Disorders of Potatoes: A Colour Handbook (Ed) Wale S, Platt HW and Cattlin N, Manson Publishing Ltd, UK, pp 16-24.

Wan J, Zhang XC, Neece D, Ramonell KM, Clough S, Kim S, Stacey MG, Stacey G (2008) A LysM receptor-like Kinase plays a critical role in chitin signalling and fungal resistance in *Arabidopsis.* Plant Cell 20, 471–481.

Williamson B, Tudzynski B, Tudzynski P, Van Kan JA (2007) *Botrytis cinerea*: The cause of grey mould disease. Mol. Plant Pathol. 8, 561-80. DOI: 10.1111/j.1364-3703.2007.00417.x.

Wise K, Woloshuk C, Freije A (2015) Diseases of wheat *Fusarium* Head Blight (Head Scab). Purdue extension. https://www.extension.purdue.edu/extmedia/BP/BP-33-W.pdf

Xie F, Wang Q, Sun R, Zhang B (2015) Deep sequencing reveals important roles of microRNAs in response to drought and salinity stress in cotton. J. Exp. Bot. 66, 789–804.

Yuriko O, Keishi O, Kazuo S, Phan TL (2014) Response of plants to water stress. Front Plant Sci. 5, DOI:10.3389/fpls.2014.00086.

Web Resources

[1] Biotic stress factors. GMO Safety. http://www.gmo-safety.eu/glossary/1278.biotische-stressfaktoren.html

[2] Viral Diseases of Plants. Texas Plant Disease Handbook. http://plantdiseasehandbook.tamu.edu/ food-crops/fruit-crops/blackberry-dewberry-and-boysenberry/viral-diseases-of-plants/

[3] Nematodes: Plant Parasitic, various. Cornell University. http://plantclinic.cornell.edu/factsheets/ nematodes.pdf

[4] Bacterial blight- *Pseudomonas syringae* pv. *Glycinea*. Field crop pathology. http:// www.fieldcroppathology.msu.edu/extension-3/soybean/soybean-foliar diseases/bacterial-blight-pseudomonas-syringae-pv-glycinea/

[5] Common diseases of some economically important crops. Diagnostic manual for plant diseases in Vietnam. 151-170. http://aciar.gov.au/files/node/8613/MN129%20part7.pdf

[6] Bacterial Wilt – *Ralstonia solanacearum*. Plant diseases. Penn State Extension. http:// extension.psu.edu/pests/plant-diseases/all-fact-sheets/ralstonia

[7] Crown Gall. Royal Horticulture Society. https://www.rhs.org.uk/advice/profile?pid=141

[8] Rice: bacterial diseases: bacterial leaf blight. Development of e courses for B.Sc (Agriculture). http://agridr.in/tnauEAgri/eagri50/PATH272/lecture01/0011.html

[9] What is *Botrytis* or Grey Mold? Alchimia. https://www.alchimiaweb.com/blogen/botrytis-gray/

[10] Descriptions. *Puccinia graminis*. Broad institute.http://www.broadinstitute.org/annotation/ genome/puccinia_group/GenomeDescriptions.html

[11] Tungro Disease. Expert system for paddy. http://agritech.tnau.ac.in/expert_system/paddy/ cpdistungro.html

[12] Root-knot nematode. Department of agriculture and fishery, Queensland Government. https:/ /www.daf.qld.gov.au/plants/fruit-and-vegetables/a-z-list-of-horticultural-insect-pests/root-knot-nematode

6

Plant Growth Promoting Microbes Potential Tool for Growth and Development of Plants in Abiotic Stress Environments

Hruda Ranjan Sahoo and Nibha Gupta

Abstract

Microbes are endowed with a high and exploitable potential of being useful associates with their surrounding environment and host in case of symbiotic relationship. Most of the microbes live alone but many of them require association of other living system for its survival. Hence, the beneficial role of microbes is always correlated with growth promotion and survivability of other associated living system. Now it is established that microbes play an important and key role in plant growth and development, even its survivability in case of symbiotic mycorrhizal association with orchids. They are nothing but potential tool protecting plants from different adverse conditions like diseased state, environmental stress like cold temperature, drought, metal contamination, salt and acidic or alkaline stress. These stress situations could be managed by various factors like deaminase properties, phytohormone metabolism, bioconversion of toxic to non-toxic forms, secretion of bioactive compounds, bioleaching and mineral solubilisation, siderophores and production of volatile compounds etc. Plant growth promoting microbes are now well known as useful agents for plant growth and enhanced production under various stressed conditions such as drought, salt, cold, metal stress etc. The details of their vital role and various factors responsible for the growth and development of plant under stress environment has been reviewed and presented here.

Keywords: Abiotic stress, Bioconversion, Growth promoting microbes, Mineral solubilization, Phytohormone, Siderophore.

1. Introduction

Microorganisms play an important role in plant growth promotion, health and enhanced productivity. They do have capability to colonize rhizosphere and phyllosphere of the host plants and help them directly and/or indirectly through various mechanisms like biodegradation of organic material and nutrient recycling. Besides, presence of microbes also influences the availability of different nutrients through chelation, oxidation, reduction and mineralization processes. Additionally, associated microbes (free living/symbiotic) also improve the nutrient uptake and the plant growth. They stimulate the growth through production of plant growth promoting substances. Antagonistic properties may also help in protecting plants from harmful effects of other microbial associates. Now it is clear that the microbes are considered to be invisible emperors of the living world and do have impact on the three 'P's important for plant living-Production, Protection and Preservation (Conservation). All beneficial activity of microbes could be imparted in the plant growth in natural and normal environmental conditions. Do they exhibit the same when plants are in the stressed condition? If yes, then what are these factors influencing their performance and up to what extent. This has been reviewed and presented here.

2. Abiotic stress in environment and soil

The term stress is nothing but any un-favourable condition or substance that affects or blocks a plant's metabolism, growth or development, which can be induced by various natural factors (Lichtenthaler 1998). Stress factors are divided into two groups- biotic (living) and abiotic (non-living) stresses. Biotic stress factors such as variety of pathogenic microorganisms, insects and higher animals including human interference and abiotic stress factors such as cold, heat, water logging, drought, wind, intense light, soil salinity and inadequate or excessive mineral content are significant in the environment (Vinocur and Altman 2005; Wahid *et al.* 2007).

Water/ osmotic stress prevails due to decreased water availability to plants and arise when water is lost due to metabolic activities and when transpiration is higher than the water available for plant absorption, due to less rainfall. Water stress also results in reduction in gaseous exchange, transpiration rate and CO_2 assimilation during photosynthesis (Cornic 2000). It is obvious to observe that drought stress alters the plant photosynthetic rate, nutrient uptake, hormone production, homeostasis, stomatal activity, enzymatic reactions etc. (Panwar *et al.* 2014). Cold stress arises due to the decrease in temperature of the surrounding environment. Therefore, only few plants having adaptation to such extreme climatic conditions are may be able to grow and survive. The temperature below 15°C results in cold shock in plants. Salinity stress is most important abiotic factor that greatly hampers proper growth and development at all stages of crop plants in agricultural land. It reduces osmotic potential of soil decreasing water availability ultimately resulting in the loss of soil aeration. Moreover, accumulation of ions in higher concentration through uptake by plant roots cause toxicity, thus it has negative effect on plant metabolism especially Ag, Cd, Hg and Pb which are nonessential and extremely toxic elements (Gadd and Griffiths 1978; Williams *et al.* 2000; Saqib *et al.* 2012; Abbas and Akladious, 2013). Accumulation of heavy metals such as Fe, Mo, Mn, Cd, Cr, Cu, Hg, Pb, and Ni in soil exhibit toxicity symptoms and redundant growth which escalate towards disruption in nutrient uptake, physiological processes and metabolic activities and may even lead to death (Wani *et al.* 2012; Khan *et al.* 2012; Selvakumar *et al.* 2012).

Abiotic stressed soils are mostly characterized by lack of organic matter, low water-holding capacity, poor soil structure and nutrient deficiency. It restricts the growth performance of the plants as there is reduction in the number, diversity and activity of soil microflora (Sgroy *et al.* 2009). However, it is observed that application of plant growth promoting microbes enhance the growth and development of plants even under stressed environment since these microbes have the mechanism to secrete different compounds under stress/derelict conditions. Microorganisms have the ability to secrete compounds that protect plants from abiotic stresses like chilling injury, drought, high temperature, metal toxicity and salinity (Panwar *et al.* 2014). Studies also prove that these microbial communities possess intrinsic property to tolerate stress thereby imparting resistance to the plants to survive in these stressed conditions. Hence, the role of microorganisms in plant growth promotion and nutrient management is well known and well established. The utilization of plant growth promoting microbes (PGPM) has become a promising alternative to alleviate plant stress and the role of microbes in the management of biotic and abiotic stresses is gaining importance (Yao *et al.* 2010).

3. Plant growth promoting microbes and stressed environment

Microorganisms colonizing the rhizosphere soil have properties of disease resistance, mineral solubilisation, nutrient acquisition and stress tolerance are also PGPM. (Kloepper and Schroth 1978; Vessey 2003). These microbes are mainly involved in production of beneficial metabolites such as antibiotics, antioxidative enzymes, exopolysaccharides, phytohormones, siderophores, volatile compounds, etc. Direct mechanisms of plant growth promotion is through secretion of metabolites like indole acetic acid (IAA), cytokinins, gibberellins, etc. and enhance the uptake of essential nutrients from the soil. Indirectly, the plant growth promotion occurs by PGPM through production of metabolites like antibiotics, hydrogen cyanide (HCN), siderophores, etc (Hemlata Chauhan *et al.* 2015). The PGPM are mostly categorized into 3 groups- bio-stimulants which promote plant growth by production of phytohormones such as auxins, gibberellic acid, cytokinins (Carmen and Roberto, 2011); biofertilizers which enable nutrient acquisition and availability through nitrogen fixation and P solubilization (Mohammadi and Sohrabi 2012; Das *et al.* 2013) and bioprotectants which provide protection to plants against phytopathogens *via* antibiotics, siderophores and induced systemic resistance (Labuschagne *et al.* 2011; Glick 2012; Figueiredo *et al.* 2011).

Several studies conducted by researchers indicate that PGPM play a definite role in improving the growth as well as yield of several agricultural and horticultural plants (Lugtenberg and Kamilova 2009; Egamberdieva *et al.* 2010; Miransari 2011). Strains which are belonging to several genera such as *Arthrobacter, Azospirillum, Azotobacter, Bacillus, Bradyrhizobium, Burholderia, Cellulomonas, Enterobacter, Pseudomonas, Rhizobium,* and *Serratia* have been reported for PGPM activity (Cakmakci *et al.* 2005; Rajasekar and Elango 2011). Plant growth promoting traits such as phosphate solubilisation activity, IAA production and siderophore secretion was observed under salt stress of up to 1.25%, 0.75% and 1% NaCl, respectively by strains of *Pseudomonas* such as *Pseudomonas aeruginosa, P. cepacia, P. fluorescens* and *P. putida* (Deshwal and Kumar 2013). Reports establish the fact that PGPM being the suitable alternative for imparting stress resistance should be exploited and their beneficial attributes should be confirmed by application through field trials.

Plant growth promoting microbes have been isolated from different stressed environments such as hilly areas, mangrove region, mines, desert, wastelands, saline lands etc. Rajasankar *et al.* (2013) isolated pesticide tolerant and phosphorus solubilizing *Pseudomonas* sp. SGRAJ09 from pesticides treated *Achillea clavennae* rhizosphere soil. Fungicide tolerant strain *Enterobacter asburiae* isolated from mustard rhizosphere had different plant growth promoting properties such as P solubilization, siderophores and IAA synthesis (Ahmed and Khan 2010a). The three PGPR isolates *Bacillus polymyxa* BcP26, *Mycobacterium phlei*MbP18 and *P. alcaligenes* PsA15 tolerating high temperature and salt concentrations have the ability to survive in arid and saline soils such as calcisol (Egamberdiyeva 2007). Metal-tolerant PGPR, NBRI K28 *Enterobacter* sp., was isolated from fly ash contaminated soils (Kumar *et al.* 2008).

Endophytic bacteria (*Arthrobacter, Bacillus, Curtobacterium, Microbacterium, Pseudomonas and Staphylococcus)* were isolated and characterized from *Alyssum bertolonii* which is nickel hyperaccumulator plant

(Barzanti *et al.* 2007). The endophytic fungus *Cuvularia* sp. isolated from *Dichathelium lanuginosum* growing on a geothermal soil was found to be thermotolerant to temperatures of 50°C to 65°C and also mitigating the effect of abiotic stress on plants (Redman *et al.* 2002). Sujatha *et al.* (2004) investigated the solubilisation potential of thermophilic bacteria. Das *et al.* (2003) reported that the cold tolerant mutants of *Pseudomonas* were more efficient phosphate solubilisers than their wild type strains. P solubilizing *Pseudomonas* and *Acinetobacter* were isolated from cold desert of Himalayas which is involved in plant growth promotion (Gulati *et al.* 2008, 2009, 2010). *Penicillium* sp. showed better efficiency of rock phosphate solubilisation in liquid culture medium isolated from chromite, iron and manganese mines of Orissa (Gupta *et al.* 2007). High P solubilisation activity of inorganic phosphate was shown by fungi *Aspergillus* sp., *Fusarium* sp. and *Penicillium* sp. isolated from soil of saline area of Purna river basin (Rajankar *et al.* 2007). The list of P solubilising fungi from stressed soil environment is given in Table1.

Table 1: Phosphate solubilizing fungi from stressed soil environment

Soil	Mineral solubilising fungi	References
Hill soil	*Penicillium citrinum, P. islandicum, P. mellini, P. olivicolor, P. restrictum, P. rugulosum*	Sharma *et al.* (2010), Sharma (2011)
Himalayan soil	*Aspergillus* sp., *A. glaucus, A. niger, A. sydowii, Penicillium* sp., *Paecilomyces hepiali*	Rinu *et al.* (2013), Chatli *et al.*(2007), Pandey *et al.* (2008), Rinu and Pandey, (2011 and 2012).
Mangrove soil	*Aspergillus niger, Penicillium nigricans, Fusarium, Helminthosporium, Cladobotrytis, Paecilomyces, Alternaria*	Vazquez *et al.* (2000), Gupta and Das (2008), Kanimozhi and Panneerselvam (2010)
Mine soil	*Candida krissii, Galactomyces geotrichum Mucor ramosissimus, Penicillium expansum, Penicillium oxalicum, Penicillium* sp.	Xiao *et al.* (2008) Yingben *et al.* (2012) Gupta *et al.* (2007) Chai *et al.* (2011)
Saline soil	*Aspergillus clavatus, A. fumigatus, A. nidulans, A. niger, A. sydowii, A. terreus, A. ustus, A. flavus, Fusarium* sp., *Penicillium* sp., *Rhizopus* sp., *Talaromyces funiculosus*	Singh *et al.* (2012), Sanjotha *et al.* (2011), Rajankar *et al.* (2007), Srinivasan *et al.* (2012), Noor *et al.* (2013), Kanse *et al.* (2015)
Volcanic soil	*Gliocladium roseum, Penicillium albidum Penicillium frequentans, Penicillium restrictum, Penicillium thomii, Myrothecium roridum, Eupenicillum javanicum*	Morales *et al.* (2011)

4. Factors influencing stress tolerance

4.1 ACC deaminase activity (Inhibition of ethylene biosynthesis)

Ethylene is a phytohormone but its excessive secretion results in root curling and shortening ultimately resulting in plant death. However, ACC deaminase property exhibited by plant growth promoters helps in combating stress by hydrolysing ACC, precursor of ethylene to α-ketoglutarate and ammonia (Das *et al.* 2013). Bacterial strains with ACC deaminase activity have been reported in several genera such as *Acinetobacter, Achromobacter, Agrobacterium, Alcaligenes, Azospirillum, Bacillus, Burkholderia, Enterobacter, Pseudomonas, Ralstonia, Serratia* and *Rhizobium* etc. (Gupta *et al.* 2015). This activity of PGPM's in various soils help in plant growth and development under different stress conditions. Bacterial ACC deaminase are important for pathogenic stress, remediation of high/heavy metal concentration, drought stress, waterlogging stress, temperature stress, flower senescence, salinity stress, Ethylene-IAA crosstalk and rhizobial infection (Saraf *et al.* 2011). Micro-organisms having ACC deaminase genes have been reported to be useful in promotion of early root development, protecting plants against a variety of environmental stress, facilitating the production of volatile organic compounds responsible for aroma formation and phytoremediation of contaminated soil (Saraf *et al.* 2011). The list of microbes with ACC deaminase activity during stress conditions is listed in Table 2.

Table 2: Microbes with ACC deaminase activity during stressed conditions

Organism	Crop Plant	References
Achromobacter piechaudii	Tomato	Mayak *et al.* (2004)
Acinetobacter, Pseudomonas	Barley and oats	Chang *et al.* (2014)
Enterobacter aerogenes, Pseudomonas fluorescens, P. syringae	Maize	Nadeem *et al.* (2009)
Pseudomonas fluorescens	Groundnut	Sarvanakumar and Samiyappan (2007)
Pseudomonas mendocina	Lettuce	Kohler *et al.* (2009)
Raoultella planticola	Cotton	Wu *et al.* (2012)
Rhizobium phaseoli, Pseudomonas fluorescens, P. syringae	Mungbean	Ahmad *et al.* (2011, 2012, 2013)

4.2 Production of phytohormones

Plant growth regulators are the substances that influence plant's physiological processes at very low concentration. Production of phytohormones by PGPM has been reported by several researchers. Microbes belonging to genera

Acinetobacter, Bacillus, Enterobacter, Klebsiella, Pseudomonas, Serratia have the ability to synthesize auxin (Zaidi *et al*. 2006; Selvakumar *et al*. 2008; Indiragandhi *et al*. 2008). Generally, soil microbes can produce phytohormones such as auxins and other compounds *via* tryptophan metabolism (Solano *et al*. 2010). L-tryptophan is the precursor for biosynthesis of auxins in plants and microbes. Auxins of microbial origin are also known to evoke physiological response in host plants. Indole acetic acid (IAA) is known for triggering both rapid responses (cell elongation and slow responses cell division and differentiation. IAA is a hormone which is responsible for root initiation and also involved in cell division and cell enlargement. IAA producing organisms are believed to increase root length which enables the plant to access more nutrients from soil due to greater root surface area (Vessey, 2003; Boiero *et al*. 2007). It also induces increased level of protection in plants against external stress conditions (Bianco and Defez, 2009; Patil *et al*. 2011).

4.3 Bioconversion

This involves the usage of plant growth promoters in conversion of toxic metals to less toxic forms so that it is accessible by plant roots. Plant-associated rhizobacteria and mycorrhizae increase the bioavailability of various heavy metal ions for their uptake by plants by catalyzing redox transformations leading to changes in heavy-metal bioavailability. They improve the metal clean up by plants through phytoremediation strategy (Varsha *et al*. 2011). Plant species such as *Thlapsi caerulescens, Seberatia acuminate, Alyssum lesbiacum, Arabidopsis halleri, Thlapsi rotundifolium, Astralagus* sp., *Pteris vittata* are capable of hyperaccumulating metals such as Zn, Ni, Cd, Pb, Se, As etc (Chaudhary and Khan, 2015). The major groups of phytoremediation strategies includes phytotransformation where cleaning of chlorinated aliphatics, herbicides present in soil occurs by means of phreatophytic trees, grasses, legumes etc; rhizosphere remediation which is through removal of biodegradable organics by grasses and phytostabilization where metal uptake and biodegradation of hydrophobic organics for soil stabilization occurs by help of trees. Besides, phytoaccumulation/phytoextraction, rhizofiltration and phytovolatilization are other techniques for removal of toxic metals from soil environment by phytoremediation technology (Schnoor, 2003). Therefore, survival of plants which are extremely tolerant to these metals is possible resulting in conversion of metals to less toxic forms by involvement of PGPR. Hence, PGPR mediated phytoremediation results in removal of metal toxicity and these microbes improve plant growth under metal stress conditions by means of biotransformation of toxic forms into less toxic forms.

4.4 Exopolysaccharide secretion

PGPM have the ability of secreting exopolysaccharides for protection against stress.Certain bacteria have the ability to synthesize several forms of multifunctional polysaccharides such as intracellular polysaccharides, extracellular polysaccharides, and structural polysaccharides. Production of Exopolysaccharides is generally important in biofilm formation; root colonization can affect the interaction of microbes with roots appendages. During salt stress, EPS form macro aggregates in the soil which can bind different cat ions and make them available to the plants (Haynes and Swift, 1990). The matrix formed by these microbes suppresses the diffusion of plant growth promoting substances from the vicinity of plant and helps in plant growth under stress.Crop inoculation with e.g. *Bacillus amylolequifaciens* leads to the production of polysaccharides (EPS) which tends to improve soil structure by facilitating the formation of macroaggregates. This in turn increases plant resistance to stress due to water shortage (Miloševiæ *et al.* 2012). *Halomonas variabilis* (HT1) and *Planococcus rifietoensis* (RT4) have capacity to form biofilm and produce exopolysaccharides under salt stress. Eventually, it can add to the plant growth and soil structure. This was proved from the pot experiment conducted by growing seedlings of *Cicer arietinum* var. CM-98 plant under salt stress by inoculation of these bacterial strains. At elevated salt stress conditions, both the strains increased plant growth and soil structure. The results suggest the feasibility of using above strains in improving plant growth and soil fertility under salinity (Qurashi and Sabri, 2012). Exopolysaccharide (EPS)-producing rhizobacterium belonging to the genus *Rhizobium* improved the physical properties of sunflower (*Helianthus annuus* L.) RAS, promoting plant growth under both normal water supply and water stress conditions (Alami *et al.* 2000).

4.5 Phosphate solubilisation

Phosphorus (P) is an essential element next only to nitrogen which is necessary for plant growth throughout the world (Vassileva *et al.* 1998). Unlike nitrogen, this element is not acquired through biochemical fixation but comes from other sources which include animal manures, chemical fertilizers, and plant residues including domestic wastes, green manures, human and industrial wastes and, native compounds of phosphorus, both organic and inorganic already present in soil to meet plant requirements (Rao 1982). Production of organic acids and acid phosphatases is the principal mechanism for solubilization of mineral phosphate.

Several soil bacteria belonging to the genera *Bacillus* and *Pseudomonas* and fungi belonging to the genera *Aspergillus* and *Penicillium* are reported to possess the ability of converting insoluble phosphates in soil into soluble forms through secretion of various organic acids such as, acetic, formic, fumaric, glycolic, lactic, propionic and succinic acids (Rashid *et al.* 2004).Phosphate solubilizing microorganisms (PSM) can grow in media with the sole phosphate source as aluminium and iron phosphate, bonemeal, rock, tricalcium phosphate and similar insoluble phosphate compounds (Gaur, 1990). *Bacillus* and *Paenibacillus* are specifically used to enhance the status P in plants (Brown, 1974). *Rhizobium* sp. such as *Allorhizobium, Azorhizobium, Bradyrhizobium, Mesorhizobium and Sinorhizobium* in combined inoculation with different phosphate solubilising bacteria, are used to enhance the plant productivity (Akhtar and Siddiqui 2009).

Vyas *et al.* (2007) showed high solubilisation of aluminium phosphate, Mussorie rock phosphate, North Carolina rock phosphate, Tri calcium phosphate by *Eupenicillium parvum* and exhibited tolerance against desiccation, salinity, acidity, aluminium and iron. *Penicillium citrinum* Thom. isolated from rhizosphere of sugarcane tested for its phosphate solubilization activity on four carbon sources–glucose, glycerol, mannitol and sucrose, as well as saline conditions – NaCl, $CaCl_2$, KCl *in vitro*. It showed maximum solubilization of TCP (461 µg/ml) in presence of glucose and minimum phosphate solubilization of 421µg/ml in presence of sucrose. In saline conditions, maximum solubilization was noticed at 1% $CaCl_2$ at pH 8. As a result, hectares of saline and alkaline land were improved by *Penicillium citrinum* (Yadav *et al.* 2011).

4.6 Siderophores

Presence of iron in the form of Fe^{3+} in soil gets limited due to accumulation of iron in the form of oxides and hydroxides. However, microbes have developed mechanism of acquiring iron by production of iron carriers/chelating agents called siderophores. They form soluble Fe^{3+} complexes taken up active transport mechanisms. Hence, siderophores can scavenge iron during iron starvation conditions prevailing in the environment. Sequestration of iron by some microbial communities is a strategy to overcome iron stress/deficient conditions and thereby adapt to such conditions in soil. Microbes belonging to *Bacillus, Enterobacter, Pseudomonas, Rhodococcus* etc. have the capacity of releasing siderophore into the soil. It is observed that majority of PGPM facilitate uptake of iron from environment and promote plant growth (Rajkumar *et al.* 2010; Ahemad and Khan 2012). However, siderophores also bind to metals like chromium, manganese, magnesium and remove these toxicants from contaminated soil and indirectly ameliorate metal stress (Mani *et al.* 2010; Akhtar *et al.* 2013).

4.7 Volatile compounds

PGPM are known to produce volatile inorganic and organic compounds such as acetoin, 2, 3-butanediol, betaineetc. which can affect plant growth and resistance against biotic and abiotic stresses. Chemicals such as acetoin and 2, 3 butanediol produced by *Bacillus sp.* are involved in Na^+ homeostasis during salt stress (Ryu *et al.* 2003) and betaine is produced during oxidative stress conditions (Farag *et al.* 2013). Thus, production of microbial volatiles modulates plant growth during stress conditions.

5. Role of PGPM on plant growth and development in stressed environment

Stressed environment often results in severe losses in crop production. Majority of economically important plants such as wheat, rice and maize are found to be stress sensitive (Bita and Gerats, 2013). Therefore, the present aim is to increase crop plant productivity and enhance resistance or tolerance against various stress factors. When plants are exposed to different environmental conditions such as extreme temperatures, toxic compounds and water deficit conditions etc., they adapt themselves by developing symbiotic associations with PGPM present in their vicinity. These microbes are important to the structure, function and health of various plant communities; hence they also help the plants to overcome environmental stresses.

5.1 Drought tolerance

Drought conditions affect at all growth stages of plant growth from morphological to molecular levels and during whichever stage the water deficit occurs (Farooq *et al.* 2009). The important effect of drought is it results in poor germination and reduction in seedling development (Kaya *et al.* 2006; Nezhadahmadi *et al.* 2013). During water stress; there is also decrease in transpiration and conductance rate in leaves. As a result, water use efficiency increases which improves plant growth in water limited conditions (Aroca and Ruiz-lozano, 2009). There is increase in hormonal levels such as ABA which induces stomatal closure minimizing water loss (Cho *et al.* 2008). Compounds like methyl jasmonate, salicylic acid and ethylene are also involved in regulating stomatal closure. However, ABA produced by PGPM enhances drought tolerance. *Arabidopsis* plants have higher ABA when inoculated with *Azotobacter brasilense and A. lipoferum* under drought stress enhanced its growth and yield (Cohen *et al.* 2008; Arzanesh *et al.* 2011).

It is observed that *Azospirillum* also improved the growth of crops such as wheat and maize during water stress conditions suggesting its potential role as plant growth promoter (Alvarez *et al.* 1996; Cohen *et al.* 2009). Various *Pseudomonas* species such as *P. entomophila, P. monteillii, P. Putida, P. Stutzei, P. Syringae* have been reported for their plant growth promotion activity under water stress (Sandhya *et al.* 2010). Similarly, *Bacillus sp.* such as *B. cereus, B. lentus* and *B. subtilis* were also observed for their role in plant growth promotion in drought stress environment (Heidari and Golpayengani, 2012; Wang *et al.* 2012). Effect of soybean seed inoculation with five *Bradyrhizobium japonicum* strains under three drought levels of conditions was studied; which showed prominent differences in the reduction of dry matter in plants. The soybean plants inoculated with the strains D 216 and 2b plants were most tolerant to soil drought and on average for all three drought levels, the lowest dry weight reduction was registered in the plants inoculated with the strain D 216 (*Miloševiœ et al.* 2012). Under drought stress, co-inoculation of bean (*Phaseolus vulgaris* L.) with *Rhizobium tropici* and two strains of *Paenibacillus polymyxa* resulted in augmented plant height; shoot dry weight and nodule number (Figueiredo *et al.* 2008).

5.2 Salt tolerance

Several studies indicated that inoculation of PGPR can mitigate the effects of salt stress in different plant species enhancing plant tolerance against salt stress. These microbes improve the growth and development of beans, lettuce, peppers and tomatoes grown in saline environments (Grover *et al.* 2010; Yildirium and Taylor 2005). In a study, it was observed that inoculation with *Azospirillum* in barley seedlings seemed to mitigate NaCl stress (Zawoznik *et al.* 2011). It is observed that salt tolerant rhizobacteria produce IAA in media containing NaCl. The presence of tryptophan stimulated auxin production of *P. aureantiaca* TSAU22 and *P. putida* TSAU1. The IAA production ability of PGPR strains at higher saline conditions balances decrease in the IAA levels of the roots and thus alleviating salt stress in plants. In other study, *Arthrobacter* species alleviate salt stress in plants by production of root associated IAA, GB and ABA (Egamberdiyeva 2013).

PGPR strains of *Pseudomonas fluorescens* 153, 169 and *P. putida* 108 have the ability to thrive in saline soil (Abbaspoor *et al.* 2009). Plants inoculated with *P. mendocina* showed greater shoot biomass than the uninoculated plants at various salinity levels (Kohler *et al.* 2010). Plant growth promoting *Pseudomonas* surviving under saline condition enhanced the plant growth in *Cicer arietinum* L., cotton, maize and wheat seedlings (Kausar and Shahzad, 2006; Yao *et al.* 2010; Mishra *et al.* 2010; Nadeem *et al.* 2013). Inoculating with PGPR, *P. mendocina,* alone or in combination with an AM fungus, *Glomus intraradices or G. mosseae* enhanced the growth and nutrient uptake and other physiological activities of *Lactuca sativa* growing under salt stress conditions (Kohler *et al.* 2009). Studies on tripartite symbiosis of bacterial-mycorrhizal-legume in saline conditions demonstrated the effects of dual inoculation with *Azospirillum brasilense* bacteria and arbuscular mycorrhizal (AM) fungus *Glomus clarum* on the host plants (*Vicia faba*) in pot cultures at five NaCl levels, resulted in decrease in salinity tolerance and higher accumulation of proline with increasing levels of salinity (Rabie and Almadini 2005). Beans, clover and corn inoculated with AM fungi improved their osmoregulation and increased proline accumulation which resulted in salinity resistance (Feng *et al.* 2002, Grover *et al.* 2010).

5.3 Cold tolerance

Microbes possess adaptations to tolerate cold stress by the synthesis of specialised molecules such as cold shock proteins, cold acclimation proteins, ice nucleators, cryoprotectors, and antifreeze molecules. This is an important adaptation exhibited by them to freezing temperatures. Microbes such as *Acinetobacter, Burkholderia, Pseudomonas, Rahnella, Rhodococcus, Serratia* etc. have cold adaptation mechanism which influence agricultural productivity in crop plants (Subramanian *et al.* 2011). *Bacillus megaterium* also stimulated the growth of crop plants such as wheat and maize under low temperature conditions (Trivedi and Pandey 2008). *Pseudomonas sp.* used as bio inoculants increased plant growth under cold stress which is observed in wheat and pea plants (Negi *et al.* 2005; Trivedi and Pandey 2007). Similarly, seed bacterization with *P. lurida* has been reported to positively influence the growth and nutrient uptake of wheat in pot culture conditions growing in controlled cold temperatures (Selvakumar *et al.* 2010). Mixed inoculation of *Pseudomonas* and *Rhizobium leguminosarum* improved the growth of lentil in cold stress environment (Mishra *et al.* 2011). Combined effect of *Exiguobacterium acetylicum* and *Pantonea dispersa* positively influenced the growth and nutrient uptake of wheat seedlings in glasshouse studies at suboptimal cold conditions (Selvakumar *et al.* 2008, 2009). Psychrotrophic bacteria such as *Mycobacterium phlei* and *Mycoplanabullata* improved the growth of winter wheat in nutrient poor calcisol soil (Egamberdiyeva and Hoflich, 2003).

5.4 Metal tolerance

Heavy metals are the main inorganic pollutants accumulating in the soil biosphere due to their non-degradable nature and consequently disturbing the environment (Rajkumar *et al.* 2010). Microbes can survive in metal stressed environment due to their ability of heavy metals uptake into their cell through acidification of the microenvironment, release of chelating substances and changes in redox potential (Whiting *et al.* 2001; Lasat 2002). Soil microbes interacting with plants which help in bioremediation have heavy metals sequestration ability (Hansda *et al.* 2014).Therefore, selection of microorganisms having metal tolerance property and efficient in producing PGPR compounds are useful in enhancing the recolonization of the plant rhizosphere in the polluted soil.

Microbes have been reported to protect the plants from metal toxicity. *Kluyvera ascorbata* protect plants from nickel, lead, and zinc toxicity (Burd *et al.* 2000). Dell'Amico *et al.* (2008) indicated that rhizobacteria (*Alcaligenes* sp. ZN4, *Mycobacterium* sp. ACC14, *Pseudomonas fluorescens*ACC9, and *Pseudomonas tolaasii* ACC23) protected plants against the inhibitory effects of cadmium. Wani *et al.* (2008) found similar results that inoculating *Rhizobium* species RP5 resulted in protection of plants against toxic effects of Ni and Zn. Ma *et al.* (2009) noted that inoculation of metal resistant strains (*Bacillus cereus* SRA10;*Bacillus* sp. SRP4; *Bacillus weihenstephanensis* SPR12 and *Psychrobacter* sp. SRA1, SRA2) showed to be very effective in protecting plants from Ni toxicity.

Plant growth promoting microbes belonging to *Bacillus, Mesorhizobium, Pseudomonas* enhanced the growth of chickpea crop under different metal stress conditions which is proved by the studies conducted by various research groups (Wani *et al.* 2008; Wani and Khan 2010; Oves *et al.* 2013). *Bacillus megaterium, Bacillus mucilaginous, Bacillus subtilis* detoxify heavy metals such as Zn, Pb, Ni and promoted growth of *Brassica juncea* (Wu *et al.* 2006; Zaidi *et al.* 2006). *Pseudomonas putida* enhanced the growth of Canola (*Brassica napus)* under Ni and Cd stress and *Vigna radiata* under Pb and Cd stress (Belimov *et al.* 2001; Tripathi *et al.* 2005; Farwell *et al.* 2014). *Pseudomonas sp.* extracted from the hydrocarbon contaminated soil possessed a major role for stimulation of root and shoot growth as well as enhances the rice yield production from hydrocarbon contaminated soil (Deepthi *et al.* 2014). It also improved the growth of various crops such as *Glycine max, Triticum vulgare* and *Vigna radiate* under Cr, Cd and Ni stress environment (Gupta *et al.* 2002).

Burkholderia sp. and *Methylobacterium oryzae* decreased nickel and cadmium stress in tomato by reducing their uptake and translocation (Marquez *et al.* 2007; Madhiyan *et al.* 2007). Consortium of growth promoting microbes such as *Azomonas* sp., *Bacillus* sp., *Pseudomonas* sp. and *Xanthomonas* sp. stimulated growth of *Brassica napus* in cadmium contaminated soil under pot culture conditions (Sheng and Xia, 2006). In *Zea mays*, mixed inoculation of microbes such as *Pseudomonas aeruginosa, Pseudomonas fluorescens and Ralstonia metallidurans* was done which increased Cr and Pb uptake enhancing its growth under stressed conditions (Braud *et al.* 2009). The biomass of *Lupines luteus* plants was improved by inoculation of consortia of *Bradyrhizobium* sp., *Pseudomonas* sp. and *Ochrobactrum cytisi* under metal stress (Dary *et al.* 2010).

It is also observed that presence of ectomycorrhizal or AM fungi associated with the roots of plants decreased metal uptake by the plants and thereby increasing their plant biomass, thus playing an important role in detoxification. Several authors have also reported the association of mycorrhizal fungal taxa such as *Glomus* and *Gigaspora* with plants growing in heavy metal contaminated sites (Khan *et al.* 2000). AM fungi improve plant tolerance growing in heavy metal polluted soil (Tonin *et al.* 2001). Mycorrhizal species such as *Suillus bovines* and *Thelephora terrestris* improves *Pinus sylvestris* growth against Cu toxicity (Hall, 2002). Vivas *et al.* (2006) showed a protective effect of the AM fungi and bacteria (*Glomus mosseae* and *Brevibacillus*) interaction against uptake of toxic metal Zn by *Trifolium repens* plants in moderately polluted soil.

Hence, inoculation of plants with microflora with growth promoting attributes may diminish toxicity of heavy metals to plants in polluted soil (Madhaiyan *et al.* 2007).The role of PGPM in improving the growth of various crop plants in different stressed environment is listed in Table 3.

Table 3: List of microorganisms improving growth of different crop plants under stressed environment

Microorganisms	Crop plant	Type of stress	References
Azospirillum	Wheat	Water stress	Alvarez et al. (1996)
Azospirillum	Maize	Water stress	Cohen et al. (2009)
Pseudomonas aeruginosa	Chick pea	Metal stress (Cr)	Oves et al. (2013)
Bradyrhizobium sp.	Green gram	Metal stress (Cd)	Wani et al. (2007)
Rhizobium sp.	Pea	Metal stress (Cu)	Wani et al. (2008)
Pseudomonas hurida	Wheat	Cold stress	Selvakumar et al. (2010)
Mesorhizobium sp.	Chick pea	Metal stress (Cr)	Wani et al. (2008)
Pseudomonas and Rhizobium leguminosarum	Lentil	Cold stress	Mishra et al. (2011)
Pseudomonas putida	Wheat	Cold stress(10-15°C)	Trivedi and Pandey, (2007)
Bacillus megaterium	Maize and wheat	Cold stress(10-15°C)	Trivedi and Pandey, (2008)
Bacillus sp.	Chick pea	Metal stress (Cr)	Wani and Khan, (2010)
Bacillus halodenitrificans and Halobacillus	Wheat	Salt stress (320 mM)	Ramadoss et al. (2013)
Pseudomonas fluorescens	Pea	Cold stress	Negi et al. (2005)
Trichoderma sp.	Cowpea (Vigna unguiculata)	Salt stress (0.25-0.65%)	Badar et al. (2015)
Mycorrhiza and Aspergillus niger	Trifolium repens	Water stress	Medina et al. (2010)
Flavobacterium sp. Rhodococcus sp. and Variovorax paradoxus	Rapeseed	Metal stress (Cd)	Belimov et al. (2005)
Burkholderia sp. and Methylobacterium oryzae	Tomato	Metal stress (Ni and Cd)	Madhaiyan et al. (2007)
Bradyrhizobium sp., Pseudomonas sp., Ochrabactrum cystisis	Lupinu sluteus	Metal stress (Cu, Cd, Pb)	Dary et al. (2010)
Bacillus thuringiensis and Pseudomonas fluorescens	Wheat	Metal stress (Cr)	Shahzadi et al. (2013)
A. brasilense and A. chroococcum	Wheat	Metal stress (Pb)	Janmohammadi et al. (2013)

(Contd.)

Microorganisms	Crop plant	Type of stress	References
Ralstonia eutropha and Chryseobacterium humi	Sunflower	Metal stress (Zn and Cd)	Marques *et al.* (2013)
Serratia sp. *and Rhizobium* sp.	Lettuce	Salt stress	Han and Lee, (2005)
R. tropici, R. etli, Sinorhizobium, Chryseobacterium balistinum	Bean and Soyabean	Salt stress (25 mM)	Estevezi *et al.* (2009)
Pseudomonas sp. *and Rhizobium* sp.	Maize	Salt stress	Bano and Fatima, (2009)
Bacillus sp., *Burkholderia* sp., *Enterobacter* sp., *Microbacterium* sp., *Paenibacillus* sp.	Wheat	Salt stress	Upadhayay *et al.* (2012)
Enterobacter cloacae, Pseudomonas putida, P. fluorescens, Serratia ficaria	Wheat	Salt stress	Nadeem *et al.* (2013)
Paenibacillus polymyxa and Rhizobium tropici	Common bean	Water stress	Figueiredo *et al.* (2008)
Pseudomonas sp.	*Asparagus officinalis*	Water stress(8 Weeks)	Liddycoat *et al.* (2009)
Pseudomonas entomophila, P. monteillii, P. putida, P.stutzei, P. syringae	Maize	Water stress	Sandhya *et al.* (2010)
B. cereus, B. subtilis and Serratia	*Cucumis sativus*	Water stress	Wang *et al.* (2012)
A. brasilense, B.lentus, Pseudomonads	Basil	Water stress	Heidari and Golpayengani, (2012)

6. Conclusion

The global recognition of the rhizosphere as a critical interface for soil and plants has led it to become promising sphere of research area at present, hence new technologies have been developed to face challenges posed by such a complex and diverse environment. In the past 10–15 years, there has been an increasing interest in the possibility of utilizing PGPM as adjuncts to agricultural and horticultural practice as well as environmental cleanup. Moreover, with this interest there has been a major effort worldwide to better understand many of the fundamental mechanisms that they use to facilitate plant growth. Therefore, it is important to study relationship between PGPM and plants before using them on a massive scale in the environment. With increasing concern about the natural environment and the understanding that the era of the large scale use of chemicals in the environment needs to come to an end, PGPM offer an attractive alternative that contains the possibility of developing more sustainable approaches to agriculture. Finally, it is efficacious to select PGPM with plant growth promoting attributes for enhancement in plant growth even under stressed conditions. Microorganisms associated with agricultural plants improve their tolerance to abiotic stresses and thereby get adapted. The complex and dynamic interactions between microorganisms and plant roots under conditions of abiotic stress affect not only the plants but also the physical, chemical, and structural properties of soil. The possibility of mitigation of abiotic stresses in plants opens a new chapter in the application of microorganisms in agriculture. Hence, microbial species and strains do play an important role in understanding plant tolerance mechanism to stressed environment, adaptation to stress, and mechanisms that develop in plants under stress conditions. Therefore, selection of microorganisms from stressed ecosystems may contribute to the concept of biotechnology application in agriculture.

References

Abbas SM and Akladious SA (2013) Application of carrot root extract induced salinity tolerance in cowpea (*Vigna sinensis* L.) seedlings. Pak. J. Bot. 45(3), 795-806.

Abbaspoor A, Zabihi HR, Movafegh S, Asl MHA (2009) The efficiency of Plant Growth Promoting Rhizobacteria (PGPR) on yield and yield components of two varieties of wheat in salinity condition. Am –Eurasian. J. Sustain. Agric. 3(4), 824-828.

Ahemad M and Khan MS (2010) Plant growth promoting activities of phosphate solubilizing *Enterobacter asburiae* as influenced by fungicides. Eurasian J. Bio. Sci. 4, 88-95.

Ahemad M and Khan MS (2012) Effect of fungicides on plant growth promoting activities of phosphate solubilising *Pseudomonas putida* isolated from mustard (*Brassica campestris*) rhizosphere. Chemosphere 86, 945-50.

Ahmad M, Zahir A, Naeem Asghar H, Asghar M (2011) Inducing salt tolerance in mung bean through coinoculation with rhizobia and plant-growth-promoting rhizobacteria containing 1-aminocyclopropane-1-carboxylate deaminase. Can. J. Microbiol. 57(7), 578-589.

Ahmad M, Zahir A, Nazli F, Akram F, Khalid M, Arshad M (2013) Effectiveness of halo-tolerant, auxin producing *Pseudomonas* and *Rhizobium* strains to improve osmotic stress tolerance in mung bean (*Vigna radiata* L.). Braz. J. Microbiol. 44 (4), 1341–1348.

Ahmad M, Zahir ZA, Asghar HN, Arshad M (2012) The combined application of rhizobial strains and plant growth promoting rhizobacteria improves growth and productivity of mung bean (*Vigna radiata* L.) under salt-stressed conditions. Ann. Microbiol. 62, 1321–1333.

Akhtar MS, Chali B, Azam T (2013) Bioremediation of arsenic and lead by plants and microbes from contaminated soil. Res. Plant Sci.1, 68-73.

Akhtar MS and Siddiqui ZA (2009) Effects of phosphate solubilising microorganisms and *Rhizobium sp.* on the growth, nodulation, yield and root-rot disease complex of chickpea under field condition. Afr. J. Biotechnol. 8, 3489-3496.

AlamiY, Achouak W, Marol C, Heulin T (2000) Rhizosphere Soil Aggregation and Plant Growth Promotion of Sunflowers by an Exopolysaccharide-Producing *Rhizobium* sp. Strain Isolated from Sunflower Roots. Appl. Environ. Microbiol. 66 (8), 3393–3398.

Alvarez MI, Sueldo RJ, Barassi CA (1996) Effect of *Azospirillum* on coleoptile growth in wheat seedlings under water stress. Cereal. Res. Commun. 24,101-107.

Aroca R and Ruíz-Lozano JM (2009) Induction of plant tolerance to semi-arid environments by beneficial soil microorganisms (a review) In: Climate Change, Intercropping, Pest Control and Beneficial Microorganisms, Sustainable Agriculture Reviews 2, (Ed) Lichtfouse E, Springer, Netherlands. pp121-135.

Arzanesh M, Alikhani H, Khavazi K, Rahimian H, Miransari M (2011). Wheat (*Triticum aestivum* L.) growth enhancement by *Azospirillum sp.* under drought stress. World J Microbiol Biotechnol. 27, 197-205.

Badar R, Batool B, Ansari A, Mustafa S, Ajmal A, Perveen S (2015) Amelioration of salt affected soils for cowpea growth by application of organic amendments. J. Pharmacogn. Phytochem. 3(6), 87-90.

Bano A, Fatima M (2009) Salt tolerance in *Zea mays* (L.) following inoculation with *Rhizobium* and *Pseudomonas*. Biol. Fertil. Soils 45, 405–413.

Barzanti R, Ozino F, Bazzicalupo M, Gabbrielli R, Galerdi F, Gonelli C, Mengoni A (2007) Isolation and characterization of endophytic bacteria from the nickel hyperaccumulator plant *Alyssum bertolonii*. Microbial Ecol. 53,306–316

Belimov AA, Hontzeas N, Safronova VI, Demchinskaya SV, Piluzza G, Bullita S, Glick BR (2005) Cadmium tolerant plant growth promoting bacteria associated with the roots of Indian mustard (*Brassica juncea* L Czern.) Soil Biol. Biochem. 37,241-50.

Belimov AA, Safronova VI, Sergeyeva TA, Egorova TN, Matveyeva VA, Tsyganoc VE, Borisov AY, Tikhonovich IA, Kluge C, Preisfeld A, Dietz KJ, Stepanok VV (2001) Characterization of plant growth promoting rhizobacteria isolated from polluted soils and containing 1-aminocyclopropane-1-carboxylatedeaminase. Can. J. Microbiol. 47, 642-652.

Bianco C and Defez R (2009) *Medicago truncatula* improves salt tolerance when nodulated by an indole-3-acetic acid-overproducing *Sinorhizobium meliloti* strain. J. Exp. Bot. 60, 3097–3107.

Bita CE and Gerats T (2013) Plant tolerance to high temperature in a changing environment: scientific fundamentals and production of heat stress-tolerant crops. Front Plant Sci. 4, 273.

Boiero L, Perrig D, Masciarelli O, Pena C, Cassa´n F, Luna V (2007) Phytohormone production by strains of *Bradyrhizobium japonicum* and possible physiological and technological implications. Appl. Microbiol Biotechnol. 74, 874–880.

Braud A, Jezequel K, Bazot S, Lebeau T (2009) Enhanced phytoextraction of an agricultural Cr-Hg and Pb-contaminated soil by bioaugmentation with siderophore producing bacteria. Chemosphere 74, 280-6.

Brown ME (1974) Seed and root bacterization. Ann Rev Phytopathol. 12, 181–97.

Burd GI, Dixon DG, Glick BR (2000) Plant growth-promoting bacteria that decrease heavy metal toxicity in plants. Can. J. Microbiol. 46, 237–245.

Cakmakci R, Donmez D, Aydýn A, Sahin F (2005) Growth promotion of plants by plant growth-promoting rhizobacteria under greenhouse and two different field soil conditions. Soil Biol. Biochem. 38, 1482–1487.

Carmen B and Roberto D (2011) Soil bacteria support and protect plants against abiotic stresses. In: Abiotic Stresses in Plants: Mechanisms and Adaptations (Ed) Shanker A, DOI: 10.57772/23310.

Chai B, Wu Y, Liu P, Liu B, Gao M (2011) Isolation and phosphate-solubilizing ability of a fungus, *Penicillium sp.* from soil of an alum mine. J. Basic Microbiol. 51(1), 5-14.

Chang P, Gerhardt KE, Huang XD, Yu XM, Glick BR, Gerwing PD, Greenberg BM (2014) Plant growth-promoting bacteria facilitate the growth of barley and oats in salt-impacted soil: implications for phytoremediation of saline soils. Int. J. Phytorem. 16 (11), 1133–1147.

Chatli AS, Beri V, Sidhu BS (2008) Isolation and characterisation of phosphate solubilising microorganisms from the cold desert habitat of *Salix alba* Linn. in trans Himalayan region of Himachal Pradesh. Ind. J. Microbiol. 48, 267–273.

Chaudhary K and Khan S (2015) Review: Plant Microbe-Interaction in Heavy Metal Contaminated Soils. Ind. Res. J. Genet. Bio. 7(2), 235 – 240.

Cho SM, Kang BR, Han SH, Anderson AJ, Park JY, Lee YH, Cho BH, Yang KY, Ryu CM , Kim YC (2008) 2R,3R-butanediol, a bacterial volatile produced by *Pseudomonas chlororaphis* O6, is involved in induction of systemic tolerance to drought in *Arabidopsis thaliana*. Mol. Plant-microbe Interact. 21, 1067-1075.

Cohen A, Bottini R, Piccoli P (2008) *Azospirillum brasilense* Sp produces ABA in chemically-defined culture medium and increases ABA content in *Arabidopsis* plants. Plant Growth Reg. 54, 97-103.

Cohen AC, Travaglia CN, Bottini R, Piccoli PN (2009) Participation of abscisic acid and gibberellins produced by endophytic *Azospirillum* in the alleviation of drought effects in maize. Botany 87, 455-62.

Cornic G (2000) Drought stress inhibits photosynthesis by decreasing stomatal aperture not by affecting ATP synthesis. Trends Plant Sci. 5, 187-188.

Dary M, Chamber-Perez MA, Palomeres AJ, Pajuela E (2010) "In situ" phytostabilisation of heavy metal polluted soils using *Lupinus luteus* inoculated with metal resistant plant-growth promoting rhizobacteria. J. Hazard Mater. 177,323–330.

Das AJ, Kumar M, Kumar R (2013) Plant growth promoting Rhizobacteria :an alternative for chemical fertilizer for sustainable, environmental friendly agriculture. Res. J. Agric. For. Sci. 1, 21-23.

Das K, Katiyar V, Goel R (2003) P solubilization potential of plant growth promoting *Pseudomonas* mutants at low temperature. Microbiol. Res. 158, 359-362.

Deepthi MS, Reena T, Deepu MS (2014)In vitro study on the effect of heavy metals on PGPR microbes from two different soils and their growth efficiency on *Oryza sativa* (L.). J. Biopest. 7(1), 64-72.

Dell'Amico E, Cavalca L, Andreoni V (2008) Improvement of *Brassica napus* growth under cadmium stress by cadmium-resistant rhizobacteria. Soil Biol. Biochem. 40, 74–84.

Deshwal VK and Kumar P (2013) Effect of Salinity on Growth and PGPR Activity of Pseudomonads. J. Aca. Indus. Res. 2(6), 353-356.

Egamberdieva D (2013) The role of phytohormone producing bacteria in alleviating salt stress in crop plants. In: Biotechnological Techniques of Stress Tolerance in Plants (Ed) Miransari M, Stadium Press, LLC USA, pp. 21-39.

Egamberdiyeva D, Hoflich G (2003) Influence of growth promoting bacteria on the growth of wheat in different soils and temperatures. Soil Biol. Biochem. 35, 973-978.

Egamberdiyeva D, Kucharova Z, Davranov K, Berg G, Makarova N, Azarova T, Chebotar V, Tikhonovich I, Kamilova F, Validov Sh, Lugtenberg B (2010) Bacteria able to control foot and root rot and to promote growth of cucumber in salinated soils. Biol. Fertil. Soils. 47, 197–205.

Egamberdiyeva D (2007) The effect of plant growth promoting bacteria on growth and nutrient uptake of maize in two different soils. Appl. Soil Ecol. 36, 184–189.

Estevezi J, Dardanellii MS, Megiase M, Rodriguez–Navarro DN (2009) Symbiotic performance of common bean and soyabean co-inoculated with rhizobia and *Chryseobacterium balustinum* Aur 9 under moderate saline conditions. Symbiosis 49, 29-36.

Farag MA, Zhang H, Ryu CM (2013) Dynamic chemical communication between plants and bacteria through air borne signals: induced resistance through bacterial volatiles. J. Chem. Ecol. 39, 1007-1018.

Farooq M, Wahid A, Kobayashi N, Fujita D, Basra SMA (2009) Plant drought stress: effects, mechanisms and management. Agron. Sustain. Dev. 29(1), 185-212.

Farwell AJ, Vessely S, Nero V, Rodriguez H, McCormack K, Shah S, Dixon DG, Glick BR (2007) Tolerance of transgenic canola plants (*Brassica napus*) amended with plant growth-promoting bacteria to flooding stress at a metal contaminated field site. Environ. Pollut. 147, 540-545.

Feng G, Zhang FS, Li XL, Tian CY, Tang C, Rengel Z (2002) Improved tolerance of maize plants to salt stress by arbuscular mycorrhiza is related to higher accumulation of soluble sugars in roots. Mycorrhiza 12, 185–190.

Figueiredo MVB, Burity HA, Martinez CR, Chanway CP (2008) Alleviation of drought stress in common bean (*Phaseolus vulgaris* L.) by co-inoculation with *Paenibacillus polymyxa* and *Rhizobium tropici*. Appl. Soil Ecol. 40, 182–188.

Figueirido MVB, Seldin L, DE Araujo FF, Mariano RdLM (2011) Plant growth promoting rhizobacteria: Fundamentals and applications. In: Maheshwari DK (ed) Plant growth and health promoting bacteria. Spinger-verlag, Berlin, Heidelberg, pp 21-42.

Gadd GM and Griffiths AJ (1978) Microorganisms and heavy metal toxicity. Microbial Ecol. 4, 303–317.

Gaur A, Rana J, Jalali B, Chand H(1990)Role of VA mycorrhizae, phosphate solubilizing bacteria and their interactions on growth and up-take of nutrients by wheat crops. In: Trends in Mycorrhiza Res, Haryana Agricultural Univ. Pub., pp 105-106.

Glick BR (2012) Plant growth promoting bacteria: mechanisms and applications. Scientifica DOI:10.6064/2012/963401.

Grover M, Ali SZ, Sandhya V, Rasul A, Venkateswarlu B (2010) Role of microorganisms in adaptation of agriculture crops to abiotic stress. World J. Microbiol. Biotechnol. 27, 1231-1240.

Gulati A, Rahi P, Vyas P (2008) Characterization of phosphate-solubilizing fluorescent Pseudomonads from the rhizosphere of seabuckthorn growing in the cold deserts of Himalayas. Curr. Microbiol. 56, 73–79.

Gulati A, Sharma N, Vyas P, Sood S, Rahi P, Pathania V, Prasad R (2010) Organic acid production and plant growth promotion as a function of phosphate solubilization by *Acinetobacter rhizosphaerae* strain BIHB 723 isolated from the cold deserts of the trans-Himalayas. Arch. Microbiol. 192, 975–983.

Gulati A, Vyas P, Rahi P, Kasana RC (2009) Plant growth-promoting and rhizosphere-competent *Acinetobacter rhizosphaerae* strain BIHB 723 from the cold deserts of the Himalayas. Curr. Microbiol. 58, 371–377.

Gupta A, Meyer JM, Goel R (2002) Development of heavy metal-resistant mutants of phosphate solubilizing *Pseudomonas* sp. NBRI4014 and their characterization. Curr. Microbiol. 45, 323–327.

Gupta G, Parihar SS, Ahirwar NK, Snehi SK, Singh V (2015) Plant Growth Promoting Rhizobacteria (PGPR): Current and Future Prospects for Development of Sustainable Agriculture. J. Microbial Biochem. Technol. 7, 96-102.

Gupta N and Das S (2008) Phosphate solubilizing fungi from mangroves of Bhitarkanika, Orissa. Hayati J Biosci. 15(2), 90–92.

Gupta N, Sabat J, Parida R (2007) Solubilization of tricalcium phosphate and rock phosphate by microbes isolated from chromite, iron and manganese mines. Acta Bot. Croat. 66, 197–204.

Hall JL (2002) Cellular mechanisms for heavy metal detoxification and tolerance. J. Exp. Bot. 53(366), 1–11.

Han HS and Lee KD (2005) Plant growth promoting rhizobacteria effect on antioxidant status, photosynthesis, mineral uptake and growth of lettuce under soil salinity. Res. J. Agric. Biol. Sci. 1, 210-215.

Hansda A, Kumar V, Usmani A, Usmani Z (2014) Phytoremediation of heavy metals contaminated soil using plant growth promoting rhizobacteria (PGPR): A current perspective. Rec. Res. Sci. Tech. 6(1), 131-134.

Haynes RJ and Swift RS (1990) Stability of soil aggregates in relation to organic constituents and soil water content. J. Soil Sci. 41, 73–83.

Heidari M and Golpayengani A (2012) Effects of water stress and inoculation with plant growth promoting rhizobacteria (PGPR) on antioxidant status and photosynthetic pigments in basil (*Ocimum basilicum* L.) J. Saudi Soc. Agric. Sci. 11, 57-61.

Hemlata Chauhan, Bagyaraj DJ, Selvakumar G, Sundaram SP (2015) Novel plant growth promoting rhizobacteria—Prospects and potential. Appl. Soil. Ecol. 95, 38-53.

Indiragandhi P, Anandham R, Madhaiyan M, Sa TM (2008) Characterization of plant growth promoting traits of bacteria isolated from larval guts of diamondback moth *Plutella xylostella* (Lepidoptera: Plutellidae). Curr. Microbiol. 56, 327-333.

Janmohammadi M, Bihamta MR, Ghasemzadeh F (2013) Influence of rhizobacteria inoculation and lead stress on the physiological and biochemical attributes of wheat genotypes. Cercet. Agron. Mold. 46, 49-67.

Kanimozhi G and Panneerselvam A (2010) Isolation and screening of phosphate solubilizing fungi from mangrove soils of muthupettai, thiruvarur district. Asian J. Microbiol. Biotechnol. Environ. Sci. 12, 359–361.

Kanse OS, Whitelaw-Weckert M, Kadam TA, Bhosale HJ (2015) Phosphate solubilisation by stress tolerant soil fungus *Talaromyces funiculosus* SLS8 isolated from neem rhizosphere. Ann. Microbiol. 65, 85-93.

Kausar R and Shahzad SM (2006) Effect of ACC-deaminase Containing Rhizobacteria on growth promotion of maize under salinity stress. J. Agri. Soc. Sci. 2(4), 216-218.

Kaya MD, Okçu G, Atak M, Çıkılı Y, Kolsarıcı Ö (2006) Seed treatments to overcome salt and drought stress during germination in sunflower (*Helianthus annuus* L.). Eur. J. Agron. 24(4), 291-295.

Khan AG, Kuek C, Chaundhry TM, Khoo CS, Hayes WJ (2000) Role of plants, mycorrhizae and phytochelators in heavy metal contaminated land remediation. Chemosphere 41, 197–207.

Khan MS, Zaidi A, Wani PA (2012) Chromium plant growth promoting rhizobacteria interactions: toxicity and management. In: Toxicity of Heavy Metals to Legumes and Bioremediation (Ed) Zaidi A, Wani PA, Khan MS, Springer, Wein, pp 67-88.

Kloepper JW and Schroth MN (1978) Plant growth-promoting rhizobacteria on radishes: Proceedings of the 4th Internatational Conf. on Plant Pathogenic Bacteria, Station dePathologie Vegetable etPhytobacteriologie, INRA, Angers, France, 879-882.

Kohler J, Caravaca F, Roldán A (2010) An AM fungus and a PGPR intensify the adverse effects of salinity on the stability of rhizosphere soil aggregates of *Lactuca sativa*. Soil Biol. Biochem. 42(3), 429-434.

Kohler J, Hernandez JA, Caravaca F, Roldan A (2009)Induction of antioxidant enzymes is involved in the greater effectiveness of a PGPR versus AM fungi with respect to increasing the tolerance of lettuce to severe salt stress. Environ. Exp. Bot. 65, 245–252.

Kumar KV, Singh N, Behl HM, Srivastava S (2008) Influence of plant growth promoting bacteria and its mutant on heavy metal toxicity in *Brassica juncea* grown in fly ash amended soil. Chemosphere 72, 678–683.

Labuschagne N, Pretorious T, Idris AH (2011) Plant growth promoting rhizobacteria as biocontrol agents against soil borne plant diseases. In: Plant growth and Health Promoting Bacteria. Microbiology Monographs (Ed) Maheshwari DK, Springer-Verlag, Berlin, Heidelberg. pp 211-230.

Lasat HA (2002) Phytoextraction of toxic metals: a review of biological mechanisms. J. Environ. Qual. 31(1), 109-120.

Lichtenthaler HK (1998). The Stress Concept in Plants: An Introduction. Ann New York Acad Sci. 851(1), 187-198.

Liddycoat SM, Greenberg BM, Wolyn DJ (2009) The effect of Plant growth promoting rhizobacteria on asparagus seedlings and germinating seeds subjected to water stress under greenhouse conditions. Can. J. Microbiol. 55, 388-94.

Lugtenberg B and Kamilova F (2009) Plant-growth-promoting rhizobacteria. Ann Rev Microbiol. 63, 541–556.

Ma Y, Rajkumar M, Freitas H (2009) Improvement of plant growth and nickel uptake by nickel resistant-plant-growth promoting bacteria. J. Hazard. Mater. 166, 1154–1161.

Madhaiyan M and Poonguzhali S, Sa T (2007) Metal tolerating methylotrophic bacteria reduces nickel and cadmium toxicity and promotes plant growth of tomato (*Lycopersicon esculentum* L.). Chemosphere 69, 220–228.

Mani R, Ae N, Prasad MNV, Freitas H (2010) Potential of siderophore producing bacteria for improving heavy metal phytoextraction. Trends Biotechnol. 28, 42-49.

Marques APGC, Moriera H, Franco AR, Rangel AOSS, Castro PML (2013) Inoculating *Helianthus annus* (sunflower) grown in zinc and cadmium contaminated soils with plant growth promoting bacteria- effects on phytoremediation strategies. Chemosphere 92, 74-83.

Marquez LM, Redman RS, Rodriguez RJ, Roosinck MJ (2007) A virus in a fungus in a plant: Three-way symbiosis required for thermal tolerance. Science 315(5811), 513–515.

Mayak S, Tirosh T, Glick BR (2004) Plant growth-promoting bacteria confer resistance in tomato plants to salt stress. Plant Physiol. Biochem. 42, 565–572.

Medina A and Azcón R (2010) Effectiveness of the application of arbuscular mycorrhiza fungi and organic amendments to improve soil quality and plant performance under stress conditions. J. Soil Sci. Plant. Nutr. 10(3), 354 – 372.

Milosevic NA, Marinkovik JB, Tintor BB (2012) Mitigating abiotic stress in crop plants by microorganisms. Proc. Nat. Sci, Matica Srpska Novi Sad 123, 17-26.

Miransari M (2011). Soil microbes and plant fertilization. Appl. Microbiol. Biotech. 92, 875–885.

Mishra M, Kumar U, Mishra PK, Prakash V (2010) Efficiency of Plant Growth Promoting Rhizobacteria for the Enhancement of *Cicer arietinum* L. growth and germination under salinity. Adv. Biol. Res. 4(2), 92-96.

Mishra PK, Bisht SC, Ruwari P, Joshi GK, Singh G, Bisht JK, Bhatt JC (2011) Bioassociative effect of cold tolerant *Pseudomonas spp.* and *Rhizobium leguminosarum*-PR1 on iron acquisition, Nutrient uptake and growth of lentil (*Lens culinaris* L.). Eur. J. Soil. Sci. 47, 35–43.

Mohammadi K and Sohrabi Y (2012) Bacterial biofertilizers for sustainable crop production: a review. ARPN J. Agric. Biol. Sci. 7, 307-316.

Morales A, Alvear M, Valenzuela E, Castillo CE, Borie F (2011) Screening, evaluation and selection of phosphate-solubilising fungi as potential biofertilizer. J. Soil Sci. Plant. Nutr. 11, 89–103.

Nadeem SM, Zahir ZA, Naveed M, Arshad M (2009) Rhizobacteria containing ACC-deaminase confer salt tolerance in maize grown on salt-affected fields. Can J Microbiol. 55, 1302–1309.

Nadeem SM, Zaheer ZA, Naveed M, Nawaz S (2013) Mitigation of salinity-induced negative impact on the growth and yield of wheat by plant growth-promoting rhizobacteria in naturally saline conditions Ann. Microbiol. 63 (1), 225–232.

Negi YK, Garg SK, Kumar J (2005) Cold tolerant fluorescent Pseudomonas isolates from garhwal Himalayas as potential plant growth promoting and biocontrol agents in pea. Curr. Sci. 89, 2151-56.

Nezhadahmadi A, Prodhan ZH, Faruq G (2013) Drought tolerance in wheat. Sci. World J. 610-721.

Noor AA, Shah FA, Kanhar NA (2013) Microbial Efficacy of Phosphate Solubilization in Agro-SalineSoils of Various Areas of Sindh Region. Mehran Uni. Res. J. Eng. Tech. 32(2), 167-74.

Oves M, Khan MS, Zaidi A (2013) Chromium reducing and plant growth promoting novel strain *Pseudomonas aeruginosa* OSG41 enhance chickpea growth in chromium amended soils. Eur. J. Soil Biol. 56, 72-83.

Pandey A, Das N, Kumar B, Rinu K, Trivedi P (2008) Phosphate solubilisation by *Penicillium sp.* isolated from soil samples of Indian Himalayan region. World J. Microbiol. Biotechnol. 24, 97-102.

Panwar M, Tewari R, Nayyar H (2014). Microbial consortium of Plant growth promoting rhizobacteria improves the performance of plants growing in stressed soils: An Overview. In: Phosphate Solubilizing Micro-organisms (Ed) Khan MS, Zaidi A, Mussarat J, Springer international publishing, Switzerland, pp 257-285.

Patil NB, Gajbhiye M, Ahiwale SS, Gunjal AB, Kapadnis BP (2011) Optimization of indole acetic acid produced by *Acetobacter diazotrophicus* L1 isolated from sugarcane. Int. J. Environ. Sci. 2, 295-303.

Qurashi AW and Sabri AS (2012) Bacterial exopolysaccharide and biofilm formation stimulate chickpea growth and soil aggregation under salt stress. Braz J Microbiol. 43 (3), 1183–1191.

Rabie GH and Almadini AM (2005) Role of bioinoculants in development of salt-tolerance of *Vicia faba* plants under salinity stress. Afr. J. Biotechnol. 4 (3), 210-222.

Rajankar PN, Tambekar DH, Wate SR (2007) Study of phosphate solubilization of fungi and bacteria isolated from saline belt Purna river basin. Res. J. Agric. Biol. Sci. 3, 701–70.

Rajasankar R, Gayathry GM, Sathiavelu A, Ramalingam C, Saravanan VS (2013) Pesticide tolerant and phosphorus solubilizing *Pseudomonas* sp. strain SGRAJ09 isolated from pesticides treated *Achillea clavennae* rhizosphere soil. Ecotoxicol. 22(4), 707-717.

Rajasekar S and Elango R (2011) Effect of Microbial consortium on plant growth and improvement of alkaloid content in *Withania somnifera* (Ashwagandha). Curr. Bot. 2(8), 27–30.

Rajkumar M, Ae N, Prasad MNV, Freitas H (2010) Potential of siderophore-producing bacteria for improving heavy metal phytoextraction. Trends Biotechnol. 28, 142-149.

Ramadoss D, Lakkineni VK, Bose P, Ali S, Annapurna K (2013) Mitigation of salt stress in wheat seedlings by halotolerant bacteria isolated from saline habitats. Springer Plus 2(6), 1–7.

Rao NSS (1982) Phosphate solubilization by soil microorganisms. In: Advances in Agricultural Microbiology (Ed) Subba Rao NS, Oxford and IBH Publishing Co, New Delhi, pp. 295-305.

Rashid M, Khalil S , Ayub N , Alam S , Latif F (2004) Organic acids production and phosphate solubilization by phosphate solubilizing microorganisms (PSM) under in vitro conditions. Pak. J. Biol. Sci. 7, 187– 196.

Redman RS, Sheehan KB, Stout RG, Rodriguez RJ, Henson JM (2002) Thermotolerance generated by plant/fungal symbiosis. Science 298, 1581.

Rinu K, Malviya MK, Sati P, Tiwari SC, Pandey A (2013) Response of cold-tolerant *Aspergillus sp.* to solubilization of Fe and Al phosphate in presence of different nutritional sources. ISRN Soil Science DOI:10.1155/2013/598541.

Rinu K and Pandey A (2010) Temperature dependent phosphate solubilisation by cold and pH tolerant species of *Aspergillus* isolated from Himalayan soil. Mycoscience 51, 263-271.

Rinu K and Pandey A (2011) Slow and steady phosphate solubilisation by a psychrotolerant strain of *Paecilomyces hepiali* (MTCC 9621).World J. Microbiol. Biotechnol. 27, 1055-62.

Ryu CM, Farag MA, Hu CH, reddy MS, Wei HX, Pare PW, Kloepper JW (2003) Bacterial volatiles promote growth in Arabidopsis. Proc. Natl. Acad. Sci. USA 100: 4927-32.

Sandhya V, Ali SKZ, Grover M, Reddy G, Venkatswarlu B (2010) Effect of plant growth promoting *Pseudomonas spp.* on compatible solutes, antioxidant status and plant growth of maize under drought stress. Plant Growth Reg. 62, 21-30.

Sanjotha P, Mahantesh P, Patil CS (2011) Isolation and screening of efficiency of phosphate solubilizing microbes. Int. J. Microbiol. Res. 3, 56–58.

Saqib ZA, Akhtar j, Ul-Haq MA, Ahmad I (2012) Salt induced changes in leaf phenology of wheat plants are regulated by accumulation and distribution pattern of Na$^+$ ion. Pak. J. Agri. Sci. 49, 141-148.

Saraf M, Jha CK, Patel D (2011). The role of ACC deaminase producing PGPR in sustainable agriculture. In: Plant Growth and Health Promoting Bacteria (Ed) Maheswari DK, Springer, Heidelberg, Berlin, pp 365-386.

Saravanakumar D and Samiyappan R (2007) ACC deaminase from *Pseudomonas fluorescens* mediated saline resistance in groundnut (*Arachis hypogea*) plants. J. Appl. Microbiol. 102, 1283–1292.

Schnoor JL (2003) Phytoremediation of soil and ground water. Technology Evaluation Report TE-02-01. USA: Ground-Water Remediation Technologies Analysis Center.

Selvakumar G, Joshi P, Suyal P, Mishra PK, Joshi GK, Bisht JK, Bhatt JC, Gupta HS (2010) *Pseudomonas lurida* M2RH3 (MTCC 9245), a psychrotolerant bacterium from the Uttarakhand Himalayas, solubilizes phosphate and seedling growth. World J. Microbiol. Biotechnol. 27, 1129-35.

Selvakumar G, Kundu S, Joshi P, Nazim S, Gupta AD, Mishra PK, Gupta HS (2008) Characterization of cold tolerant plant growth promoting bacterium *Pantoea dispersa* IA isolated from a sub-alpine soil in the North western Indian Himalayas. World J. Microbiol. Biotechnol. 24, 955-960.

Selvakumar G, Panneerselvam P, Ganeshamurthy AN (2012) Bacterial mediated alleviation of abiotic stress in crops. In: Bacteria in Agrobiology: Stress Management (Ed) Maheshwari DK, Springer, New York USA, pp 205-224.

Selvakumar G, Joshi P, Nazim S, Mishra PK, Kundu S, Gupta HS (2009) *Exiguobacterium acetylicum* strain 1P (MTCC 8707) a novel bacterial antagonist from the north western Indian Himalayas. World. J. Microbiol. Biotechnol. 25, 131-37.

Sgroy S, Cassan F, Masciarelli O (2009) Isolation and characterization of endophytic plant growth promoting (PGPB) or stress homeostasis regulating bacteria associated to be halophyte *Prosopis strombulifera*. Appl. Microbiol. Biotechnol. 85, 371-81.

Shahzadi I, Khalod A, Mahmood S, Arshad M, Mahmood T, Aziz I (2013) Effect of bacteria containing ACC deaminase on growth of wheat seedlings grown with chromium contaminated water. Pak. J. Bot. 45, 487-94.

Sharma K (2011) Inorganic phosphate solubilization by fungi isolated from agriculture soil. J. Phytol. 3, 11–12.

Sharma K, Pandey K, Dubey S (2010) Exploitation of phosphate solubilization potential of certain species of *Penicillium* for agriculture. Biosci. Biotech. Res. Comm. 2, 209–210.

Sheng XF and Xia JJ (2006) Improvement of rape (*Brassica napus*) plant growth and cadmium uptake by cadmium-resistant bacteria. Chemosphere 64, 1036–1042.

Singh R, Algh GS, Singh S (2012) Screening of phosphorus solubilizing Aspergilli from unusual habitats of Agra region. Ind. J. L. Sci. 2,165-168.

Solano RB, Garcia JAL, Garcia-Villaraco A, Algar E, Garcia-Christobal J, Manero FJG (2010) Siderophore and chitinase producing isolates from the rhizosphere of Nicotiana glauca Graham enhances growth and induce systemic resistance in Solanum lycopersicum L. Plant Soil. 334, 189-197.

Srinivasan R, Yandigeri MS, Kashyap S, Alagawadi AR (2012) Effect of salt on survival and P-solubilization potential of phosphate solubilizing microorganisms from salt affected soils. Saudi J. Biol. Sci. 19 (4), 427–434.

Subhramanian P, Joe MM, Yim W, Hang B, Tipayao SC, Saravanan VS, Yoo J, Cheng J, Sultana T, Sa T (2011). Psychrotolerance mechanism in cold adapted bacteria and their perspectives as plant growth promoting bacteria in temperate agriculture. Korean J. Soil. Sci. Fert. 44(4), 625-636.

Sujatha E, Grisham S, Reddy SM (2004) Phosphate solubilization by thermophillic microorganisms. Ind. J. Microbiol. 44, 101–104.

Tonin C, Vandenkoornhuyse P, Joner EI, Straczek J, Leyval C (2001) Assesment of arbucular mycorrhizal fungi diversity in the rhizosphere of *viola calaminaria* and effect of these fungi on heavy metal uptake by clover. Mycorrhiza 10, 161–168.

Tripathi M, Munot HP, Shouch YJ, Meyer M, Goel R (2005) Isolation and functional characterization of siderophore-producing lead and cadmium resistant *Pseudomonas putida* KNP9. Curr. Microbiol. 5, 233–237.

Trivedi P and Pandey A (2007) Application of immobilized cells of *Pseudomonas putida* to solubilize insoluble phosphate in broth and soil conditions. J. Plant Nutr. Soil Sci. 170, 629–631.

Trivedi P and Pandey A (2008) Plant growth promotion abilities and formulation of *Bacillus megaterium* strain B 388 isolated from a temperate Himalayan location. Ind. J. Microbiol.48, 342–347.

Upadhayay SK, Singh JS, Saxena AK, Singh DP (2012) Impact of PGPR inoculation on growth and antioxidant status of wheat under saline conditions. Plant Biol. 14, 605-611.

Varsha YM, Naga Deepthi CH, Chenna S (2011) An emphasis on xenobiotic degradation in environmental cleanup. J Bioremed Biodeg. S11, 001-010. DOI: 10.4172/2155-6199.S11-0011.

Vassileva M, Azcon R, Barea JM, Vassilev N (1998). Application of an encapsulated filamentous fungus in solubilization of inorganic phosphate. J. Biotechnol. 63, 67– 72.

Vazquez P, Holguin G, Puente ME, Lopez-Cortes A, Bashan Y (2000) Phosphate solubilizing microorganisms associated with the rhizosphere of mangroves in a semi-arid coastal lagoon. Biol. Fertil. Soils. 30, 460–468.

Vessey JK (2003) Plant growth promoting rhizobacteria as biofertilizers. Plant Soil 255, 571–586.

Vinocur B and Altman A (2005) Recent advances in engineering plant tolerance to abiotic stress: achievements and limitations. Curr. Opin. Biotechnol. 16(2), 123-132.

Vivas A, Biro B, Ruiz-Lozano JM, Barea IM, Azcon R (2006) Two bacterial strains isolated from a Zn-polluted soil enhance plant growth an mycorrhizal efficiency under Zn-toxicity. Chemosphere 62, 1523–1533.

Vyas P, Rahi P, Chauhan A, Gulati A (2007) Phosphate solubilization potential and stress tolerance of *Eupenicillium parvum* from tea soil. Mycol. Res. 111, 931-938.

Wahid A, Gelani S, Ashraf M and Foolad MR (2007). Heat tolerance in plants: An overview. Environ. Exp. Bot. 61(3), 199-223.

Wang CJ, Yang W, Wang C, Gu C, Niu DD, Liu H, Wang P, Guo JH (2012) Induction of drought tolerance in cucumber plants by consortium of three plant growth promoting rhizobacteria strains. PLoS ONE 7, 1-10.

Wani PA and Khan MS (2010) *Bacillus* species enhance growth parameters of chickpea (*Cicer arietinum* L.) in chromium stressed soils. Food Chem. Toxicol. 48, 3262–3267.

Wani PA, Khan MS, Zaidi A (2007) Effect of metal tolerant plant growth promoting *Bradyrhizobium sp.* (Vigna) on growth, symbiosis, seed yield and metal uptake by green gram plants. Chemosphere 70, 36–45.

Wani PA, Khan MS, Zaidi A (2008) Effect of metal-tolerant plant growth-promoting *rhizobium* on the performance of pea grown in metal-amended soil. Arch. Environ. Contam. Toxicol. 55, 33–42.

Wani PA, Khan MS, Zaidi A (2012) Toxic effects of heavy metals on germination and physiological processes of plants. In: Toxicity of Heavy Metals to Legumes and Bioremediation (Ed) Zaidi A, Wani PA, Khan MS, Springer, Wein, pp 45-66.

Whiting SN, de Souza MP, Terry N (2001) Rhizosphere bacteria mobilize Zn for hyperaccumulation by *Thlaspi caerulescens*. Environ. Sci Technol. 35(15), 3144–3150.

Williams LE, Pittman JK, Hall J (2000) Emerging mechanisms for heavy metal transport in plants. Biochem. Biophys. Acta 1465, 104–126.

Wu CH, Wood TK, Mulchandani A, Chen W (2006) Engineering plant-microbe symbiosis for rhizoremediation of heavy metals. Appl. Environ. Microbiol.72, 1129–1134.

Wu Z, Haitao Y, Lu J, Li C (2012) Characterization of rhizobacterial strain Rs-2 with ACC deaminase activity and its performance in promoting cotton growth under salinity stress. World J. Microbiol. Biotechnol. 28(6), 2383-93.

Xiao CQ, Chi RA, Huang XH, Zhang WX, Qiu GZ, Wang DZ (2008) Optimization for rock phosphate solubilization by phosphate-solubilizing fungi isolated from phosphate mines. Ecol. Eng. 33,187–193.

Yadav J, Verma JP, Yadav SK and Tiwari KN (2011) Effect of salt concentration and pH on soil inhabiting fungus *Penicillium citrinum* Thom. for solubilization of tricalcium phosphate. Microbiol. J. 1, 25-32.

Yao L, Wu Z, Zheng Y, Kaleem I, Li C (2010) Growth promotion and protection against salt stress by *Pseudomonas putida* Rs-198 on cotton. Eur. J. Soil Biol. 46, 49–54.

Yildirium E and Taylor AG (2005) Effect of biological treatment on growth of bean plants under salt stress. Ann. Rep. Bean Improv. Coop. 48, 176–177.

Yingbean W, Yeulin H, Hongmei Y, Wei C, Zhen W, Liguan X, Aiqua Z (2012) Isolation of phosphate solubilising fungus and its application in solubilisation of rock phosphates. Pak. J. Biol. Sci. 15(23), 1144-1151.

Zaidi S, Usmani S, Singh BR, Musarrat J (2006) Significance of *Bacillus subtilis* strain SJ-101 as a bioinoculant for concurrent plant growth promotion and nickel accumulation in *Brassica juncea*. Chemosphere 64, 991-997.

Zawoznik MS, Ameneiros M, Benavides MP, Vázquez S, Groppa MD (2011) Response to saline stress and aquaporin expression in *Azospirillum*-inoculated barley seedlings. Appl. Microbiol. Biotechnol. 90(4), 1389-1397.

7

Endophytes and Their Possible Roles in Plant Stress Management

Sanjay K. Singh

Abstract

Variously defined endophytes are highly diverse in plants in natural ecosystems, and most of them are symbiotic with of them are Asymptomatic presence of endophytic fungi in internal tissues on healthy host plants are able to play vital roles in physiology and biochemical processes. As a result of this interaction endophyte produces chemical substances having various applications. This chapter provides a brief review of certain grass and non-grass endophytes capable to extend beneficial impacts on host plants to sustain the unfavourable conditions of abiotic stresses. However, biotic stresses are also a matter of great concern, where endophytes can play important roles in combating the unfavourable situation. Endophytic fungi used in bio-control and other agriculture related sectors are also highlighted in this review. Endophytic fungi producing volatile organic compounds reported from all over the world and their various applications find place in this article which has opened new frontiers of bioprospecting of endophytic fungi. Present article reflects various benefits of endophytes to their host plant like improvement in plant growth, development and productivity. As such fungal endophytes provide ample scope as one of the untapped resources for their documentation, and applications in various fields.

Keywords: Abiotic stress, Biotic stress, Biocontrol, Biofumigation, Endophyte fungi, Plant-microbe interaction

1. Introduction

Fungi being the second largest group after insects from key component of ecosystems throughout world. They are ubiquitous in diverse habitats ranging from psychrophilic to thermophilic and remarkably play a vital role in every ecosystems. Being heterotrophic, they are usually saprophytes and parasites. During evolution when plants colonized the land successfully, fungi developed different types of relationship with them. The group 'endophytes' form one of these associations and their existence found in the fossil records suggest that endophyte-host association may have evolved about > 400 million years ago and these endophytes including mycorrhizal fungi were found initially associated with higher plants (Redecker *et al.* 2000; Krings *et al.* 2007).

Literature reveals the existence of endophytic fungi inside the organs of asymptomatic plants since the end of 19th century (Guerin 1898; Vogl 1898). Except for a few fragmentary report (Sampson 1933), work on endophytic fungi began to receive attention by the end of 20th century, when concerted efforts were made by researchers to understand the biology of grass endophytes. Most endophyte and grass associations were reported to be mutualistic and were co-evolved (Arachivaleta *et al.* 1989). In this regard, discovery of cause of fescue toxicosis by Bacon *et al.* (1977) has been considered as landmark in the history of grass endophyte research. Cattles fed on infected grass, *Festuca arundinacea* suffered a lot due to toxicoses and expressed syndrome like symptomsin United States, while in New Zealand endophyte infected ryegrass *Lolium perrenne* caused ryegrass staggers in sheep. Later on by research it was unveiled that although pasture grasses did not show any symptoms, leaves and stems of most of the plants of *Festuca arundinacea* found systemically infected by *Neotyphodium coenophialum* (*F. Clavicipitaceae*), and the infected plants contained several ergot alkaloids, in addition to other metabolites. *Neotyphodium* species and its *Epichloë* sexual morph (teleomorph) has been highly studied group, and widely known species are *N. coenophialum*, *N. loli*, and *N. uncinatum*. They colonize tall fesue, perennial rye-grass, and meadow fescue, respectively. Later on concept of mutualistic symbioses between endophytic *Neotyphodium* spp. and their grass hosts were further revised and defined as 'asymptomatic symbioses' (Wilkinson and Schardl 1997). As such biology of grass endophytes has seen a tremendous change and is known for more than a century. Researches on grass endophytes, led to increased recognition due to enormous beneficial impacts they conferred to host plant communities in various ways, like through tolerance to biotic and a biotic stress (Schardl and Phillips 1997), increasing biomass and decreasing water consumption, etc.

2. Definition of endophytes

From time to time term endophyte has been defined in various ways by researchers. De Barry (1866) coined the term 'endophyte' to detect fungi that reside within host tissues, from epiphytes, and were first characterized in depth from north-temperate Poaceous member by Sampson (1933). Carroll (1986) used the term for those organisms causing asymptomatic infections within plant tissues, excluding pathogenic and mycorrhizal fungi. Petrini (1991) expanded Carroll's definition to include all the organisms that colonize a plant without causing apparent harm to the host at any stage in their life cycle. However, invisible pathogens in this category were also included. Wilson (1995) further expanded definition and included both fungi and bacteria, which for all or part of their life cycle invade living plants and cause asymptomatic infections entirely within tissues but cause no disease. Later on, all microorganisms residing inside plants without producing visible symptoms were categorized as endophytes (Azevedo *et al.* 2000). According to Stone *et al.* (2000), the term 'endophyte' is an all-encompassing topographical term which includes all organisms that during a variable period of their life occupy living tissues of their hosts without any symptom. However, Bacon and White (2000) more precisely defined that an endophytic fungus lives in a mycelial form in biological association with the living plant, at least for some period. Although, by definition, endophytes do not cause disease in their hosts, they are obligatory heterotrophic and are often congeneric with a taxonomically diverse range of fungal pathogens. The majority of plant species examined to date harbour endophytic fungi inside aerial tissues becoming a vital component of terrestrial plant communities (Arnold and Herre 2003; Gonthier *et al.* 2006). These endophytes were considered to be among the least-known groups of plant associated fungi, differentiated from pathogenic fungi based on asymptomatic growth under different conditions (Verhoeff 1974), and from mycorrhizal fungi, on the basis of taxonomy (Bills and Polishook 1991) and tissue specificity (Carlile and Watkinson 1989; Agrios 1997).

3. Group, diversity and distribution of fungal endophytes

According to Stone *et al.* (2000) endophytic microbes in relation to their plant organ belong to distinct classes. Major groups are : 1) Endophytic Clavicipitaceae, 2) Fungal endophytes of dicots, 3) Other systemic fungal endophytes, 4) Fungal endophytes of lichens, 5) Endophytic fungi of bryophytes and ferns, 6) Endophytic fungi of tree bark, 7) Fungal endophytes of xylem, 8) Fungal endophytes of roots, 9) Fungal endophytes of galls and cysts, 10) Prokaryotic endophytes of plants (includes endophytic bacteria and actino). Such ubiquitous endophytes dominated their assemblages of distinct hosts suggesting that certain genera of fungi are well adapted to make an endophytic way of life (Bills *et al.*

2004). Rodriguez *et al.* (2008) provided comprehensive accounts on different functional groups broadly classified into Clavicipitaceous and Nonclavicipitaceous endophytes. They are further divided into class 1 to 4, based on host colonization, mode of transmission, in plant diversity and fitness conferred to hosts. This information is highly important for application oriented research work. Literature reveals that endophytic fungi are mostly documented from grasses (Bacon *et al.* 1977; Clay 1988), conifers (Petrini and Fisher 1988; Wang and Guo 2007; Hormazabal and Piontelli 2009), marine algae (Hawksworth 1988), lichens (Li *et al.* 2007), mosses and ferns (Fisher 1996), palms (Rodrigues 1994; Fröhlich *et al.* 2000), liverworts (Boullard 1988), pteridophytes (Dhargalkar and Bhat, 2009) etc Endophytes are largely confined to gymnosperms in temperate regions (Bernstein and Carroll, 1977; Petrini and Fisher 1988). It is, however evident that endophytic fungal diversity peaks in tropical forests where woody angiosperm diversity is also higher (Lodge *et al.* 1996; Gamboa and Bayman 2001; Banerjee 2011). These reports also states that most studies on endophytes have been carried out in the northern hemisphere and tropical and subtropical regions.

4. Plant stresses and endophytes

Plant stresses are broadly categorised as follows:

(i)   Abiotic stresses consist of non-organismal, non-pathogenic factors that affect plant function includes drought, flood, high soil salinity, heat, cold, oxidative stress, heavy metal toxicity and nutrient deficiency, etc. which lead to stress in the plant resulting in lower or no yield.

(2)   Biotic stress consists of pathogenic or organism factors that inhibit plant function, resulting in lower or no yield.

In addition to systematics, evolutionary biology, ecology, there has been an increasing interest in applied research of endophytic fungi. This ranges from biological control to bioprospecting based on pioneering work done by different workers worldwide on this cryptic guild of fungi (Spurr and Welty 1975; Carroll and Carroll 1978; Petrini 1986, Sieber 1989; Rodrigues and Samuels, 1990; Espinosa-Garcia & Langenheim 1990; Stierle *et al.* 1993; Backman and Sikora 2008). Studies on methods of detection, taxonomy, species composition, distribution, biological, ecological and physiological aspects of endophytes of woody plants in Europe and North America has been extensively carried out by Carroll (1995), Petrini (1986, 1996) and Bills (1996). In addition, interactions and mutualistic symbiosis between endophytes and host plants has also been studied in detail. Tropical and temperate rainforests are considered as most biologically diverse terrestrial ecosystems on earth (Mittermeier *et al.* 1999).

Host response to endophyte is a rather complex phenomenon involves physiological and biochemical processes. It appears from different reports that during course of evolution grasses have developed symbiotic relationship with fungi including mycorrhizal one that are able to grow in or on the roots (Smith and Read 1997) and other fungi that infect grass shoots systemically form non-pathogenic, systemic, and intercellular associations (Bacon and Battista 1991). Many grass endophytes being symbiotic in association extends broad spectrum of benefits to their hosts, e.g., *Neotyphodium coenophialum* exerts several benefits to its host, *Lolium arundinaceum.* Though, actual mechanisms by which this symbiotic relationship imparts such benefits are not well studied, it has been reported that host plant persistence and vigour were enhanced when some novel endophytes were transferred into elite tall fescue germplasm. This phenomenon has opened a new avenue to exploit the agronomic benefits in general.

Endophyte and host plant interaction play vital roles in biology of both endophytes as well as host plants. This interaction involves number of plant physiological and ecological processes in various ways. Similarly biotic and abiotic stresses can interact to influence host-endophyte responses. The fossils and molecular studies suggest that arbuscular mycorrhiza fungi (AMF) assisted plants in their colonization of land, and endophyte can also improve tolerance to abiotic stress, in addition to providing mineral nutrients. Toxic alkaloids produced by endophyte (*Neotyphodium*) confer resistance to herbivory by insects and mammals, and ultimately allows plants to survive stressful conditions. Among best known grass endophytes of family Clavicipitaceae (Ascomycetes) include *Atkinsonella, Balansia, Balansiopsis, Echinodothis, Epichloe, Myriogenospora.* Later on during detailed study two other group of endophytes were reported. These were *Gliocladium*-like endophytes (An *et al.* 1993) in perennial ryegrass, and *Phialophora*-like endophytes in meadow fescue grass (Schmidt 1993). An endophytic *Curvularia* species has been demonstrated to confer adaptation of the *Dichanthelium* grass to extreme heat condition. A root-endophytic *Piriformospora indica* has been reported to play important roles in regulation of systemic defence responses by inducing resistance in barley plants against both necrotrophic root parasites and biotrophic leaf pathogen, *Blumeria graminis* (Erysiphaceae).

A comprehensive review has also been provided by Belesky and West (2009) on attributes of host plant-endophyte interactions with emphasis on tall fescue responses to abiotic stresses associated with drought and soil mineral deficiencies. The endophyte and grass association play important roles during extreme water stress condition. This association is reported to maintain higher level of water content at tiller bases than those in non-infected plants during drought condition due to enhanced accumulation of solutes in tissues of endophyte-infected plants. This may also happen due to slow-down of the transpiration stream, or thicker cuticle (Elbersen and West 1996; Buck *et al*.1997).This situation compliment to the survival of plant in drought conditions. Mechanisms of drought tolerance is generally divided into three major groups: (a) accumulation and translocation of assimilates, (b) osmotic adjustment, and (c) maintenance of cell wall elasticity (Arraudeau 1989). However, accumulation of carbohydrates which are osmotically active is reported to play important role in endophyte-related osmotic adjustment in tall fescue. Accumulation of solutes in tissues helps to maintain turgor and facilitates the biological processes, and different solutes are considered to contribute to osmotic adjustment in endophyte-infected grasses e.g., water-soluble sugars, fungal mannitol and arabitol, amino acid proline and other amino acids, loline alkaloids, etc. (Richardson *et al*. 1992; Bush *et al*. 1993).

5. Endophytic fungi producing VOCs and their various applications

Fungi are known to produce low-molecular weight volatile substances since long (Dennis and Webster 1971) but spectrum of their potential in agriculture, and allied sectors are quite recent. However, detailed analysis of volatiles of fungi was never attempted earlier, except by Wheatley *et al.* (1997), who identified VOCs and their inhibitory activity. Later on number of endophytic fungi was reported to produce VOCs having promising applications in agriculture and allied sectors, which has opened new frontiers of bioprospecting.

Among endophytes, *Muscodor* is a genus placed under Xylariaceae, erected by Worapong, Strobel & W.M. Hess (Worapong *et al*. 2001), for accommodating a sterile/nonsporulating fungus isolated from *Cinnamomum zeylanicum*, collected from Honduras. To date, 18 species have been described based on ITS sequencing and volatile gas composition analysis and hyphal structures using a scanning electron microscope (SEM). *Muscodor* species has been reported from different parts of the world. Due to its remarkable ability to produce arrays of volatile organic compounds exhibiting broad spectrum of bioactivity, they are being extensively used as a model organisms in agriculture and allied sectors primarily in controlling plant pathogens, nematodes, insects pest, and thereby indirectly improving the plant health and yields. The biofumigation by *M. albus* has been reported to control blue and gray molds of apple, *Botrytis cinerea*; brown rot of peaches, *Monilinia fructicola* (Mercier and Jiménez 2004), green mold, *Penicillium digitatum,* and sour rot of lemons, *Geotrichum candidum* (Mercier and Smilanick 2005). Considering an important carrier of many devastating pathogens from one generation to another, biofumigation of infected seed materials may be helpful in reducing/controlling the incidence of different diseases to a significant level. Four days biofumigation treatment of barley seed contaminated with *Ustilago hordei* by agar culture of *M. albus* is reported control 100% covered smut (Strobel *et al.* 2001). As such *Muscodor albus* and *M. roseus* were used in controlling plant pathogenic fungi (Stinson *et al.* 2003; Banerjee *et al.* 2010), nematodes (Riga *et al.* 2008), and potato tuber moth (PTM), *Phthorimaea operculella,* a serious pest of stored potato of most of the potato growing countries (Lacey and Neven 2006). A naphthalene like compound producing *Muscodor vitigenus* has been reported as an effective insect repellent (Daisy *et al.* 2002a, b). In addition to plant pathogens, insects and pests, antibiotic effect of *M. albus* was demonstrated to control human pathogenic microbes, like *Yersinia pestis*, *Mycobacterium tuberculosis* and *Staphylococcus aureus.*

Most recently, number of endophytic species of *Muscodor* has also been reported to show broad spectrum of bioactivities against selected fungal and bacterial pathogens, suggesting inherent potential of this genus to be exploited for various purposes (Meshram *et al.* 2013, 2014, 2015; Banerjee *et al.* 2014; Saxena *et al.* 2014, 2015). As such gaseous products of *Muscodor* spp. have attracted great attention of researchers as 'mycofumigant' for various applications in the fields of agriculture, medicine, and industry (Mitchell *et al.* 2008). Therefore, considering its significance details of 18 different species with their properties, and reported from all over world are summarised in Table 1. Except *Muscodor* and its species, there are number of other endophytic fungi reported to produce volatiles with various application. Singh *et al.* (2011) reported an endophytic *Phomopsis* sp., from *Odontoglossum* sp. (Orchidaceae) producing a unique mixture of volatile organic compounds (VOCs) inhibitory to range of plant pathogens, like *Pythium, Sclerotinia, Rhizoctonia, Fusarium, Botrytis, Colletotrichum* including sabinene (a monoterpene with a peppery odour). The *Phoma* sp. isolated from *Larrea tridentata* (creosote bush) growing in the desert region of southern Utah, USA was reported to produce a unique mixture of volatile organic compounds including a series of sesquiterpenoids, some alcohols and several reduced naphthalene derivatives. The gases of *Phoma* sp. possess antifungal properties markedly similar to that of a methanolic extract of the host plant. Some of the test organisms with the greatest sensitivity to the VOCs *Phoma* sp. were *Verticillium, Ceratocystis, Cercospora* and *Sclerotinia* while those being the least sensitive were *Trichoderma, Colletotrichum* and *Aspergillus* (Strobel *et al.* 2011). This report also discussed the possible involvement of VOC production by the fungus and its role in the biology/ecology of the fungus/plant/environmental relationship with implications for utilization as an energy source.

Table 1: Biological and culture characteristic comparison of *Muscodor* species reported till date.

Muscodor spp. & Geographic origin	Host & Family	Mycelium pigment production	Hyphal & Mycelial Growth Characteristics	Major Compounds Produced	Bioactivity	Reference
Muscodor albus [Honduras]	*Cinnamomum zeylanicum* [Lauraceae]	White	Straight; Rope-like	2-Methylpropanoic acid; naphthalene and azulene derivatives	Antifungal and antibacterial	Worapong et al. (2001)
Muscodor vitigenus [Peru]	*Paullinia paullinoide* [Sapindaceae]	White	Straight Rope-like	Naphthalene	Anti-insect	Daisy et al. (2002)
Muscodor crispans [Bolivia]	*Ananas ananassoides* [Bromeliaceae]	White in the dark and pink in the light	Wavy growing Rope-like with cauliflower like bodies	2-Methylpropanoic acid	Antifungal and antibacterial	Mitchell et al. (2008)
Muscodor yucatanensis [Quintana Roo]	*Bursera simaruba* [Burseraceae]	White	Cottony-like pattern; Rope-like with coils structure and swollen cell	2-Methylpropanoic acid	Antifungal and antibacterial	González et al. (2009)
Muscodor fengyangensis [Southeast China]	*Cinnamomum zeylanicum* [Lauraceae]	White	Straight; Rope-like with coils structure	2-Methylpropanoic acid	Antifungal and antibacterial	Zhang et al. (2010)
Muscodor cinnamomi [Thailand]	*Cinnamomum bejolghota* [Lauraceae]	White in the dark and pale orange in the light	Rope-like with cauliflower-like bodies	Ethyl 2-methylpropanoate	Antifungal and antibacterial	Suwannarach et al. (2011)
Muscodor sutura [Colombia]	*Prestonia trifida* [Apocynaceae]	White in the light and reddish in the dark	Suture-like pattern Rope-like nondescript extracellular bodies	2-Methylpropanoic acid	Antifungal	Kudalkar et al. (2012)
Muscodor equiseti [Thailand]	*Equisetum debile* [Equisetaceae]	White	Cottony-like pattern Rope-like with coils structure and swollen cell	2-Methylpropanoic acid	Antifungal and antibacterial	Suwannarach et al. (2013)
Muscodor musae [Thailand]	*Musa acuminate* [Musaceae]	White	Straight and hairy-like Mycelium; Rope-like with coils structure	2-Methylpropanoic acid	Antifungal and antibacterial	Suwannarach et al. (2013)
Muscodor oryzae [Thailand] [Poaceae]	*Oryza rufipogon*	Pale orange	Straight Rope-like with coils structure	3-Methylbutan-1-ol	Antifungal and antibacterial	Suwannarach et al. (2013)
Muscodor roseus [Northern Territory]	*Grevillea pteridifolia* [Proteaceae]	Lightly rose	Felt-like mycelium; Forming rope-like stands and coils structure	Ethyl 2-butenoate and 1,2,4- trimethylbenzene	Antifungal	Worapong et al. (2002)

Fungus	Host plant	Colour	Structure	Compounds	Activity	Reference
Muscodor suthepensis (2013) [Thailand]	Cinnamomum bejolghota [Lauraceae]	White in the dark and pale pink in the light	Straight Rope-like with coils structure	2-Methylpropanoic acid	Antifungal and antibacterial	Suwannarach et al. (2013)
Muscodor kashayum [India]	Aegle marmelos [Rutaceae]	White	Thick & coiled appear fused rope-like hyphal strands with branching at right angle	3-cyclohexen-1-ol, 1-(1,5-dimethyl-4-hexenyl)-4-methyl	Antifungal and antibacterial	Meshram et al.(2013)
Muscodor albus [India]	Piper nigrum [Piperaceae]	White	Mcelium with heavily intertwining hyphae	acetic acid, ethyl ester; propanoic acid, 2-methyl-	Antifungal	Banerjee et al. (2014)
Muscodor darjeelingensis [India]	Cinnamomum camphora [Lauraceae]	White	Thick sterile mycelium with cauliflower-like structures	2, 6-Bis (1, 1-dimethylethyl)-4-(1-oxopropyl) phenol	Antifungal and antibacterial	Saxena et al. (2014)
Muscodor strobelii [India]	Cinnamomum zeylanicum [India]	Pale yellow (in light)	Straight ; Rope-like, slimy; Zinnia- & bud-like bodies	4-octadecyl-morpholine, Tetraoxa-propellan, Aspido-fractinine-3-methanol	Antifungal and antibacterial	Meshram et al. (2014)
Muscodor tigerii [India]	Cinnamomum camphora [Lauraceae]	White turns to brown after 15 days	Swollen; branches at a certain angle & terminates into coils later fuse to form thick ropy mycelium	4-Octadecylmorpholine, 1-Tetradecanamine, N,N-dimethyl	Antifungal and antibacterial	Saxena et al. (2015)
Muscodor ghoomensis [India]	Cinnamomum camphora [Lauraceae]	Pale yellow	Straight appears as fused rope with branching at right angle; mycelium flat radial crevices with rough margins	4-octadecylmorpholine; 1-nonadecamine-N.N dimethyl	Antifungal and antibacterial	Meshram et al. (2015)
Muscodor indica [India]	Cinnamomum camphora [Lauraceae]	No pigment; whitish turning pale colored when old	Coiled, appear like fused rope-like hyphal strands with branching at right angle	1, 6-dioxacyclododecane-7, 12-dione; 4-octadecy lmorpholine	Antifungal and antibacterial	Meshram et al. (2015)

6. Conclusion and future outlook

Endophytic fungi are poorly researched group of organism. Current trends are towards exploring diversity of endophytic fungi for their exploitation and becoming increasingly difficult to define and delimit the fungal taxa on group level for technological developments. Fungi, as a whole have established their credentials as promising goldmines for a variety of bioactive metabolites and enzymes. It is well documented that less than 1% of the fungal diversity available in nature has received serious attention for biotechnology investigations (Heywood 1995; Staley *et al*.1997). However, dimension of fungal biology is changing tremendously and fungi can play important role in mainstream biology as model organism. Increased incidences of new diseases, drug resistance in microorganisms require effective strategies to explore new and useful bioactive compounds to combat this situation. Various studies have established that the endophytic fungi isolated from various tissues, seed, stem, leaf impacts the ecology of the host plants in different ways, often increasing fitness traits and are capable of producing plant metabolites independently. However, it is required to understand the actual mechanism of action to achieve a range of applications associated with endophyte-plant interaction. Recent report on production of low molecular weight volatile organic compounds (VOCs) by endophytic fungi reflects wide spectrum of antibiotic effects of VOCs for application in various fields. Also it has been suggested that volatiles produced by endophytes may be useful as an alternatives of a wonder agro-chemical 'methyl bromide' to be phased out due to environmental concern. The genome sequencing of certain fungi has reflected the fact that number of secondary metabolites, encoded by the gene clusters, far exceeds the actual number of metabolites produced by them (Hertweck 2009; Brakhage *et al*. 2009). This untapped potential has led the researchers to explore the potential of filamentous fungi through various advanced techniques leading to the activation of silent gene clusters. Epigenetic modulations have been used in order to enhance the production of cryptic and bioactive compounds in endophytic fungi (Sun *et al*. 2012; Hassan *et al*. 2012). This work may serve as pointer for future direction.

Endophytes are the most biologically diverse in the regions where plant life is the most unusual and diverse. Certainly, India is one among these regions. Considering such a vast plant diversity and wide range of vegetation distributed along with the 'Western Ghats' and Eastern Himalayan regions, it is presumed that a high degree of endophytic fungal diversity may be present, which may serve as valuable germplasm for exploration of biologically active compounds. Therefore, target oriented different strategies need to be conceptualized for systematically exploring germplasm of bioactive endophytes from plants of different ecological settings, including rain forests which in other parts of world are reported to posses high endophytic microbial diversity. Strobel and Daisy (2003) emphasized to understand the rationale of plant selection strategies which is governed by various hypotheses.

Similarly, several other endophytes are reported to produce enzymes which can degrade carbon sources into compounds that can be used as biofuels. Endophytic *Gliocladium roseum* has been reported to produce a series of volatile hydrocarbons, some of them are similar to diesel fuel termed as mycodiesel (Strobel *et al.* 2008). An endophytic *Phomopsis* sp., was reported to produce sabinene (a monoterpene with a peppery odour (Singh *et al.* 2011). Sabinene along with other monoterpenes such as pinene are being explored as the components for the next generation aircraft fuel (Reninger *et al.* 2008; Rude and Schirmer, 2009). A fungal source of this monoterpene has the genetic potential to facilitate its production via microbial fermentation technology.

Acknowledgement

Author thanks Director, MACS' Agharkar Research Institute, Pune for extending all facility, and to Department of Science and Technology (DST), New Delhi for providing financial support for setting up National Facility for Culture Collection of Fungi [No. SP/SO/PS-55/2005] at this Institute. Technical help by Mr. D.K. Maurya is acknowledged.

References

Agrios G (1997) Plant Pathology. 4ᵗʰ edition. Academic Press, San Diego, California.

An ZQ, Siegel MR, Hollin W, Tsai HF, Schmidt D, Bunge G, Schardl CL (1993) Relationships among non-*Acremonium* sp. fungal endophytes in five grass species. Appl. Environ. Microbiol. 59, 1540–1548.

Arachivaleta M, Bcon CW, Hoverland CS, Radeliffe DE (1989) Effect of the tall fescue endophyte on plant response to environmental stress. Agron. J. 81, 83-90.

Arraudeau MA (1989) Breeding strategies for drought resistance. In: Drought Resistance in Cereals (Ed) Baker FWC, International, Wallingford, UK, pp 107–116.

Arnold AE, Herre EA (2003) Canopy cover and leaf age affect colonization by Baihua mountain of Beijing, China. Fungal Divers. 25, 69–80.

Azevedo JL, Maccheroni W-Jr, Pereira JO, de Araujo WL (2000) Endophytic microorganisms: A review on insect control and recent advances on tropical plants. Electron. J. Biotechnol. 3(1), 40-65.

Backman PA, Sikora RA (2008) Endophytes: An emerging tool for biological control. Biol. Control 46, 1-3.

Bacon CW, Porter JK, Robins JD, Luterell EJ (1977). *Epichloë typhina* from 10 toxic tall fescue grass. Appl. Env. Microbiol. 34, 576-581.

Bacon CW, White JF (2000) Physiological adaptations in the evolution of endophytism in the Clavicipitaceae. In: Microbial Endophytes (Ed) Redlin SC, and Carris LM, Marcel Dekker, New York, pp 237–261.

Bacon CW, Battista J De (1991) Endophytic fungi of grasses. In: Handbook of applied mycology, Vol.1. (Ed) Arora DK, Rai B, Mukerji KG and Knudsen RG, Marcel Dekker, New York, pp 231–256.

Banerjee D (2011) Endophytic fungal diversity in tropical and subtropical plants. Res. J. Microbiol. 6, 54–62.

Banerjee D, Pandey A, Jana M, Strobel G (2014) *Muscodor albus* MOW12 an endophyte of *Piper nigrum* L. (Piperaceae) collected from North-East India produces volatile antimicrobials. Indian J. Microbiol. 54(1), 27–32.

Banerjee D, Strobel G, Geary B, Sears J, Ezra D, Liarzi O, Coombs J (2010) *Muscodor albus* strain GBA, an endophytic fungus of *Ginkgo biloba* from United States of America, produces volatile antimicrobials. Mycology 1, 179–186.

Belesky DP, West CP (2009) Abiotic stresses and endophyte effects. In: Tall Fescue for the Twenty-First Century (Ed) Fribourg HA, Hannaway DB and West CP, Agronomy Monograph 53, pp 49-64.

Bernstein ME, Carroll G (1977) Internal fungi in old-growth Douglas fir foliage. Can. J. Bot. 55, 644–653.

Bills GF (1996) Isolation and analysis of endophytic fungal communities from woody plants. In: Endophytic Fungi in Grasses and Woody Plants (Ed) Redlin SS and Carris LM, APS Press, Saint Paul, pp 121–132.

Bills GF, Polishook JD (1991) Microfungi from *Carpinus caroliniana*. Can. J. Bot. 69, 1477–1482.

Bills JK, Christensen M, Powell M, Thorn G (2004) Endophytic Fungi. In: Biodiversity of Fungi: Inventory and Monitoring Methods (Ed) Mueller GM, Bills GF and Foster M, Elsevier Academic Press, CA, USA, pp 241–270.

Boullard B (1988) Observations of the coevolution of fungi with hepatics. In: Coevolution of Fungi with Plants and Animals (Ed) Pirozynski KA and Hawksworth DL, Academic Press, London, UK, pp 107–124.

Brakhage AA, Bergmann S, Schuemann J, Scherlach K, Schroeckh V, Hertweck C (2009) Fungal genome mining and activation of silent gene clusters In: The Mycota XV (Ed) Esser K., Springer-Verlag, Heidelberg, Germany, pp 297-303.

Buck GW, West CP, Elbersen HW (1997) Endophyte effect on drought tolerance in diverse *Festuca* species In: *Neotyphodium*/Grass Interactions (Ed) Bacon CW and Hill NS, Plenum Press, New York, pp 141–143.

Bush LP, Fannin FF, Siegel MR, Dahlman DL, Burton HR (1993) Chemistry, occurrence and biological effects of pyrrolizidine alkaloids associated with endophyte-grass associations. Agric. Ecosys. Environ. 44, 81–102.

Carlile M, Watkinson SC (1989) The fungi. Academic Press 6, 983–987.

Carroll GC (1986) The biology of endophytism in plants with particular reference to woody perennials. In: Microbiology of the Phyllosphere (Ed) Fokkema NJ, and vanden Huevel J, Cambridge University Press, Cambridge, pp 205–222.

Carroll GC (1995) Forest endophytes: pattern and process. Can. J. Bot.73, 1316–1324.

Carroll GC, Carroll FE (1978) Studies on the incidence of coniferous needle endophytes in the Pacific Northwest. Can. J. Bot. 56, 3032–3043.

Clay K (1988) Fungal endophytes of grasses: A defensive mutualism between plants and fungi. Ecology 69, 10–16.

Daisy B, Strobel G, Ezra D, Castillo U, Bairn G, Hess WM (2002a) *Muscodor vitigenus* anam. sp. nov., an endophyte from *Paullinia paullinioides*. Mycotaxon 81, 463–475.

Daisy BH, Strobel GA, Castillo U, Ezra D, Sears J, Weaver DK, Runyon JB (2002b) Naphthalene, an insect repellent, is produced by *Muscodor vitigenus*, a novel endophytic fungus. Microbiology 148, 3737–3741.

De Bary A (1866) Morphologie und Physiologie der Pilze, Flechten, und Myxomyceten, (Vol. II), Hofmeister's Handbook of Physiological Botany, Leipzig, Germany.

Dennis C, Webster J (1971) Antagonistic properties of species of *Trichoderma*.II. Production of volatiles antibiotics. Trans. Br. Mycol. Soc. 57, 41-48.

Dhargalkar S, Bhat DJ (2009) *Echinosphaeria pteridis* sp. nov. and its *Vermiculariopsiella* anamorph. Mycotaxon 108, 115–122.

Elbersen HW, West CP (1996) Growth and water relations of field grown tall fescue as influenced by drought and endophyte. Grass Forage Sci. 51, 333–342.

Espinosa-Garcia F, Lagenheim JH (1990) The leaf fungal endophytic community of a coastal redwood population diversity and spatial patterns. New Phytol. 116, 89–97.

Fisher PJ (1996) Survival and spread of the endophytic *Stagonospora pteridiicola* in *Pteridium aquilinum*, other ferns and some flowering plants. New Phytol.132, 199–122.

Fröhlich J, Hyde KD, Petrini O (2000) Endophytic fungi associated with palms. Mycol. Res. 104, 1202–1212.

Gamboa MA, Bayman P (2001) Communities of endophytic fungi in leaves of a tropical timber tree (*Guarea guidonia*: Meliaceae). Biotropica 33, 352–360.

Gonthier P, Massimo G, Nicolotti G (2006) Effects of water stress on the endophytic mycota of *Quercus robur*. Fungal Divers. 21, 69–80.

González MC, Anaya AL, Glenn AE, Macías-Rubalcava ML, Hernández-Bautista BE, Hanlin RT (2009) *Muscodor yucatanensis*, a new endophytic ascomycete from Mexican chakah, *Bursera simaruba*. Mycotaxon 110, 363–372.

Guerin P (1898) Sur la présence d'un champignon dans l'ivraie. J. Botanique. 12, 230-238.

Hawksworth DL (1988) The variety of fungal-algal symbioses, their evolutionary significance and the nature of lichens. Bot. J. Linn. Soc. 96, 3–20.

Hertweck C (2009) Hidden biosynthetic treasures brought to light. Nat. Chem. Biol. 7, 450-452.

Heywood VH (1995). Global biodiversity assessment. United Nations Environment Programme, Cambridge University Press, Cambridge, UK.

Hormazabal E, Piontelli E (2009) Endophytic fungi from Chilean native gymnosperms: antimicrobial activity against human and phytopathogenic fungi. World J. Microbiol. Biotechnol. 25, 813–819.

Hassan SRU, Strobel G, Booth E, Kingston B, Floerchinger C, Sears J (2012) Epigenetic modulation of volatile organic compound formation in the mycodiesel producing endophyte- *Hypoxylon* sp. CI-4. Microbiology 158, 465–473.

Krings M, Taylor TN, Hass H, Kerp H, Dotzler N, Hermsen EJ (2007) Fungal endophytes in a 400-million-yr-old land plant: infection pathways, spatial distribution, and host responses. New Phytol. 174, 648–657.

Kudalkar P, Strobel G, Hassan S, Geary B, Sears J (2012) *Muscodor sutura*, a novel endophytic fungus with volatile antibiotic activities. Mycoscience 53(4), 319-325.

Lacey LA, Neven LG (2006) The potential of the fungus, *Muscodor albus*, as a microbial control agent of potato tuber moth (Lepidoptera: Gelechiidae) in stored potatoes. J. Invertebr. Pathol. 91, 195–198.

Li WC, Zhou J, Guo LD (2007) Endophytic fungi associated with lichens in Baihua mountain of Beijing, China. Fungal Divers. 25, 69–80.

Lodge DJ, Fisher PJ, Sutton BC (1996) Endophytic fungi of *Manilkara bidentata* leaves in Puerto Rico. Mycologia 88, 733–738.

Mercier J, Jiménez JI (2004) Control of fungal decay of apples and peaches by the biofumigant fungus *Muscodor albus*. Postharvest Biol. Technol. 31, 1-8.

Mercier J, Smilanick JL (2005) Control of green mold and sour rot of stored lemon by biofumigation with *Muscodor albus*. Biological Control 32, 401-407.

Meshram V, Gupta M, Saxena S (2015) *Muscodor ghoomensis* and *Muscodor indica*: new endophytic species based on morphological features, molecular and volatile organic analysis from Northeast India. Sydowia 67, 133–146.

Meshram V, Kapoor N & Saxena S (2013) *Muscodor kashayum* sp. nov. a new volatile anti-microbial producing endophytic fungus. Mycology 4(4), 196-204.

Meshram V, Saxena S, Kapoor N (2014) *Muscodor strobelii*, a new endophytic species from South India. Mycotaxon 128, 93-104.

Mitchell AM, Strobel GA, Hess WM, Vargas PN, Ezra D (2008) *Muscodor crispans*, a novel endophyte from *Ananas ananassoides* in the Bolivian Amazon. Fungal Divers. 31, 37–44.

Mittermeier RA, Myers N, Gil PR Mittermeier CG (1999) Hotspots:Earth's Biologically richest and most endangered ecoregions, CEMEX Conservation International, Washington, DC.

Petrini O (1986) Taxonomy of endophytic fungi of aerial plant tissues. In Microbiology of the Phyllosphere (Ed) Fokkema NJ, Van Den Heuvel J, Cambridge University Press, Cambridge, pp 175–187.

Petrini O (1991) Fungal endophytes of tree leaves. In: Microbial Ecology of Leaves (Ed) Andrews JH and Monano SS, Springer-Verlag, New York, pp 179–197.

Petrini O (1996) Ecological and physiological aspects of host specificity in endophytic fungi. In: Endophytic Fungi in Grasses and Woody Plants (Ed) Redlin SC and Carris LM, APS Press, St Paul, pp 87–100.

Petrini O, Fisher PJ (1988) A comparative study of fungal endophytes in xylem and whole stem of *Pinus sylvestris* and *Fagus sylvatica*. Trans. Br. Mycol. Soc. 91: 233–238.

Redecker D, Kodner R, Graham LE (2000) Glomalean fungi from the Ordovician. Science 289, 1920–1921.

Renninger NS, Ryder JA, Fischer KJ (2008) Jetfuel compositions and methods of making and using same.WO200813049217.

Richardson MD, Chapman GW-Jr, Hoveland CS, Bacon CW (1992) Sugar alcohols in endophyte-infected tall fescue. Crop Sci. 32, 1060–1061.

Riga E, Lacey LA, Guerra N (2008) *Muscodor albus*, a potential biocontrol agent against plant-parasitic nematodes of economically important vegetable crops in Washington State, USA. Biol. Control 45, 380–385.

Rodrigues KF (1994) The foliar fungal endophytes of the Amazonian palm *Euterpe oleracea*. Mycologia 86, 376–385.

Rodrigues KF, Samuels G (1990) Preliminary study of endophytic fungi in a tropical palm. Mycol. Res. 94, 827–830.

Rodriguez RJ, White Jr, JF, Arnold AE, Redman RS (2008) Fungal endophytes: diversity and functional roles. New Phytol. 182(2), 314-330.

Rude M, Schirmer A (2009) New microbial fuels: a biotech perspective. Curr. Opin. Microbiol. 12, 274–281.

Sampson K (1933) The systemic infection of grasses by *Epichloë typhina*. Trans. Brit. Mycol. Soc. 196B, 1-27.

Saxena S, Meshram V, Kapoor N (2015) *Muscodor tigerii* sp. nov. volatile antibiotic producing endophytic fungus from the northeastern Himalayas. Ann. Microbiol. 65, 47–57.

Saxena S, Meshram V, Kapoor N (2014) *Muscodor darjeelingensis*, a new endophytic fungus of *Cinnamomum camphora* collected from northeastern Himalayas. Sydowia 66 (1), 55–67.

Schardl CL, Phillips TD (1997) Protective grass endophytes: where are they from and where are they going? Plant Dis. 81, 430–438.

Schmidt D (1993) Effects of *Acremonium uncinatum* and a *Phialophora*-like endophyte on vigour, insect and disease resistance of meadow fescue. In: Proc 2nd Intl Symposium on *Acremonium/Grass* Interactions (Ed) Hume DE, Latch GCM and Easton HS, AgResearch, Palmerston North, pp 185–188.

Sieber TN (1989) Endophytic fungi in twigs of healthy and diseased Norway spruce and white fir. Mycol. Res. 92, 322–326.

Singh SK, Strobel GA, Knighton B, Geary B, Sears J & Ezra D (2011) An endophytic *Phomopsis* sp. possessing bioactivity and fuel potential with its volatile organic compounds. Microbial Ecol. 61, 729–739.

Smith SE, Read DJ (1997) Mycorrhizal symbiosis. Second edition. Academic Press, San Diego, CA.

Spurr Jr HW, Welty RE (1975) Characterization of endophytic fungi in healthy leaves of *Nicotiana* spp. Phytopathology 65, 417– 422.

Stierle A, Strobel GA, Stierle D (1993) Taxol and taxane production by *Taxomyces andreanae*, an endophytic fungus of Pacific yew. Science 260, 214–216.

Stinson AM, Zidack NK, Strobel GA, Jacobsen BJ (2003) Mycofumigation with *Muscodor albus* and *Muscodor roseus* for control of seedling diseases of sugar beet and *Verticillium* wilt of eggplant. Plant Dis. 87, 1349-1354.

Stone JK, Bacon CW, White JF (2000) An overview of endophytic Microbes: Endophytism defined. In: Microbial Endophytes (Ed) Bacon C and White J, Marcel Dekker, New York, pp 3–30.

Strobel GA, Daisy B (2003) Bioprospecting for microbial endophytes and their natural products. Microbiol. Mol. Biol. Rev. 67(4), 491-502.

Strobel GA, Dirkse E, Sears J, Markworth C (2001) Volatile antimicrobials from *Muscodor albus* a novel endophytic fungus. Microbiology 147, 2943- 2950.

Strobel GA, Knighton B, Kluck K, Ren Y, Livinghouse T, Griffin M, Spakowicz D, Sears J (2008) The production of mycodiesel hydrocarbons and their derivatives by the endophytic fungus *Gliocladium roseum* (NRRL 50072). Microbiology 154, 3319–3328.

Strobel GA, Singh SK, Hassan SR, Mitchell AM, Geary B, Sears J (2011) An endophytic/ pathogenic *Phoma* sp. from creosote bush producing biologically active volatile compounds having fuel potential. FEMS Microbiol. Lett. 320, 87- 94.

Sun J, Awakawa T, Noguchi H, Abe I (2012) Induced production of mycotoxins in an endophytic fungus from the medicinal plant *Datura stramonium* L. Bioorg. Med. Chem. Lett. 22, 6397–6400.

Suwannarach N, Bussaban B, Hyde KD, Lumyong S (2010) *Muscodor cinnamomi*, a new endophytic species from *Cinnamomum bejolghota*. Mycotaxon 114, 15–23.

Suwannarach N, Kumla J, Bussaban B, Hyde KD, Matsui K, Lumyong S (2013) Molecular and morphological evidence support four new species in the genus *Muscodor* from northern Thailand. Ann. Microbiol. 63, 1341-1351.

Verhoeff K (1974) Latent infection by fungi. Annu Rev Phytopathology 12:99–110.

Vogl AK (1898) Mehl und die anderen Mehl Produkte der Cerealien und legumonosen. Nahrungsm.Unters Hyg Warenk 12, 25-29.

Wang Y, Guo LD (2007) A comparative study of endophytic fungi in needles, bark, and xylem of *Pinus tabulaeformis*. Can. J. Bot. 85, 911–917.

Wheatley R, Hackett C, Bruce A, Kundzewicz A (1997) Effect of substrate composition on production of volatile organic compounds from *Trichoderma* spp. inhibitory to wood decay fungi. Int. Biodeterior. Biodegrad. 39, 199-205.

Wilkinson HH, Schardl CL (1997) The evolution of mutualism in grass-endophyte associations. In: *Neotyphodium*/grass interactions (eds.) Bacon CW, Hill NS, Plenum Press, New York, pp 13–25.

Wilson D (1995) Endophyte–the evolution of a term and clarification of its use and woody perennials. In: Microbiology of the phyllosphere (Ed) Fokkema NJ, vanden Huevel J, Cambridge University Press, Cambridge, pp 205–222.

Worapong J, Strobel GA, Daisy B, Castillo UF, Baird G, Hess WM (2002) *Muscodor roseus* anam. sp. nov., an endophyte from *Grevillea pteridifolia*. Mycotaxon 81, 463–475.

Worapong J, Strobel GA, Ford EJ, Li JY, Brird G, Hess WM (2001) *Muscodor albus* anam. nov., an endophyte from *Cinnamomum zeylanicum*. Mycotaxon 79, 67–79.

Zhang CL, Wang GP, Mao LJ, Komon-Zelazowska M, Yuan ZL, Lin FC, Druzhinina IS, Kubicek CP (2010) *Muscodor fengyangensis* sp. nov. from South-East China: morphology, physiology and production of volatile compounds. Fungal Biol. 114, 797–808. http://dx.doi.org/10.1016/j.funbio.2010.07.0

8

Application of Endophytic Microorganisms for Alleviation of Abiotic Stresses in Crop Plants

Kamal K. Pal and Rinku Dey

Abstract

A majority of the plants studied in natural ecosystems are symbiotic with microorganisms that either reside entirely (endophytes) or partially within plants. These microorganisms express different associations ranging from mutualism to parasitism. These symbiotic relationships appear to impart tolerance to various types of abiotic stresses such as heat, drought, salinity, heavy metals, etc. and sometimes may be responsible for the survival of both plant hosts and microbial symbionts in high stress habitats. The amelioration of the abiotic stresses by the endophytes assumes increasing significance in the light of rapidly changing global climate, which is likely to face frequent incidences of extreme weather conditions like high temperature, droughts, etc. To compensate the loss in crop productivity due to vagaries of nature and depleting areas of cultivable land, the application of endophytic microorganisms in agriculture is seen as a potential and ecologically sound means of maintaining profitability and sustainability in crop production. Apart from providing tolerance to abiotic stresses, majority of the endophytes are also known to confer tolerance to biotic stresses such as diseases, pests, etc. The endophytic microorganisms have also found application in remediation of heavy metal contaminated sites or polluted soils through phytoremediation. Here, we describe the role of endophytic microorganisms in alleviation of abiotic stresses in plants and the different ways by which this symbiosis can potentially mitigate the impacts of climate change and anthropogenic activities on crop plants.

Keywords: Abiotic stress, Endophyte, Phytoremediation, Stress tolerance, Symbiosis

1. Introduction

Throughout evolutionary time plants have been confronted with changing environmental conditions, forcing them to adapt or succumb to selective pressures such as extreme temperatures, insufficient water and toxic chemicals (Rodriguez *et al.* 2004). During recent years, the pace of climate change has accelerated resulting in elevated levels of CO_2, temperature, ultraviolet radiations, etc. The incidences of extreme drought may increase in the future due to global warming (IPCC 2014), and predicted increases in drought- and temperature- related stresses are expected to reduce crop productivity (Ciais *et al.* 2005, Larson 2013). These changes adversely affect the rainfall pattern and distribution, photosynthetic activities of plants, utilization of water, etc. Added to this is the rapid expansion of human population, causing changes in the natural and agricultural ecosystems. The agricultural ecosystems are facing limitations of decreasing area and availability of water. The rapid changes at the global level are causing abiotic and biotic stresses to crop plants. In future, local climate changes may require crops to have a greater tolerance of stresses such as drought, salinity, high temperature, etc. in order to produce an economic crop. While plant-breeding programmes and genetic modification can produce crop cultivars with much improved drought tolerance (Cattivelli *et al.* 2008), supplementary techniques and practices using micro-organisms may help to alleviate the worst effects of drought (Reynolds and Tuberosa 2008; Coleman-Derr and Tringe 2014). The same is true for other stresses like salinity, high temperature, water-logging, etc. Providing food security to an ever-growing population is a major challenge which will require cultivation of additional farmland along with improvements in crop yield. This will increase the burden of farming on arid and semi-arid lands resulting in heat- and drought-related stresses on crops.

Plants show, to a certain extent, some degree of adaptation to the different stresses. Due to lack of locomotion plants have to depend on complex physiology in order to escape the different types of stresses or to mitigate their effects to certain extent. There may be several reasons for the tolerance shown by the plants to different stresses. One of these reasons could be the symbiotic associations of plants with microorganisms – bacteria, fungi, etc. Endophytes are microorganisms (bacteria, fungi and unicellular eukaryotes) which can live at least part of their life cycle inter- or intracellularly inside of plants usually without inducing pathogenic symptoms. This can include competent, facultative, obligate, opportunistic and passenger endophytes. Endophytes can have several functions and/or may change function during their life cycle (Murphy *et al.* 2014a, b). Under stress conditions, endophytic plants exhibit different types of stress responses including the production of osmolytes, altering water movement, scavenging reactive oxygen species, etc.

The endophytic microorganisms are ubiquitous in the plants (Table 1). It has been reported that all plant life on Earth is symbiotic with fungi (Rodriguez *et al.* 2004), which contribute to and may be responsible for the adaptation of plants to environmental stresses (Clay and Holah 1999; Morton 2000; Redman *et al.* 2002a). The fungal symbiosis with plants is of two categories – 1) Fungal endophytes reside inside the plant tissues of roots, stems, leaves, etc, and 2) Mycorrhizal fungi residing only in the plant roots, but may extend into the rhizosphere. In natural ecosystems, symbiosis plays an important role in plant adaptation and survival. Endophytic fungi may help plants to adapt to high stress environments and mitigate the effects of such stresses. Fungal endophytes confer tolerance to drought, heavy metals, high temperature, besides promoting growth and nutrient acquisition sometimes. Recent developments in this field are the aspects of symbiotic lifestyle switching (Redman *et al.* 2001) and the phenomenon of habitat-adapted symbiosis (Rodriguez *et al.* 2008), which allows the plants to establish in high stress habitats. Similar is the case with bacterial endophytes, which are also widely present in the plant kingdom. The bacterial endophytes may also confer tolerance to a wide range of biotic and abiotic stresses in plants.

Another type of stress faced by plants growing in metal polluted sites or mines is the heavy metal stress. Man-made activities such as industrial productions, mining activities, agrochemicals usage, biosolids waste disposals, etc. are primary sources of heavy metal contaminations. Phytoremediation is suggested to be an environment-friendly approach for the clean- up of metal contaminated sites. Endophytes can also play an important role in accelerating phytoremediation (Table 2). The plant-endophyte partnership can accelerate the phytoremediation process; improve the plant growth by sequestration of heavy metals.

Table 1: Non-exhaustive list of endophytic bacteria associated with crop plants

Endophyte	Plant species	Endophyte	Plant species
Azorhizobium caulinodans	Rice	*Pantoea* sp.	Rice, soybean
Azospirillum brasilense	Banana	*Pantoea agglomerans*	Citrus plants, sweet potato
Azospirillum amazonense	Banana, pineapple	*Pseudomonas chlororaphis*	Marigold (*Tagetes* spp.), carrot
Bradyrhizobium japonicum	Rice	*Pseudomonas fluorescens*	Carrot
Methylobacterium extorquens	Scots pine, citrus plants	*Pseudomonas citronellolis*	Soybean
Rhizobium leguminosarum	Rice	*Pseudomonas synxantha*	Scots pine
Rhizobium (Agrobacterium) radiobacter	Carrot, rice	*Serratia* sp.	Rice
Gluconacetobacter diazotrophicus	Sugarcane	*Bacillus* spp.	Citrus
Sinorhizobium meliloti	Sweet potato	*Bacillus megaterium*	Maize, carrot, citrus plants
Azoarcus sp.	Callar grass	*Clostridium*	Grass *Miscanthus sinensis*
Burkholderia sp.	Banana, pineapple, rice	*Paenibacillus odorifer*	Sweet potato
Herbaspirillum seropedicae	Sugarcane, rice, maize, sorghum, banana	*Staphylococcus saprophyticus*	Carrot
Herbaspirillum rubrisulbalbicans	Sugarcane	*Kocuria varians*	Marigold
Citrobacter sp.	Banana	*Streptomyces* spp.	Wheat
Enterobacter spp.	Maize, citrus	*Pseudomonas fluorescens*	Carrot
Enterobacter cloacae	Citrus	*Burkholderia cepacia*	Lupine, citrus
Enterobacter agglomerans	Soybean	*Arthrobacter globiformis*	Maize
Enterobacter asburiae	Sweet potato	*Curtobacterium flaccumfaciens*	Citrus plants
Klebsiella sp.	Wheat, sweet potato, rice	*Microbacterium esteraromaticum*	Marigold
Klebsiella pneumoniae	Soybean	*Microbacterium testaceum*	Maize
Klebsiella terrigena	Carrot	*Nocardia* sp.	Citrus plants
Klebsiella oxytoca	Soybean	*Streptomyces*	Wheat
Serratia sp.	Rice	*Stenotrophomonas*	Dune grasses

Source: Rosenblueth and Martinez-Romero, 2006

Table 2: Application of endophytes in phytoremediation

Compound	Plant association	Organism
Mono- and dichlorinated	Wild rye	*Pseudomonas aeruginosa* strain R75 and
benzoic acids	(*Elymus dauricus*)	*Pseudomonas savastanoi* strain CB35
2,4-D	Poplar and willow	*P. putida* VM1450
Methane	Poplar	*Methylobacterium populi* BJ001
TNT, RDX, HMX	Poplar	*Methylobacterium populi* BJ001
MTBE, BTEX, TCE	Poplar	*Pseudomonas* sp
Toluene	Poplar	*B. cepacia* Bu61
TCP and PCB	Wheat	*Herbaspirillum* sp. K1
Volatile organic	Lupine	*Burkholderia cepacia* G4
Compounds and toluene		

Source: Newman & Reynolds 2005

2. Fungal endophytes and stress tolerance

The fungal endophytes reside entirely within the host tissues and comprise a phylogenetically different group that are members of the dikarya (Girlanda *et al.* 2006; Arnold and Lutzoni 2007). Though most of the fungal endophytes belong to the Ascomycota clade, some belong to the Basidiomycota. Fungal endophytes are widespread in occurrence and reportedly found in major taxonomic groups of plants thriving under various environments. The endophytic fungi express a range of symbiotic lifestyles spanning from mutualism to parasitism and this is known as symbiotic continuum (Schardl and Leuchtmann, 2005).

Endophytic fungi provide tolerance to plants to many types of abiotic stresses such as drought, high temperature, heavy metals, etc. Mutualistic fungi may confer several benefits to plants such as tolerance to drought. The stress tolerance provided to the plants due to the symbiotic association with endophytic fungi may involve two mechanisms: 1) rapid activation of host stress response systems upon exposure to stress (Redman *et al.* 1999), or 2) synthesis of anti-stress biochemicals by the fungus (Bacon and Hill 1996). The mechanisms by which endophytes activate host stress response systems are not known.

2.1 High temperature stress

In response to high temperature stress, plants are known to initiate complex biosynthetic responses involving heat shock proteins, antioxidant systems, and adjustments in osmotic potential, and membrane lipids (Iba 2002). The stress tolerance conferred by some endophytes involves habitat-specific fungal adaptations. Some plants have been reported to thrive in geothermal soils tolerating high temperature and drought stress, due to the symbiotic association with endophytic fungi. Studies conducted by Rodriguez *et al.* (2004) indicated that the endophytic fungus *Curvularia protuberata* conferred thermotolerance to *Dichanthelium lanuginosum* and this symbiosis was responsible for the survival of both the species in geothermal soils of Yellowstone National Park. It was also observed that the individual partners could not tolerate the stress, when exposed to heat stress >38°C, but the symbiotic association provided the tolerance to the habitat-specific stress. The ability of the endophyte to confer heat tolerance was due to the presence of a fungal RNA virus, which provides biochemical functionality to the fungus for conferring heat tolerance (Ma´rquez *et al.* 2007).

2.2 Moisture-deficit stress

To address the issues of providing food security to an ever growing human population, efforts are needed for enhancement in crop yield and bringing additional farmland under cultivation. For this, we may need to bring marginal, arid and semi-arid lands under cultivation, resulting in the crops facing more stresses like heat and drought related stresses. These stresses will reduce the crop productivity, with strong adverse effects on regional, national, and household livelihood and food security (IPCC 2014). Plants need to show adaptation to drought stress in order to survive (Seki *et al.* 2007).

Plants respond to moisture-deficit stress conditions by several mechanisms including osmotic adjustments, production of antioxidants, altered transcriptional and translational regulation, and altered stomatal activity (Griffiths and Parry 2002). The physiological and biochemical responses to drought stress include stomatal closure (Roelfsema and Hedrich 2005), reduction of growth and photosynthesis (Flexas *et al.* 2004), and activation of respiration (Rennenberg *et al.* 2006).

Very few plant species show drought-tolerance, though all the plants respond to moisture-deficit stress to certain extent. Some fungal and bacterial functional groups have been reported to enhance drought tolerance in a variety of crop hosts (Table 3), including mycorrhizal (Boyer *et al.* 2014) and endophytic fungi (Oberhofer *et al.* 2014). There are reports suggesting fungal symbionts conferring drought-tolerance to some plants (Clay and Schardl 2002), the mechanisms of which may involve osmotic adjustments and/or altered stomatal activity. The mutualistic *Colletotrichum magna* mutants have been reported to provide drought tolerance to watermelon plants. Interestingly, both the non-pathogenic mutants as well as the wild type *C. magna* asymptomatically colonize non-cucurbit hosts including tomato and pepper (Redman *et al.* 2001), which allows these plants to survive desiccation for longer duration as compared to the non-symbiotic plants. The host genotype plays a very important role in

Table 3: Non-exhaustive list of beneficial endophytes capable of alleviating abiotic stress and plant growth promotion

Endophyte	Crop/plant	Function(s)	Reference
Pseudomonas pseudoalcaligenes	Rice	Salinity tolerance	Jha *et al.* 2010
Penicillium minioluteum LHL09	Soybean	Salinity	Khan *et al.* 2011
Pseudomonas aeruginosa PW9	Cucumber	Abiotic stress tolerance	Pandey *et al.* 2012
Clavibacter sp.	*Chorispora bungeana*	Chilling stress tolerance	Ding *et al.* 2011
Piriformospora indica	*Prosopis juliflora, Z. nummularia,* Wheat, mustard, tomato, cabbage	Salinity, moisture & biotic stress tolerance and plant growth, nutrient uptake, etc.	Walker *et al.* 2005

determining the expression and magnitude of benefits of endophytic fungi. It is now thought that the host range of these fungi may be wider and provide benefits to unrelated plant species also (Rodriguez *et al.* 2004).

Fungal root endophytes isolated from a wild barley species (*Hordeum murinum* subsp. *murinum*) induced significant improvements in agronomic traits for a severely drought-stressed barley cultivar grown in a controlled environment, including number of tillers, grain yield, and shoot biomass (Murphy *et al.* 2015). This group studied the inoculation effects of five endophytes, and the trait that showed maximum significant difference was the number of tillers per plant that was more in all the inoculated treatments.

An endophytic fungus *Piriformospora indica* has been widely studied and used worldwide for alleviation of abiotic stresses. This fungus belongs to the Sebacinaceae family, which colonizes the roots of many plant species, and has been reported to impart benefits to plants under drought-stress conditions (Sahay and Varma 1999; Shahollari *et al.* 2005, 2007). Besides, the fungus has also been reported to promote nutrient uptake; helps the plants to circumvent moisture, temperature and salt stress, confers resistance to toxins, heavy metals, pathogens; and also promote plant growth and seed production (Verma *et al.* 1998; Waller *et al.* 2005, 2008; Oelmüller *et al.* 2009). *P. indica* was isolated from the roots of *Prosopis juliflora* and *Zizyphus nummularia* plants grown in the Thar desert of India (Verma *et al.* 1998), and root colonization and association of fungal hyphae with roots have been found to result in promotion of plant growth and higher seed yield under drought-stress conditions. This fungus colonises the roots by growing inter- and intracellularly and forms pear shaped spores within the cortex, but does not invade the aerial parts of the plants. The fungus does not show host specificity. It has a wide host range including bryophytes (*Aneura pinguis*), pteridophytes (*Pteris ensiormis*), gymnosperms (*Pinus halepensis*), and a large number of angiosperms (Varma *et al.* 2001; Shahollari *et al.* 2005; Waller *et al.* 2005; Serfling *et al.* 2007). This includes the cereal crops like rice, wheat, barley, etc. Some drought-inducible genes have been identified (Seki *et al.* 2002), which are mainly classified into two major groups: proteins that function directly in abiotic stress tolerance and regulatory proteins, which are involved in signal transduction or expression of stress-responsive genes (Shinozaki *et al.* 2003). Under drought-stress conditions, many genes for signaling components themselves are upregulated. Using *Arabidopsis* as a model system, Sherameti *et al.* (2008) demonstrated that *P. indica* conferred drought tolerance by priming the aerial parts of the plants for the expression of a set of quite diverse (drought) stress-related genes. *P. indica* also promotes the expression of the two genes *MDAR2* (At3g09940) and *DHAR5* (At1g19570) in the aerial parts of drought-stressed seedlings. These two genes are crucial for the beneficial interaction between *P. indica* and *Arabidopsis*, because reduced ascorbate maintains a constant redox balance in the cytoplasm (Vadassery *et al.* 2009b). The mutualistic interactions between microorganisms and plants have been utilized for enhancing growth, biomass and seed production, often in poor soils, with little extraneous application of chemical fertilizers and pesticides. *P. indica* has the advantage of rapid and large scale propagation for field level applications.

Arbuscular-mycorrhizal (AM) fungi are important group of endophytic microorganisms playing significant role in sustainable agriculture. They provide multiple benefits to their plant hosts by increasing drought resistance (Allen and Allen 1986; Nelsen 1987), mineral uptake and providing resistance to diseases. The symbiotic relationship between AM fungi and their host plants is generally nonspecific in nature. There are considerable differences at the species level and within the geographic isolates, and thus great biodiversity exists for the AM endophytes (Bethlenfalvay *et al.* 1989). Suitable endophytes need to be selected for a particular environmental condition based on the specific plant host and compatibility with the environmental conditions. AM endophytes may alleviate the effects of reduced photosynthesis due to reduction in leaf water status and stomatal closure by altering the physiological parameters of the plants so that the plant can adapt to low soil moisture content (Ruiz-Lozano *et al.* 1995). AM fungi mediated improved host nutrition, particularly phosphorus nutrition has also been reported to improve the water balance of plants (Giovannetti and Mosse 1980; Graham and Syvertson 1984). The several mechanisms by which AM fungi improve drought-tolerance in their host plants, are increased CO_2 assimilation, changes in stomatal conductance (Auge'and Duan 1991) and transpiration. Potassium nutrition plays a significant role in stomatal movement with any changes in leaf water status. Accumulation of proline in the leaves has been reported to enhance osmotic adjustments. Ruiz-Lozano *et al* (1995) reported higher proline content in drought-stressed plants as compared to well-watered plants and a lower proline content was stated to be an indication of better tolerance to drought. AM fungal isolates need to be carefully selected based on the situation or the problems for which they are intended. Suitable AM fungal isolates have great potential for restoring drought-prone degraded lands.

2.3 Salinity stress

The problem of high soil salinity, particularly in arid and semi-arid regions, is a limiting factor for agricultural productivity in such areas (Flowers *et al.* 1977). Accumulation of salts in the rhizosphere zone reduces the water potential and its availability to the plants (Heyster and Nabors, 1982). Salinity stress affects the metabolic processes of plants. But, the plants growing in saline environments have developed mechanisms to circumvent the detrimental effects of high salt concentrations that include compartmentalization or translocation of salt, exclusion, cellular osmotic adjustments, and/or antioxidant systems (Gilbert *et al.* 2002, Yoshida *et al.* 2003). They may produce amino acids such as proline and accumulate inside the cells to maintain osmotic balance. Most of the crops of arid and semi-arid regions are mycorrhizal. Some mycorrhizal fungi have been reported to confer salt tolerance through symbiosis (Yano-Melo *et al.* 2003). The fungus-plant symbiotic association confers salinity tolerance in some plant species including banana, tomato and lettuce. In plants such as tomato (Al-Karaki 2006) and soybean (Sharifi *et al.* 2007) increased growth under saline conditions was observed when their roots were colonized by AM fungi.

Salinity stress affects the growth parameters of plants, irrespective of their being mycorrhizal or non-mycorrhizal. However, the reduction in biomass production is observed to be more in non-mycorrhizal plants. The improvement in growth of salinity stressed plants inoculated with AM fungi was reported by Tain *et al.* (2004). According to Zandavalli *et al.* (2004), better nutritional status of plants may be the reason for enhancement in growth and salt tolerance of mycorrhizal peanuts. It is a well known fact that beneficial effects of AM fungi on plant growth is to a large extent due to higher uptake of P. Rabie and Almadini (2005) reported enhancement in contents of N and K as a result of AM association. Investigations conducted by Al-Khaliel (2010) revealed that *Glomus mosseae* could improve growth of peanuts under salinity stress through enhanced nutrient absorption and photosynthesis. This association helped in the establishment of peanuts under salinity and phosphorus deficiency conditions.

There are reports of alleviation of moderate salt stress by *P. indica* (Waller *et al.* 2005). Barley plants exposed to moderate (100 mM NaCl) salt concentrations in hydroponic culture showed leaf chlorosis and reduced growth. The harmful effects of moderate salt stress were completely alleviated by *P. indica*, as shown by the higher biomass than non-stressed control plants.

Antioxidants are molecules that function to reduce oxidative stress by scavenging or quenching ROS and these molecules are crucial for beneficial plant/microbe interactions (Alguacil *et al.* 2003). Generally, ROS and H_2O_2 are produced as a result of abiotic and biotic stresses. The fungus *P. indica* has been found to enhance the antioxidative capacity of plants. Baltruschat *et al.* (2008) studied the *P. indica* - mediated salt tolerance in barley. It was reported that the fatty acid composition, lipid peroxidation, ascorbate concentration and activities of catalase, ascorbate peroxidase, dehydroascorbate reductase (DHAR), monodehydroascorbate reductase (MDHAR) and glutathione reductase enzymes were strongly influenced by the fungus under salt-stress condition.

Also, the endophytic fungus significantly elevated the amount of ascorbic acid and increased the activities of antioxidant enzymes in barley roots under salt stress conditions. Fungal endophyte induced increase in antioxidants was reported in barley grown under salt and pathogen stress (Harrach *et al.* 2013).

There are instances of cross-species alleviation of abiotic stresses in literature. Recently, Khan *et al.* (2011) demonstrated that the endophytic fungus *Exophiala* sp. LHL08 isolated from cucumber roots could confer salinity and drought stress tolerance in rice seedlings by modulating stress responses.

2.4 Habitat-adapted symbiosis

A new term was coined by Rodriguez *et al.* (2008) which defines habitat-specific, symbiotically-conferred stress tolerance as habitat-adapted symbiosis. This was hypothesized to be responsible for the establishment of plants under high-stress habitats. It was observed that grass endophytes from coastal areas conferred salt tolerance, geothermal endophytes conferred heat tolerance and endophytes from agricultural crops conferred disease tolerance, respectively, to plants under the respective habitats. But, the same fungal species isolated from plants in habitats devoid of salt or heat stress did not confer these stress tolerances. Further studies also showed that the agricultural, coastal and geothermal plant endophytes also colonized tomato (a model eudicot) and conferred disease, salt and heat tolerance, respectively. Strengthening the idea of habitat-adapted symbiosis was the observation that the coastal plant endophyte colonized rice (a model monocot) and conferred salt tolerance. These endophytes showed a broad host range covering both monocots and eudicots.

2.5 Heavy metal stress

Fast paced industrialization has resulted in widespread environmental pollution. The use of chemicals such as pesticides, agrochemicals, industrial solvents, etc., have gone into the environment polluting soil, air and water. Among the pollutants, heavy metals are of particular concern for human health because of their cytotoxicity and carcinogenicity. The use of plants for remediation of polluted soils (phytoremediation) is an eco-friendly and cost-effective option. Many plants are known to be capable of hyperaccumulating heavy metals in their tissues. The association of metal tolerant endophytes with hyperaccumulating plants results in enhanced uptake of heavy metals by plants and increase in biomass of the plants.

Fungal endophytes also show instances of imparting tolerance to their host plants against heavy metals or trace metals. The AM fungus *Glomus intraradices* was shown to enhance growth of *Helianthus annuus* in Ni contaminated soil (Ker and Charest, 2010) along with significantly increased activity of glutamine synthetase, indicating an enhanced Ni tolerance. AM fungal symbiosis may enhance plant growth and tolerance to trace elements like Ni, As, etc. in hyperaccumulating plants growing in metal contaminated sites.

3. Bacterial endophytes and stress tolerance

Like their fungal counterparts, endophytic bacteria are also widely present in the plant kingdom colonizing internal tissues of their host plants and forming relationships like symbiotic, mutualistic, commensalistic and trophobiotic (Ryan et al. 2008). Endophytic bacteria may benefit their plant hosts by improving growth and development. This may involve the facilitating of primary and secondary nutrient uptake by processes such as atmospheric nitrogen fixation, siderophore production, and solubilization of minerals such as phosphate (PO_4^{3-}), potassium (K^+) and zinc (Zn^{2+}). Endophytic bacteria may also supply plant roots with phytohormones like auxin, cytokinin and gibberellins (Table 4). Some endophytic bacteria provide tolerance to stresses such as pathogen infections, drought, soil salinity, etc. by inhibiting the production of the plant hormone ethylene (Siddikee et al. 2010).

Table 4: Production of phytohormones by beneficial bacterial endophytes

Endophyte	Plant origin	Beneficial action
Azospirillum sp. BS10	Rice	ACC deaminase, BNF, IAA, Siderophore
Burkholderia phytofirmans	Onion	ACC deaminase, IAA, Siderophore
Enterobacter sp. 638	Poplar	IAA, Siderophore, Volatiles
Gluconacetobacter diazotrophicus Pal5	Sugarcane	Gibberellins, IAA and Volatiles
Herbaspirillum seropedicae SmR1	Sorghum	ACC deaminase, IAA
Pseudomonas putida W619	Poplar	IAA
Bacillus sp., *Micrococcus* sp., *Pseudomonas* sp., *Serratia* sp.	Legume	IAA, Gibberellins, Cytokinins

Source: Sturz et al. 2000.

3.1 Moisture-deficit stress

In the interaction of plants with bacterial endophytes, production and modulation of hormones such as auxins and ethylene play an important role in plant development and stress (e.g. drought) tolerance. Stress tolerance in such interactions is reportedly influenced by endophyte-derived hormones. The hormones abscisic acid and gibberellins produced by the endophyte *Azospirillum lipoferum* was found to alleviate drought stress symptoms in maize (Cohen et al. 2009).

3.2 Salinity stress

Salinization of agricultural soils is a serious threat to crop production. It is estimated that by the year 2050 approximately 50% of the arable land will be affected by salinity (Munns and Tester, 2008). Not only climatic conditions but also human activities are responsible for soil salinization. Such conditions limit the growth of vegetation and the number of plant species growing in these situations. However, in saline conditions also certain plant populations are successfully adapted and exhibit strategies of salt tolerance. One of the mechanisms of coping with salt stress or other abiotic stresses is the association of plants growing under such conditions with the endophytic microorganisms. These microorganisms along with rhizospheric microorganisms may be involved in the biogeochemical cycling of nutrients under saline conditions. Cassán *et al.* (2009) coined the term plant stress homeoregulating bacteria (PSHB), that can either directly or indirectly facilitate the plant growth in optimal, biotic, or abiotic stress conditions. The PSHB may benefit plants by providing stress-related phytohormones, like abscisic acid (Cohen *et al.* 2008); plant growth regulators like cadaverine (Cassán *et al.* 2009); and catabolism of ACC deaminase. Many plant species have been reported to thrive under saline environments by developing effective associations with endophytic bacteria. A halophyte *Prosopis strombulifera* is reported to be naturally associated with different endophytes having plant growth-promoting physiological and biochemical capabilities (Sgroy *et al.* 2009). These workers mentioned the production of phytohormones by many isolates and also production of ABA and ACC deaminase activity as homeostasis regulation mechanisms by some isolates. Some endophytic bacteria can produce the enzyme 1-aminocyclopropane-1-carboxylic acid (ACC) deaminase that breaks down ACC, the direct precursor of ethylene, into ammonia and á-ketobutyrate. ACC is used by bacteria as a source of nitrogen and carbon and thereby the deleterious effect of ethylene on plant tissues is reduced (Glick *et al.* 2007). Under environmental stress conditions, the availability of auxin, ACC deaminase and the nutrients produced by endophytic bacteria is important to minimize the deleterious effects of physiological stress and to support the level of growth and development required to complete the lifecycle of the plants (Timmusk *et al.* 2011).

Yaish *et al.* (2015) isolated and characterized endophytic bacteria from date palm (*Phoenix dactylifera* L.) seedling roots, and tested for their ability to help plants grow under saline conditions. Molecular characterization showed that the majority of these strains belonged to the genera *Bacillus* and *Enterobacter* . These endophytic bacteria had a likely role to play in salinity tolerance due to their ability to produce the growth regulator IAA and reduce the production of stress hormone ethylene through the production of ACC deaminase, besides providing essential nutrients such as ammonia, K^+, Fe^{3+}, PO_4^{3-}, and Zn^{2+}.

Certain endophtyic bacteria can also bring about cross-species alleviation of abiotic stresses. A salt-tolerant endophytic bacteria *Pseudomonas aeruginosa* PW09 isolated from wheat stem could alleviate salinity stress in cucumber (Pandey *et al*. 2012). There was enhanced biomass accumulation under salinity stress along with higher antioxidant activities and proline accumulation in cucumber plants inoculated with *P. aeruginosa* PW09. This endophyte could also reduce plant mortality in cucumber due to *Sclerotium rolfsii* infection.

3.3 Heavy metal stress

Contamination of soil with heavy metals has become a serious environmental problem and threat to the delicate ecological balance. Phytoremediation is the use of plants to extract pollutants from soil in an eco-friendly way. However, many metal accumulating plants exhibit slow growth and often are inhibited by high concentrations of heavy metals. Studies involving plant-associated microorganisms have shown that such microorganisms can enhance seedling emergence and growth of plants under metal polluted conditions (Chen *et al*. 2010). Recent studies have demonstrated that endophytic bacteria improve the tolerance of plants to heavy metals and increase heavy metal translocation factors, biomass, and trace element concentrations of hyperaccumulators (He *et al*. 2009). Endophytic bacteria are influenced by the physicochemical properties of the soil and have evolved with the progress of heavy metal contamination (Chen *et al*. 2012). It is important to study the abundance and composition of bacterial endophytes in the metal contaminated sites for understanding their interactions with the environment, and also for exploring the possible uses of these bacterial species for the bioremediation of heavy metals.

Phytolacca americana is reported to be a Mn-hyperaccumulating plant that has great potential for remediation of Mn-contaminated soils. Wei *et al*. (2014) investigated the diversity of the endophytic bacterial populations in the tissues of *P. americana* growing in Mn mine by PCR-DGGE. Phylogenetic analyses of the recovered DNA sequences classified the bacteria into 10 different divisions, indicating a high level of diversity amongst the endophytic bacterial species of *P. americana*. From the sequencing results it was observed that Proteobacteria, specifically the γ, δ and α subclasses, may be the dominant endophytic bacterial genera of *P. americana*. These endophytic bacteria may have an important role in assisting *P. americana* in phytoremediation.

Endophytic bacteria can improve trace element uptake by plants and the efficiency and rate of phytoextraction. Zinc-resistant endophytes were isolated from Zn-accumulating willows (*Salix caprea*), comprised mostly *Sphingomonas* spp., *Methylobacterium* spp. and various actinobacteria (Kuffner *et al.* 2010). Endophytic bacteria isolated from both the root and shoot tissues of the Cd/Zn-hyperaccumulator *Sedum alfredii* were closely related to *Pseudomonas*, *Bacillus*, *Stenotrophomonas* and *Acinetobacter* (Long *et al.* 2011). Inoculation with these endophytes resulted in increase in plant growth and biomass, along with increase in concentration of metals. Inoculation of *Nicotiana tabacum* with Cd-resistant seed endophyte *Sanguibacter* sp. S_d2 increased shoot Cd concentrations compared to non-inoculated plants (Mastretta *et al.* 2009). Pea plants inoculated with an endophyte *Pseudomonas* sp., capable of degrading the organochlorine herbicide, 2,4-dichlorophenoxyacetic acid (2,4-D), showed no accumulation of the herbicide in the tissues and no signs of phytotoxicity when exposed to 2,4-D (Germaine *et al.* 2006). Thus, endophytes can alleviate the toxic effects of agrochemicals and reduce the impact of hazardous chemicals.

Certain endophytic bacteria are capable of metal biosorption and result in phytostabilization. These endophytes show increased biosorption and bioaccumulation of the metals. Inoculation with endophytic bacteria, *Magnaporthe oryzae* and *Burkholderia* sp. increased plant growth but reduced the Ni and Cd accumulation in roots and shoots of tomato and also their availability in soil (Madhaiyan *et al.* 2007). Certain endophytes produce metabolites such as siderophores, biosurfactants and organic acids and alter the availability/toxicity of the heavy metals to the plant (Sheng *et al.* 2008). A plant growth promoting endophyte (PGPE) *Pseudomonas* sp. A3R3 effectively promoted the phytoremediation of both host (*Alyssum serpyllifolium*) and non-host (*Brassica juncea*) plants by improving either the Ni accumulation or biomass production (Ma *et al.* 2011). Metal tolerant endophytic bacteria have great potential for reclamation of heavy metal and chemically polluted sites.

Conclusion

Many beneficial endophytic fungi and bacteria have been identified in the past and the basis for their molecular interaction with plants has been studied. The primary interest of these endophytic microorganisms is to gain access to photoassimilates by the plants. The endophytic microorganisms often take shelter inside the plant tissues to escape the harsh environment outside and also to get access to carbon compounds. However, the mutualistic association involves many recognition and signaling processes between the endophytes and the host plants.

Though not much is known about the biochemical basis of stress tolerance in plants mediated by endophytes, in near future it may be possible to develop effective symbiosis between specific fungi and plants to achieve stress tolerance for a particular geographic region. Such symbiosis may be effective and environment friendly strategy for mitigating the adverse effects of stresses on plant communities. Though much has been studied on the symbiotic association between endophytes and their plant hosts but some questions still need answers as to why only few plant species are adapted to severe stress conditions and whether plants can adapt to such habitats without the endophytic association. The effectiveness and performance of each endophyte may differ from each other, depending on environmental factors. Endophytes need to be selected for a particular crop growing conditions. Endophytes mediated drought-tolerance will have far-reaching consequences in light of current and future climate change brought by global warming. This will help the marginal farmers to raise crops in vast areas of arid land and help the crop to adapt to drier areas. It is known that majority of plants have endophytes residing in them, but their diversity in the different plant species is yet to be discovered. Thus, scope exists for finding many potentially beneficial endophytes from the different plant species thriving in different ecosystems. Sustainable agricultural practices must exploit the potential of endophytes for mitigating the impacts of abiotic stresses on crop plants. The endophytes can also have biotechnological potential in large-scale remediation of polluted sites and sustainable production of bioenergy and biofuel crops.

References

Alguacil MM, Hernande JA, Caravaca F, Portillo B, Roldan A (2003) Antioxidant enzyme activities in shoot from three mycorrhizal shrub species afforested in a degraded semiarid soil. Physiol. Plant. 118, 562–570.

Al-Karaki GN (2006) Nursery inoculation of tomato with arbuscular mycorrhizal fungi and subsequent performance under irrigation with saline water. Sci. Hortic. 109, 1–7.

Al-Khaliel AS (2010) Effect of salinity stress on mycorrhizal association and growth response of peanut infected by *Glomus mosseae*. Plant Soil Environ. 56, 318-324.

Allen EB, Allen MF (1986) Water relations of xeric grasses in the field: interactions of mycorrhizas and competition. New Phytol. 104, 559–571.

Arnold AE, Lutzoni F (2007) Diversity and host range of foliar fungal endophytes: are tropical leaves biodiversity hotspots? Ecology 88, 541–549.

Auge RM, Duan X (1991) Mycorrhizal fungi and non-hydraulic root signals of soil drying. Plant Physiol. (Bethesda) 97, 821–824.

Bacon CW, Hill NS (1996) Symptomless grass endophytes: products of coevolutionary symbioses and their role in the ecological adaptations of grasses. In: Endophytic Fungi in Grasses and Woody Plants (Ed) Redkin SC and Carris LM, APS Press, St. Paul, pp 155–178.

Baltruschat H, Fodor J, Harrach BD, Niemczyk E, Barna B, Gullner G, Janeczko A, Kogel KH, Schäfer P, Schwarczinger I, Zuccaro A, Skoczowski A (2008) Salt tolerance of barley induced by the root endophyte *Piriformospora indica* is associated with a strong increase in antioxidants. New Phytol. 180, 501–510.

Bethlenfalvay GJ, Brown MS, Franson R, Mihara KL (1989) The Glycine-Glomus-Bradyrhizobium symbiosis. IX. Nutritional, morphological and physiological responses of nodulated soybean to geographic isolates of the mycorrhizal fungus *Glomus mosseae*. Physiol. Plant. 76, 226–232.

Boyer LR, Brain P, Xu X-M, Jeffries P (2014) Inoculation of drought-stressed strawberry with a mixed inoculum of two arbuscular mycorrhizal fungi: effects on population dynamics of fungal species in roots and consequential plant tolerance to water deficiency. Mycorrhiza 25, 215–227.

Cassán F, Maiale S, Masciarelli O, Vidal A, Luna V, Ruiz O (2009) Cadaverine production by *Azospirillum brasilense* and its possible role in plant growth promotion and osmotic stress mitigation. Eur. J. Soil Biol. 45, 12–19.

Cattivelli L, Rizza F, Badeck F-W, Mazzucotelli E, Mastrangelo AM, Francia E, Mare C, Tondelli A, Stanca M (2008) Drought tolerance improvement in crop plants: an integrated view from breeding to genomics. Field Crop Res. 105, 1–14.

Chen L, Luo SL, Chen JL, Wan Y, Liu CB, Liu YT, Pang XY, Lai C, Zeng GM (2012) Diversity of endophytic bacterial populations associated with Cd hyperaccumulator plant *Solanum nigrum* L. grown in mine tailings. Appl. Soil Ecol. 62, 24-30.

Chen L, Luo SL, Xiao X, Guo HJ, Chen JL, Wan Y, Li B, Xu TY, Xi Q, Rao C, Liu CB, Zeng GM (2010) Application of plant growth-promoting endophytes (PGPE) isolated from *Solanum nigrum* L. for phytoextraction of Cd-polluted soils. Appl. Soil Ecol. 46, 383-389.

Ciais P, Reichstein M, Viovy N, Granier A, Ogee J, Allard V, Aubinet M, Buchmann N, Bernhofer C, Carrara A, Chevallier F, De Noblet N, Friend AD, Friedlingstein P, Grünwald T, Heinesch B, Keronen P, Knohl A, Krinner G, Loustau D, Manca G, Matteucci G, Miglietta F, Ourcival JM, Papale D, Pilegaard K, Rambal S, Seufert G, Soussana JF, Sanz MJ, Schulze ED, Vesala T, Valentini R (2005) Europe-wide reduction in primary productivity caused by the heat and drought in 2003. Nature 437, 529–533.

Clay K, Holah J (1999) Fungal endophyte symbiosis and plant diversity in successional fields. Science 285, 1742–1744.

Clay K, Schardl C (2002) Evolutionary origins and ecological consequences of endophyte symbiosis with grasses. Am. Nat. 4, S99-S127.

Cohen A, Bottini R, Piccoli P (2008) *Azospirillum brasilense* Sp 245 produces ABA in chemically defined culture medium and increases ABA content in Arabidopsis plants. Plant Growth Regul. 54, 97–103.

Cohen AC, Travaglia CN, Bottini R, Piccoli PN (2009) Participation of abscisic acid and gibberellins produced by endophytic *Azospirillum* in the alleviation of drought effects in maize. Botany 87, 455-462.

Coleman-Derr D, Tringe SG (2014) Building the crops of tomorrow: advantages of symbiont-based approaches to improving abiotic stress tolerance. Front. Microbiol. 5, 1–6.

Ding SI, Huang CL, Sheng HM, Song CL, Li YB, An LZ (2011) Effect of inoculation with the endophyte *Clavibacter* sp. strain Enf12 on chilling tolerance in *Chorispora bungeana*. Physiol. Plant. 141(2), 141-151.

Flexas J, Bota J, Loreto F, Cornic G, Sharkey TD (2004) Diffusive and metabolic limitations to photosynthesis under drought and salinity in C(3) plants. Plant Biol. 6, 269-279.

Flowers TJ, Torke PF, Yeo AR (1977) The mechanism of salt tolerance in halophytes. Ann Rev Plant Physiol. 28, 89–121.

Germaine K, Liu X, Cabellos G, Hogan J, Ryan D, Dowling DN (2006) Bacterial endophyte enhanced phyto-remediation of the organochlorine herbicide 2,4-dichlorophenoxyacetic acid. FEMS Microbiol. Ecol. 57, 302–310.

Gilbert GS, Mejia-Chang M, Rojas E (2002) Fungal diversity and plant disease in mangrove forests: salt excretion as a possible defense mechanism. Oecologia 132, 278–285.

Giovannetti M, Mosse B (1980) An evaluation of techniques for measuring vesicular arbuscular infection in roots. New Phytol. 84, 489–500.

Girlanda M, Perotto S, Luppi AM (2006) Molecular diversity and ecological roles of mycorrrhiza-associated sterile fungal endophytes in mediterranean ecosysems. In: Microbial Root Endophytes (Ed) Boyle CJC, Sieber TN, Springer-Verlag, Berlin, pp 207–226.

Glick BR, Cheng Z, Czarny J, Duan J (2007) Promotion of plant growth by ACC deaminase producing soil bacteria. In: New perspectives and Approaches in Plant Growth-Promoting Rhizobacteria Research (Ed) Bakker PAHM, Raaijmakers JM, Bloemberg G, Hçfte M, Lemanceau L and Cooke BM, Springer, Heidelberg, pp 329–339.

Graham JH, Syvertson JP (1984) Influence of vesicular arbuscular mycorrhiza on the hydraulic conductivity of roots of two citrus rootstocks. New Phytol. 97, 277–284.

Griffiths H, Parry MAJ (2002) Plant responses to water stress. Ann. Botany 89, 801–802.

Harrach BD, Baltruschat H, Barna B, Fodor J, Kogel K-H (2013) The mutualistic fungus Piriformospora indica protects barley roots from a loss of antioxidant capacity caused by the necrotrophic pathogen Fusarium culmorum. Mol. Plant Microbe Interact. 26, 599–605.

He LY, Chen ZJ, Ren GD, Zhang YF, Qian M, Sheng XF (2009) Increased cadmium and lead uptake of a cadmium hyperaccumulator tomato by cadmium resistant bacteria. Ecotoxicol Environ. Saf. 72, 1343-1348.

Heyster JW, Nabors MW (1982) Growth, water content and solute accumulation of two tobacco cells cultured on sodium chloride, dextran, and polyethylene glycol. Plant Physiol. 68, 1454–1459.

Iba K (2002) Acclimative response to temperature stress in higher plants: approaches of gene engineering for temperature tolerance. Ann. Rev. Plant Biol. 53, 225–245.

IPCC (2014) Summary for policymakers. In: Climate Change 2014: Impacts, adaptation, and vulnerability. Part A: Global and sectoral aspects. Contribution of Working Group II to the Fifth Assessment Report of the Intergovernmental Panel on Climate Change (Ed) Field CB, Barros VR, Dokken DJ, Mach KJ, Mastrandrea MD, Bilir TE, Chatterjee M, Ebi KL, Estrada YO, Genova RC, Girma B, Kissel ES, Levy AN, MacCracken S, Mastrandrea PR, White LL, Cambridge University Press, Cambridge, UK and New York, NY, USA, pp. 1–32.

Ker K, Charest C (2010) Nickel remediation by AM-colonized sunflower. Mycorrhiza 20, 399-406.

Khan AL, Hamayun H, Ahmad N, Waqasa M, Kanga SM, Kima YH, Lee IJ (2011) Exophiala sp. LHL08 reprograms Cucumis sativus to higher growth under abiotic stresses. Physiol Plant. 143, 329–343.

Khan AL, Muhammad H, Ahmad N, Hussain J, Kang S, Kim YH, Muhammad A, Tang DS, Muhammad W, Radhakrishnan R, Hwang YH, Lee IJ (2011) Salinity stress resistance offered by endophytic fungal interaction between Penicillium minioluteum LHL09 and Glycine max L. J. Microbiol. Biotechnol. 21(9), 893–902.

Kuffner M, Maria D, Puschenreiter S, Fallmann M, Wieshammer K, Gorfer G, Strauss M, Rivelli J, Sessitsch AAM (2010) Bacteria associated with Zn and Cd-accumulating Salix caprea with differential effects on plant growth and heavy metal availability. J. App. Microbiol. 108, 1471-1484.

Larson C (2013) Losing arable land, China faces stark choice: adapt or go hungry. Science 339, 644–645.

Long XX, Chen XM, Chen YG, Woon-Chung WJ, Wei ZB, Wu QT (2011) Isolation and characterization endophytic bacteria from hyperaccumulator Sedum alfredii Hance and their potential to promote phytoextraction of zinc polluted soil. World J. Microbiol. Biotechnol. 27, 1197-1207.

Ma Y, Rajkumar M, Luo Y, Freitas H (2011) Inoculation of endophytic bacteria on host and non-host plants – Effects on plant growth and Ni uptake. J. Hazard. Mater. 195, 230-237.

Ma´rquez LM, Redman RS, Rodriguez RJ, Roossinck MJ (2007) A virus in a fungus in a plant three way symbiosis required for thermal tolerance. Science 315, 513–515.

Madhaiyan M, Poonguzhali S, Sa T (2007) Metal tolerating methylotrophic bacteria reduces nickel and cadmium toxicity and promotes plant growth of tomato (*Lycopersicon esculentum* L.). Chemosphere 69, 220–8.

Mastretta C, Taghavi S, van der Lelie D, Mengoni A, Galardi F, Gonnelli C, Barac T, Boulet J,Weyens N, Vangronsveld J (2009) Endophytic bacteria from seeds of *Nicotiana tabacum* can reduce cadmium phytotoxicity. Int. J. Phytoremed. 11, 251-267.

Morton JB (2000) Biodiversity and evolution in mycorrhizae in the desert. In: Microbial Endophytes (Ed) Bacon CW and White J, CRC Press, New York, pp 3–30.

Munns R, Tester M (2008) Mechanisms of salinity tolerance. Annu. Rev. Plant Biol. 59, 651–681.

Murphy BR, Doohan FM, Hodkinson TR (2014a) Fungal endophytes of barley roots. J. Agric. Sci. 152, 602–615.

Murphy BR, Doohan FM, Hodkinson TR (2014b) Persistent fungal root endophytes isolated from a wild barley species suppress seed-borne infections in a barley cultivar. Biocontrol 60, 281–292.

Murphy BR, Martin Nieto L, Doohan FM, Hodkinson TR (2015) Fungal endophytes enhance agronomically important traits in severely drought-stressed barley. J. Agro. Crop Sci. DOI:10.1111/jac.12139

Nelsen CE (1987) The water relations of vesicular-arbuscular mycorrhizal systems. In: Ecophysiology of VA Mycorrhizal Plants (Ed) Safir GR, CRC Press, Boca Raton, pp 71–79.

Newman LA, Reynolds CM (2005) Bacteria and phytoremediation: new uses for endophytic bacteria in plants. Trends Biotechnol. 23(1): 6-8.

Oberhofer M, Güsewell S, Leuchtmann A (2014) Effects of natural hybrid and non-hybrid Epichloë endophytes on the response of *Hordelymus europaeus* to drought stress. New Phytol. 201, 242–253.

Oelmüller R, Sherameti I, Tripathi S, Varma A (2009) *Piriformospora indica*, a cultivable root endophyte with multiple biotechnological applications. Symbiosis 49, 1-17.

Pandey PK, Yadav SK, Singh A, Sarma BK, Mishra A, Singh HB (2012) Cross-species alleviation of biotic and abiotic stresses by the endophyte *Pseudomonas aeruginosa* PW09. J. Phytopathol. 160, 532-539.

Rabie GH, Almadini AM (2005) Role of bioinoculants in development of salt-tolerance of *Vicia faba* plants under salinity stress. Afr. J. Biotechnol. 4, 210–222.

Rangarajan S, Saleena LM, Nair S (2002) Diversity of Pseudomonas spp. isolated from rice rhizosphere populations grown along a salinity gradient. Microbial Ecol. 43(2), 280-289.

Redman RS, Dunigan DD, Rodriguez RJ (2001) Fungal symbiosis: from mutualism to parasitism, who controls the outcome, host or invader? New Phytol. 151, 705–716.

Redman RS, Freeman S, Clifton DR, Morrel J, Brown G, Rodriguez RJ (1999) Biochemical analysis of plant protection afforded by a nonpathogenic endophytic mutant of *Colletotrichum magna*. Plant Physiol. 119, 795–804.

Redman RS, Sheehan KB, Stout RG, Rodriguez RJ, Henson JM (2002a) Thermotolerance conferred to plant host and fungal endophyte during mutualistic symbiosis. Science 298, 1581.

Rennenberg H, Loreto F, Polle A, Brilli F, Fares S, Beniwal RS, Gessler A (2006) Physiological responses of forest trees to heat and drought. Plant Biol. 8, 556-571.

Reynolds M, Tuberosa R (2008) Translational research impacting on crop productivity in drought-prone environments. Curr. Opin. Plant Biol. 11, 171–179.

Rodriguez RJ, Henson J, Van Volkenburgh E, Hoy M, Wright L, Beckwith F, Kim Y, Redman RS (2008) Stress tolerance in plants via habitat-adapted symbiosis. ISME J. 2, 404-416.

Rodriguez RJ, Redman RS, Henson JM (2004) The role of fungal symbioses in the adaptation of plants to high stress environments. In: Mitigation and Adaptation Strategies for Global Change, Kluwer Academic Publishers, Netherlands, pp 261-271.

Roelfsema MR, Hedrich R (2005) In the light of stomatal opening: New insights into the Watergate. New Phytol. 167, 665-691.

Rosenblueth, M & Martínez-Romero E (2006) Bacterial endophytes and their interactions with hosts. Mol. Plant-Microb. Interact. 19(8), 827-837.

Ruiz-Lozano JM, Azcon R, Gomez M (1995) Effects of arbuscular-mycorrhizal Glomus species on drought-tolerance: physiological and nutritional plant responses. Appl. Environ. Microbiol. 61, 456-460.

Ryan RP, Germaine K, Franks A, Ryan DJ, Dowling DN (2008) Bacterial endophytes: recent developments and applications. FEMS Microbiol Lett. 278, 1-9.

Sahay NS, Varma A (1999) *Piriformospora indica*: A new biological hardening tool for micropropagated plants. FEMS Microbiol Lett. 181, 297-302.

Schardl C, Leuchtmann A (2005) The epichloe endophytes of grasses and the symbiotic continuum. In: The Fungal Community: Its Organization and Role in the Ecosystem (Ed) Dighton J, White JF, Oudemans P. Taylor & Francis CRC Press, Boca Raton, pp 475–503.

Seki M, Narusaka M, Ishida J, Nanjo T, Fujita M. Oono Y, Kamiya A, Nakajima M, Enju A, Sakurai T, Satou M, Akiyama K, Taji T, Yamaguchi-Shinozaki K, Carninci P, Kawai J, Hayashizaki Y, Shinozaki K (2002) Monitoring the expression profiles of 7000 Arabidopsis genes under drought, cold and high-salinity stresses using a fulllength cDNA microarray. Plant J. 31, 279–292.

Seki M, Umezawa T, Urano K, Shinozaki K (2007) Regulatory metabolic networks in drought stress responses. Curr. Opinion Plant. Biol. 10, 296-302.

Serfling A, Wirsel SGR, Lind V, Deising HB (2007) Performance of the biocontrol fungus *Piriformospora indica* on wheat under greenhouse and field conditions. Phytopathology 97, 523–531.

Sgroy V, Cassán F, Masciarelli O, Del Papa MF, Lagares A, Luna V (2009) Isolation and characterization of endophytic plant-growth promoting (PGPB) or stress homeostasis-regulating (PSHB) bacteria associated to the halophyte *Prosopis strombulifera*. Appl. Microbiol. Biotechnol. 85, 371-381.

Shahollari B, Vadassery J, Varma A, Oelmüller R (2007) A leucine- rich repeat protein is required for growth promotion and enhanced seed production mediated by the endophytic fungus *Piriformospora indica* in *Arabidopsis thaliana*. Plant J. 50, 1-13.

Shahollari B, Varma A, Oelmüller R (2005) Expression of a receptor kinase in Arabidopsis roots is stimulated by the basidiomycete *Piriformospora indica* and the protein accumulates in Triton X-100 insoluble plasma membrane microdomains. J. Plant Physiol. 162, 945-958.

Sharifi M, Ghorbanli M, Ebrahimzadeh H (2007) Improved growth of salinity-stressed soybean after inoculation with salt pre-treated mycorrhizal fungi. J. Plant Physiol. 164, 1144–1151.

Sheng XF, Xia JJ, Jiang CY, He LY, Qian M (2008) Characterization of heavy metal-resistant endophytic bacteria from rape (*Brassica napus*) roots and their potential in promoting the growth and lead accumulation of rape. Environ. Pollut. 156, 1164–1170.

Sheramati I, Tripathi S, Varma A, Oelmüller R (2008) The root-colonizing endophyte *Piriformospora indica* confers drought tolerance in Arabidopsis by stimulating the expression of drought stress-related genes in leaves. Mol. Plant-Microbe Interact. 21, 799-807.

Shinozaki K, Yamaguchi-Shinozaki K, Seki M (2003) Regulatory network of gene expression in the drought and cold stress responses. Curr. Opin Plant. Biol. 6, 410-417.

Siddikee M, Chauhan P, Anandham R, Han G-H, Sa T (2010) Isolation, characterization, and use for plant growth promotion under salt stress, of ACC deaminase-producing halotolerant bacteria derived from coastal soil. J. Microbiol. Biotechnol. 20, 1577–1584.

Sturz AV, Christie BR, Nowak J (2000) Bacterial Endophytes: Potential Role in Developing Sustainable Systems of Crop Production. Crit. Rev. Plant Sci. 19(1), 1-30.

Tain CY, Feng G, Li XL, Zhang FS (2004) Different effects of arbuscular mycorrhizal fungal isolates from saline or non-saline soil on salinity tolerance of plants. Appl. Soil Ecol. 26, 143–148.

Timmusk S, Paalme V, Pavlicek T, Bergquist J, Vangala A, Danilas T, Nevo E (2011) Bacterial distribution in the rhizosphere of wild barley under contrasting microclimates. PLoS ONE DOI: 10.1371/journal.pone.0017968.

Vadassery J, Tripathi S, Prasad R, Varma A, Oelmüller R (2009b) Monodehydroascorbate reductase 2 and dehydroascorbate reductase 5 are crucial for a mutualistic interaction between *Piriformospora indica* and Arabidopsis. J. Plant Physiol. 166, 1263–1274.

Varma A, Singh A, Sudha, Sahay N, Sharma J, Roy A, Kumari M, Rana D, Thakran S, Deka D, Bharti K, Franken P, Hurek T, Blechert O, Rexer K-H, Kost G, Hahn A, Hock B, Maier W, Walter M, Strack D, Kranner I (2001) *Piriformospora indica*: A cultivable mycorrhiza-like endosymbiotic fungus. In: Mycota IX, Springer Series, Germany, pp 123–150.

Verma SA, Varma A, Rexer K-H, Hassel A, Kost G, Sarbhoy A, Bisen P, Bütehorn B, Franken P (1998) *Piriformospora indica*, gen. et sp. nov., a new root-colonizing fungus. Mycologia 90, 898-905.

Waller F, Achatz B, Baltruschat H, Fodor J, Becker K, Fischer M, Heier T, Hückelhoven R, Neumann C, von Wettstein D, Franken P, Kogel KH (2005) The endophytic fungus *Piriformospora indica* reprograms barley to salt-stress tolerance, disease resistance, and higher yield. Proc. Natl. Acad. Sci. USA 102, 13386–13391.

Waller F, Mukherjee K, Deshmukh SD, Achatz B, Sharma M, Schäfer P, Kogel KH (2008) Systemic and local modulation of plant responses by *Piriformospora indica* and related Sebacinales species. J. Plant Physiol. 165, 60–70.

Wei Y, Hou H, ShangGuan Y, Li J, Li F (2014) Genetic diversity of endophytic bacteria of the manganese-hyperaccumulating plant *Phytolacca americana* growing at a manganese mine. Eur. J. Soil Biol. 62, 15-21.

Yaish MW, Antony I, Glick BR (2015) Isolation and characterization of endophytic plant-growth promoting bacteria from date palm tree (*Phoenix dactylifera* L.) and their potential role in salinity tolerance. Antonie Leeuwenhoek 107, 1519-1532.

Yano-Melo AM, Saggin OJ, Maia LC (2003) Tolerance of mycorrhized banana (*Musa* sp. cv. Pacovan) plantlets to saline stress. Agric. Ecosyst. Environ. 95, 343–348.

Yoshida K, Kaothien P, Matsui T, Kawaoka A, Shinmyo A (2003) Molecular biology and application of plant peroxidase genes. Appl. Microbiol. Biotechnol. 60, 665–670.

Zandavalli RB, Dillenburg LR, Paulo VD (2004) Growth responses of *Araucaria angustifolia* (Araucariaceae) to inoculation with the mycorrhizal fungus *Glomus clarum*. Appl. Soil. Ecol. 25, 245–255.

9

The Role of Arbuscular Mycorrhizal Fungi in Salt and Drought Stresses

Padmavathi Tallapragada

Abstract

Drought and salinity are two major abiotic stresses that affect various aspects of human lives of one third world population including human health and agricultural productivity. Abiotic stresses such as salinity, drought, nutrient deficiency or toxicity, and flooding limit crop productivity world-wide. A living organism is considered resistant if it tolerates a given physicochemical stress. The presence of microorganisms within the soil, specially a main group known as arbuscular mycorrhizal fungi (AMF), is a key factor in the adaptation of plants to the different ecosystems. The AMF colonizing within the plant roots can determine the success of plants in salt- and/or drought-affected areas. The symbiosis with AMF has been proposed as one of the mechanisms of salinity and water stress avoidance. Studies have shown greater drought and salinity tolerance in AMF-colonized plants by a number of mechanisms. These mechanisms include enhanced water and nutrient uptake directly by extraradical hyphae, higher leaf stomatal conductance and/or better root system architecture, higher capacity of osmotic adjustment and antioxidant defense systems, better soil structure due to higher glomalin and over expression of genes encoding antioxidant enzymes. This is due to complex mechanism of abiotic stress tolerance, which is controlled by the expression of several minor genes. The techniques employed for selecting tolerant plants are time consuming and consequently expensive. Understanding molecular mechanisms of abiotic stress tolerance as well as inducing stress tolerance in some potential crops show some promising results. The future research should focus on molecular, physiological and metabolic aspects of stress tolerance to facilitate the development of crops with an inherent capacity to withstand abiotic stresses. This chapter provides an overview of the physiological mechanisms by which growth and development of crop plants are affected by salinity and drought stress. We will also revise how the presence of AMF may affect tolerance of plants to these stresses and the different mechanisms involved.

Keywords: Abiotic stresses, Antioxidant enzymes, AMF, Glomalin, Salinity, Water stress,

1. Introduction

As the world population continues to increase, more food needs to be grown to feed the people. This can be achieved by an increase in cultivated land and by an increase in crop productivity per area. Salinity and drought are the major factors that will trigger crop development and yield. The basic physiology of the different stresses overlaps with each other and are interconnected, affecting plants mainly by disturbing their water balance and/or inducing photoinhibition and photooxidative stress. Plants will try to reduce their water loss by closing the stomata (Yang et al. 2005; Schachtman and Goodger 2008) and adjusting their root hydraulic conductance (Maurel et al. 2008; Aroca et al. 2012). Both drought and salinity generate low water potentials within the soil that lead to a primary osmotic stress in the plants. Furthermore, the presence of high salt concentrations within the soil will also cause an ionic and toxic effect in the plant tissues. Drought and high salt levels will affect the water available for plants by reducing the osmotic potential of the soil solution. In addition to the intrinsic protective system of plants against stress, a number of soil microorganisms have been proved to alleviate the stress symptoms. Mycorrhiza is an association or symbiosis between the roots of most land plants and many soil fungi that colonize the cortical tissue of roots during periods of active plant growth, from which both partners benefit. The aim of the present review is to have further exploitation to the effects of AMF on host plant and its mechanism based on previous works.

2. Strategies for improving plants against water and salt stresses

Plant stress tolerance is defined in terms of yield stability under abiotic stress conditions. However, yield losses due to abiotic stresses vary depending on timing, intensity and duration of the water stress, coupled with other environmental factors such as high light intensity and temperature (Parry et al. 2005; Reynolds et al. 2005; Neumann 2008). To overcome this, the following strategies were suggested:

a) Water management practices that save irrigation water

b) Agronomic practices such as addition of AMF by which plants can perform well under water and salinity stress conditions

c) Selection of crop cultivars that require relatively lower quantity of water for their growth and crop productivity.

2.1 Salinity stress

Salinity can occur naturally when normal life is disturbed the natural ecosystem changed the hydrology of landscapes, the movement of salts into lands and into rivers have been accelerated. This leads to affect the natural environment. Salinity can be categorized based on different ways irrigation, dry land, urban, river and industrial salinity. Dry land salinity causes reduction in the agricultural production, reduced yield, low profitability due to mitigation changed the land use and extreme conditions removes the land from the agriculture production. Salinity affects plants in different ways such as osmotic effects, specific-ion toxicity and/or nutritional disorders (Läuchli and Epstein 1990). The extent by which one mechanism affects the plant over the others depends upon many factors including the species, genotype, plant age, ionic strength and composition of the salinizing solution and the organ in question. Plants undergo characteristic changes from the time salinity stress is imposed until they reach maturity (Munns 2002).

Salinization of soil is a serious problem and is increasing steadily in many parts of the world, in particular in arid and semiarid areas (Giri *et al.* 2003; Al-Karaki 2006). Salinity in soil or water is of increasing importance to agriculture because it causes a stress condition to crop plants. Salt-affected soil is one of the serious abiotic stresses that cause reduced plant growth, development and productivity worldwide. The salt-affected soils occupy approximately 7% of the global land surface (Sheng *et al.* 2008). Salt stress reduces plant growth, leaf expansion and induces mineral deficiency, like other stresses, destabilizes cell membranes, alters selective permeability (leakage of cell solutes), fluidity, micro viscosity, and affects the solubility of many essential substrates and ions. The direct effects of salt on plant growth may involve:

(a) Reduction in the osmotic potential of the soil solution that reduces the amount of water available to the plant causing physiological drought – to counteract this problem plants must maintain lower internal osmotic potentials in order to prevent water movement from roots into the plant and soil (Feng *et al.* 2002; Jahromi *et al.* 2008).

(b) Toxicity of excessive Na_2 and Cl_2 ions towards the cell – the toxic effects include disruption to the structure of enzymes and other macromolecules, damage to cell organelles and plasma membrane, disruption of photosynthesis, respiration and protein synthesis (Juniper and Abbott 1993; Feng *et al.* 2002); and

(c)   Nutrient imbalance in the plant caused by nutrient uptake and/or transport to the shoot leading to ion deficiencies (Marschner 1995; Adiku *et al.* 2001).

Salinity affects both vegetative and reproductive development which has profound implications depending on whether the harvested organ is a stem, leaf, root, shoot, fruit, fiber or grain (Läuchli and Epstein 1990). Reduction in shoot growth due to salinity is commonly expressed by a reduced leaf area and stunted shoots (Läuchli and Epstein 1990). Final leaf size depends on both cell division and cell elongation. AMF have been shown to promote plant growth and salinity tolerance by many researchers. They promote salinity tolerance by employing various mechanisms, such as enhancing nutrient acquisition (Al-Karaki and Al-Raddad 1997), producing plant growth hormones, improving rhizospheric and soil conditions (Lindermann 1994) altering the physiological and biochemical properties of the host (Smith and Read 1995) and defending roots against soil-borne pathogens (Dehne 1982).

2.2 Drought stress

In many arid and semiarid regions of the world, drought limits crop productivity. The incorporation of factors enabling plants to withstand drought stress would be helpful to improve crop production under drought conditions. Three main mechanisms reduce crop yield by water stress:

(i)   Reduced canopy absorption of photosynthetically active radiation

(ii)   Lessened radiation-use efficiency and

(iii)   Reduced harvest index.

The reproducibility of drought stress treatments is very cumbersome, which significantly impedes research on plant drought tolerance (Farooq *et al.* 2009). Although different plant species can vary in their sensitivity and responses to the decreased water potential caused by water deficit, it is assumed that all plants have an encoded capability for stress perception, signaling and response (Bohnert *et al.* 1995). Plants can respond to drought stress at morphological, metabolic and cellular levels with modifications that allow the plants to avoid the stress or to increase its tolerance (Bray 1997). Plants using drought avoidance mechanism have deeper and dense root system, greater root penetration ability, higher stomatal conductance, and higher cuticular resistance to prevent water loss, higher pre-drawn leaf water potential, and avoid leaf rolling for longer intervals (Peng and Ismail, 2004). Although mechanism of drought tolerance is poorly understood, osmotic adjustment is considered to be associated with dehydration tolerance. Osmotic adjustment is the accumulation of organic or inorganic solutes in response to water stress thereby maintaining tissue turgor potential. In addition to the intrinsic protective system of plants against stress, a number of soil microorganisms have been proved to be able to alleviate the stress symptoms. Arbuscular mycorrhiza is an association or symbiosis between the roots of most land plants and many soil fungi that colonize the cortical tissue of roots during periods of active plant growth, from which both partners benefit. An efficient organism is an organism which shows physiological and biochemical processes such that it can successfully cope with limiting environmental conditions (Smith and Gianinazzi-Pearson. 1988). Generally, the decline in CO_2 assimilation rate associated with a reduction in leaf water status has been attributed primarily to stomatal closure and the resulting increase in leaf epidermal resistance. Transpiration is normally suppressed by water stress concurrently with the suppression of photosynthesis. Thus, particular abilities of AM endophytes to alter physiological plant parameters that enhance adaptation to low soil water content can provide suitable criteria for the selection of inoculants (Ruiz-lozano et al.1995) and AMF could alter water relations and played a great role in the growth of host plant in the condition of drought stress.

3. Arbuscular mycorrhizal fungi

AMF are group of obligate biotrophs, to the extent that they must develop a close symbiotic association with the roots of a living host plant in order to grow and complete their life cycle (Parniske 2008). They are found in the roots of about 80-90% of plant species (mainly grasses, agricultural crops and herbs) and exchange benefits with their partners, as is typical of all mutual symbiotic relationships (Wang and Qiu 2006). They represent an interface between plants and soil, growing their mycelia both inside and outside the plant roots. The process of AM colonization of host plant root is characterized by distinct stages involving a series of morphogenetic changes in the fungus; spore germination hypal differentiation, appresorium formation, root penetration, intracellular growth, arbuscule formation and nutrient transport (Bagyaraj and Padmavathi 1993).

AMF provide the plant with water, soil mineral nutrients (mainly phosphorus and nitrogen) and pathogen protection. In exchange, photosynthetic compounds are transferred to the fungus (Bonfante and Genre 2010). The most agriculturally significant and frequently investigated is AMF, from both the ecological and physiological points of view (Smith and Smith, 2011), is their positive effect on plant nutrition and, consequently, on plant fitness. In particular, they play an important role in helping the plant uptake phosphorus from the soil (Padmavathi *et al.*2015). Without AMF, it is rather difficult for the plant to absorb this macro element from the soil, since it is mainly available in its insoluble organic or inorganic form. Besides phosphorus, AMF can also translocate water and other mineral nutrients (in particular nitrogen) from the soil to the plant. These nutritional exchanges are bidirectional. As a consequence, particularly efficient symbiotic associations have been demonstrated to stabilize through unknown mechanisms, with the plant selecting the most cooperative fungal partners (Kiers *et al.*2011). The AMF-inducible recovery of plant nutritional deficiency can inevitably lead to an improvement in plant growth, with a potential positive impact on productivity. AMF are also responsible for other functions that favour the plants they colonize:

(a)   They positively affect plant tolerance towards both biotic (e.g., pathogens) and abiotic stresses (i.e., drought and soil salinity) by acting on several physiological processes, such as the production of antioxidants, the increment of osmolyte production or the improvement of abscisic acid regulation (Ruiz-Lozano 2012) and the enhancement of plant tolerance to heavy metals (Leyval *et al.* 2002)

(b)   They help plants become established in harsh/degraded ecosystems, such as desert areas and mine spoils (Requena *et al.* 2001)

(c)   They increase the power of phytoremediation (the removal of pollutants from the soil by plants) by allowing their host to explore and depollute a larger volume of soil (Göhre and Paszkowski 2006).

Another crucial ecological role played by AMF is their capacity to directly influence the diversity and composition of the aboveground plant community. Several studies have confirmed that plant species richness can be altered not only by climatic and edaphic factors, but also by soil microbial assemblages (Facelli *et al.*2010). The underlying mechanism is not completely understood, but could be related to the promotion of seedling establishment of secondary plant species (Hart *et al.*2003). AMF can also negatively affect the diversity and growth of plants, which is particularly significant for the management of weeds (Veiga 2011) AMF also plays a critical role in soil aggregation, their thick extraradical hyphal network, which envelops and keeps the soil particles compact. It has been suggested that glycoproteins (glomalin and glomalin related proteins) secreted by AMF into the soil could exert a key role in this process (Padmavathi and Ranjini 2011). These proteins are exuded in great quantities into the soil, and could have implications on carbon sequestration. This potential capability of AMF is likely to contribute to a great extent to the soil ecosystem carbon dioxide (CO_2) sequestration process. AM symbiosis has been shown to increase tolerance to biotic and abiotic stresses. Regarding abiotic stress, several studies for years have demonstrated that AM symbiosis confers tolerance to drought (Miransari 2010), heat (Compant *et al.* 2010), salinity (Miransari 2010) or osmotic stress (Ruiz-Lozano 2003).

Extensive studies have demonstrated AM-mediated plant resistance to drought and salinity conditions, but the underlying mechanisms have not yet been clearly elucidated. Our incomplete understanding of how AM symbiosis affects the ability of plants to withstand conditions of limited water and soil salinity represents an important challenge to meet the goal of improved plant productivity globally.

4. Arbuscular mycorrhizal plant strategies to cope with salinity and drought

Plants growing in saline soil are subjected to three distinct physiological stresses. First, the toxic effects of specific ions such as sodium and chloride, prevalent in saline soils, disrupt the structure of enzymes and other macromolecules, damage cell organelles, disrupt photosynthesis and respiration, inhibit protein synthesis, and induce ion deficiencies (Juniper and Abbott 1993) (Fig.1). Colonization of plant roots by some AMF is reduced in the presence of NaCl (Sheng *et al.* 2008) probably due to the direct effect of NaCl on the fungi (Juniper and Abbott 2006) indicating that salinity can suppress the formation of arbuscular mycorrhiza. The rate of germination and maximum germination of AMF spores may also

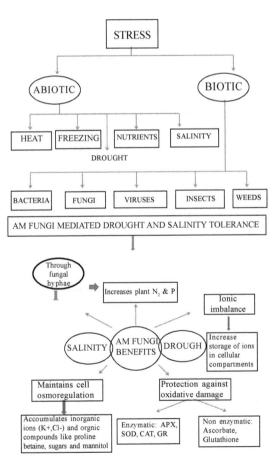

Fig. 1: Benefits of arbuscular mycorrhizal symbioses and its contribution to the host plants under stress conditions.

depend on the salt type. Contarary to the above reports increased AMF sporulation and colonization under salt stress conditions has also been reported (Aliasgharzadeh *et al*. 2001). Colonization rates were not reduced in all AMF present in coastal vegetation on Okinawa Island, Japan even when treated with high salinity of 200 mM reported byYamato *et al*. (2008). The mycorrhizal association is well known to increase host nutrient acquisition, particularly P (Smith and Read 1997). The improved growth of mycorrhizal plants in saline conditions is primarily related to mycorrhiza-mediated enhancement of host plant P nutrition (Al-Karaki 2000). Giri *et al*. (2007) showed that *Acacia nilotica* plants colonized by *G. fasciculatum* had a higher concentration of K^+ in root and shoot tissues at all salinity levels assayed. Plant growth and biomass suffered a lot under salt stress. There is considerable evidence that AMF can enhance plant growth and vigor under salt stress conditions. Phosphorus (P) is the macronutrient with the lowest mobility in soil and thus often limiting plant growth, particularly when soil water potential and P diffusion rate is lowered in dry or saline soils. However, mycorrhization was found to increase the fitness of the host plant by enhancing its growth and biomass. Several researchers have reported that AMF-inoculated plants grow better than non-inoculated plants under salt stress (Zuccarini and Okurowska, 2008). Ying-Ning (2011) reported markedly increase both plant performance (leaf number, leaf area, shoot and root dry weights) and leaf relative water content of citrus seedlings in AM association when exposed to salt stress. AM symbiosis plays a vital role in improving the P nutrition of the host plants under salt stress conditions. It has been seen that external hyphae of AMF deliver upto 80% of a plants P requirements (Marschner and Dell 1994). This is probably due to the extended network of AM fungal hyphae that allow them to explore more soil volume than non-mycorrhizal plants (Ruiz-lozano and Azcon 2000).

Plants can be classified as drought avoiders or as drought tolerant based on the absolute value of leaf water potential. Drought avoidance allows the plant to withstand water-limiting conditions by maintaining a higher water status, mainly through enhanced water uptake and/or minimized water loss; tolerance to dehydration is associated with survival and sustained physiological activity when the leaf water potential is low, resulting in the ability of leaves to endure dehydration. AM symbiosis protects host plants against the detrimental effects of drought stress through mechanisms of drought avoidance (Ruiz-Sanchez *et al*. 2010). Drought avoidance in mycorrhizal plants rely on the ability to maintain an adequate hydration status on the level of whole plants as characterized by relative water content, although a thorough review of the literature indicates that leaf water potential was not measured in some experiments (Augé and Moore 2005). The improved capability of drought avoidance mediated by AM colonization has often been associated with the AM promotion of plant growth through enhanced nutrition. The influence of AM symbiosis on leaf hydration, mainly via the increased water uptake characteristic of mycorrhizal plants, may be the basis for their improved drought resistance. Mycorrhizal plants have also been characterized as drought tolerant, mainly because of more improved osmotic adjustment, which allows the hydration and turgor of leaves to be sustained when leaf water potentials are low. The role of AM symbiosis in ameliorating plant responses to drought stress has suggested the up-regulation and down-regulation of several physiological and biochemical processes (Fig 1).

1. The direct uptake and transfer of water and nutrients by AMF

2. Increased osmotic adjustment

3. Enhanced gas exchange and water use efficiency, and

4. Better protection against oxidative damage when water is limiting may ameliorate, mitigate, and compensate the negative impacts of water stress in mycorrhizal plant.

The main absorption apparatus of mycorrhizal extension hyphae with a diameter of 2-5mm can penetrate soil pore inaccessible to root hairs (10- 20mm) and so absorb water that is not available to non-mycorrhizal plants (Gong *et al.*,2000). Because the number of extension hyphae is greater than that of root hairs, the area of surface where plants with soil and AM interaction increased greatly. In addition, colonized with AMF might change the architecture of root, which may be used to increase the interaction of roots in soil (Atkinson, 1994). Li *et al.* (1994) proved that AM hyphae were able to absorb phosphorus from the dense soil location where the host roots could not access. Several factors such as host plant, AM fungal isolate, and soil environment can influence effectiveness of root- AMF symbioses. It is important to understand and manipulate these factors to optimize plant growth responses to AMF. It may also be necessary to select AM fungal isolates best adapted to the environment in which a plant species is to be grown. Isolates of AMF differ in ability to enhance plant growth (Ruiz- Lozano *et al.*1995). Fungal isolates have been reported to differ in their ability to ameliorate plant water stress. Isolates of *G. monosporum* have been shown to be less effective in relieving water stress on wheat in comparison to *G. mosseae* (Al-Karaki *et al.* 1998). Moreover, *G. fasciculatum* has been reported to increase drought resistance in several plant species (Ellis *et al.* 1985). The improved growth, yield, and nutrient uptake in wheat plants reported here demonstrate the potential of mycorrhizal inoculation to reduce the effects of drought stress on wheat grown under field conditions in semiarid areas of the world.

5. Biochemical responses of AM plants to salinity and drought

5.1 Osmoregulation and Osmoprotectants

Drought and high salt levels will affect the water available for plants by reducing the osmotic potential of the soil solution The best characterized biochemical response of plant cells to osmotic stress is accumulation of some inorganic ions such as Na^+ and compatible organic solutes like proline, glycine betaine, and soluble sugars (Flowers and Colmer 2008). A major category of organic osmotic solutes consists of simple sugars (mainly fructose and glucose), sugar alcohols (glycerol and methylated inositols) and complex sugars (trehalose, raffinose and fructans) (Bohnert and Jensen 1996). Others include quaternary amino acid derivatives (proline, glycine betaine, â-alanine betaine, proline betaine), tertiary amines 1, 4, 5, 6-tetrahydro-2-mehyl-4-carboxyl pyrimidine), and sulfonium compounds (choline osulfate, dimethyl sulfonium propironate) (Nuccio et al. 1999). Many organic osmolytes are presumed to be osmoprotectants, as their levels of accumulation are insufficient to facilitate osmotic adjustment. Proline and sugars accumulation within the plant cells has been recognized as a response to several environmental stresses (Samaras et al. 1995). Proline accumulates within the cell cytoplasm without interfering with the metabolism and regulating its osmotic potential. Proline can also work as a sink for energy to regulate redox potentials, as a hydroxyl radical scavenger, and as a solute that protects macromolecules against denaturation (Kishor et al. 1995). The presence of AMF under drought and high salt levels may modify the production and accumulation of different osmoregulators within the plant, mainly sugars and amino acids (proline). In several studies, while proline content increased in response to water deficit, a lower accumulation of proline has been observed in mycorrhizal plants relative to nonmycorrhizal counterparts (Asrar et al. 2012). Proline could also be considered as a marker of the potential injury caused by water deficit, indicating that mycorrhizal plants, characterized by lower proline accumulation, were less stressed than the nonmycorrhizal plants. Furthermore, proline can act as an effective scavenger of ROS in the protection against denaturation and in the stabilization of membranes and subcellular structures (Kishor et al. 2005). The levels of free polyamines, other soluble nitrogenous compounds, increased in the leaves of drought-stressed mycorrhizal plants, and this increase was interpreted as indicating that free polyamines could serve as osmoprotectants under drought conditions, conferring drought resistance to mycorrhizal plants (Goicoechea et al. 1998). The increase in sugar content is found to be positively correlated with mycorrhization of the host plant as reported by Thomson et al. (1990). Porcel and Ruiz-Lozano (2004) also reported increased sugar concentrations in soybean roots colonized by G. intraradices and subjected to drought stress. Similar to proline, there were positive correlations between the increase on root sugar content and the degree of mycorrhization under salt (Feng *et al.* 2002) and drought conditions (Porcel and Ruiz-Lozano 2004), although the opposite has also been reported (Sheng *et al.* 2011). This higher accumulation of soluble sugars in mycorrhizal plant tissue, especially in roots, could make mycorrhizal plants more resistant to osmotic stresses.

5.2 Ion homeostasis – compartmentalization

Since NaCl is the principal soil salinity stress, a research focus has been on the transport systems that are involved in utilization of Na^+ as an osmotic solute (Blumwald *et al.* 2000). Na^+ competes with K^+ for uptake through common transport systems and does this effectively since the (Na^+)ext in saline environments is usually considerably greater than (K^+)ext. Ca^{2+} enhances K^+/Na^+ selective intracellular accumulation (Maathuis *et al.*1996). A low K^+: Na^+ ratio will disrupt the ionic balance in the cytoplasm as well as several metabolic pathways (Giri *et al.* 2007). H^+ pumps in the plasma membrane and tonoplast energize solute transport necessary to compartmentalize cytotoxic ions away from the cytoplasm and to facilitate the function of ions as signal determinants (Maeshima 2001). These pumps provide the driving force (H^+ electrochemical potential) for secondary active transport and function to establish membrane potential gradients that facilitate electrophoretic ion flux. The plasma membrane localized H^+ pump is a P-type ATPase and is primarily responsible for the large (pH and membrane potential gradient across this membrane (Morsomme and Boutry 2000). The increase of the K^+:Na^+ ratio is regulated through the compartmentalization of Na^+ within the plant tissues and the regulation of Na^+ uptake through the Na^+/H^+ antiporter system (Olýfas *et al.* 2009) the Na^+ influx by HKT transporters family, and the tonoplast Na^+/H^+ by antiporter family NHX (Ruiz-Lozano *et al.* 2012). AMF exclude Na^+ during its transfer to the plants or by discriminating its uptake from the soil (Hammer *et al.* 2011), maintaining a fine balance of K^+: Na^+ and Ca^{2+}: Na^+ ratios. This higher K^+: Na^+ will allow plants to maintain a fine balance in their nutrient status, preventing the disruption of metabolic processes and the absorption and translocation of Na^+ to the shoot tissues. Not much information is available on AM fungal K^+ transporters or how the presence of AMF may affect K^+ plant transporters; however, Corratge' *et al.* (2007) described K^+ transporters on ectomycorrhizal fungi *Hebeloma cylindrosporum*. These transporters may be modified by the presence of AMF which could be critical for the resistance of plants to the transport of Na^+. The description of these transporters may also play a key role in the understanding of K^+ and Na^+ transport in the fungi–plant symbiosis and open a new aspect in the molecular study of AMF. Chloride ion is an essential micronutrient that regulates enzyme activities and photosynthesis, helps maintain membrane potential, and is involved in cell turgor and cytoplasmatic pH regulation (Xu *et al.* 2000). Many plant species are not able to effectively regulate Cl entry into the shoot, and they accumulate Cl in greater amounts than Na^+ (Tavakkoli *et al.* 2010). Chloride ions reach the xylem by the symplastic pathway by ion channels or carriers as an active process (Tyerman and Skerrett 1999). Chloride ion levels within the plant tissues can be alleviated with the presence of AMF (Copeman *et al.* 1996; Zuccarini and Okurowska 2008), as Cl ions can be compartmentalized into the fungal vacuolar membranes. However, AMF have been also shown to increase the Cl concentration in citrus seedlings (Graham and Syvertsen 1989). $[Ca^{2+}]$ ext enhances salt tolerance and salinity stress elicits a transient $[Ca^{2+}]$ cyt increase, from either an internal or external source, which has been implicated in adaptation (Läuchli 1990). In the presence of AMF, $Ca^{2+}:Na^+$ ratios have been shown to increase (Evelin *et al.* 2009), although no changes in Ca^{2+} tissue concentrations have been observed (Giri and Mukerji 2004). The involvement of AMF in these processes is not clear.

5.3 Plant–water relations

The physiological effects of AM symbiosis include aboveground modifications of water relations and physiological status in terms of leaf water potential, relative water content, stomatal conductance, CO_2 assimilation, and efficiency of photosystem II as compared to nonmycorrhizal plants (Barzana *et al.* 2012). The mechanism by which AM symbiosis affects these physiological parameters is still unclear. The role played by abscisic acid ABA has been suggested as one of the nonnutritional mediated mechanisms by which AM symbiosis influence stomatal conductance and other physiological traits when plants are drought stressed (Ludwig-Müller 2010). Recent studies have shown that ABA levels increased in response to water deficit and increased more in nonmycorrhizal plants than in mycorrhizal plants, suggesting that AM plants experience less intense drought stress (Doubková *et al.* 2013). Several researchers have reported that gas exchange in host plants is often related to effect of AM symbiosis on the hydration of leaves (Augé 2001). Numerous findings show that the positive effects of AM symbiosis on foliar gas exchange, the influence of these processes on leaf water potential in mycorrhizal plants subjected to drought is still unclear. Recent studies have demonstrated a higher (less negative) leaf water potential in mycorrhizal plants in water-limited conditions, which was interpreted as an AM-mediated mechanism of avoidance to mitigate the negative impact of drought on plant growth (Asrar *et al.* 2012). Leaf water potential is recognized as an index of the water status of an entire plant and hence represents a fundamental trait revealing a potentially improved resistance of plants to drought through better hydration. Hence, measurements of water use efficiency (WUE) provide an integrated measure of plant water use and thus allow a further dissection of the plant–water relations of mycorrhizal plants when water is limiting. AM symbiosis under drought conditions enhances the photochemical efficiency of photosystem II, given by Fv/Fm assessed by chlorophyll fluorescence in rice plants (Ruiz-Sánchez *et al*. 2011) and in woody tree nut species (Yooyongwech *et al*. 2013). Such results indicate the improved performance of the photosynthetic machinery and the absence of photoinhibition when mycorrhizal plants were exposed to water deficit.

5.4 Root systems and AMF

Improved nutrient uptake by AMF is a fundamental mechanism that can alleviate the adverse effects of water stress on plant growth. One of the most common explanations for the improved nutrient status in mycorrhizal plants is the enhanced absorbing surface provided by the hyphae in the soil together with the ability of fungi to take up water from soil with low water potential (Augé 2001; Ruiz-Lozano 2003). The diameter size of hyphae (2–5 μm) is one or two times smaller than the diameter size of roots (10–20 μm), a trait conferring the ability to access very small soil pores that retain water and nutrients as soil dries. This allows to bypass the zones of water and nutrient depletion around the roots and, thus, a more extensive exploration of the soil (Miransari *et al*. 2007; Smith *et al*. 2010, 2011) that in turn may induce dense growth of roots (Miransari *et al*. 2007). AM symbiosis is considered the most common strategy for enhancing P availability in the soil or P uptake capacity (Smith *et al*. 2011). The fundamental contribution of P nutrition in the promotion of plant growth by AM symbiosis is well documented, but little information is available on the role of nitrogen (N) nutrition in the AM-mediated responses of plants to environmental limiting conditions, including drought. Even though few studies have investigated N uptake, an increased uptake of ammonium by fungal hyphae and the significant transfer of N from the fungus to the roots have been demonstrated (He *et al*. 2003), especially under drought conditions (Subramanian and Charest 1999). Addition to the effects of AM symbiosis on plant–water relations where AMF act independently and directly on nutrient and water uptake, AM symbiosis could increase drought resistance in plants through secondary actions such as the improvement of soil structural stability that in turn increases the retention of soil water (Augé 2001; Ruiz-Lozano 2003). AMF hyphae can enhance soil structure through the entanglement of soil particles to form aggregates and through the production of the glycoprotein glomalin (Rillig and Mummey 2006; Singh *et al*. 2011). AMF, in part due to their filamentous structure, also influence the development of soil structure both in the rhizosphere and in bulk soil (Miransari *et al*. 2007). Augé *et al*. (2001) reported that the soil in which mycorrhizal plants were grown was characterized by more water-stable aggregates and substantially higher extraradical hyphal densities than the soils of nonmycorrhizal plants, and this pattern correlated well with the improved retention of moisture of the mycorrhizal soil. By binding roots to the soil, fungal hyphae may even maintain liquid continuity and limit the loss of hydraulic conductivity caused by air gaps (Augé 2001).

5.5 Aquaporins

Aquaporins are membrane intrinsic proteins present in all living organisms, including fungi (Agre *et al*. 1993), that facilitate the transport of water and another small uncharged molecules across cell membranes following a gradient (Maurel 2007; Maurel *et al*. 2008). In plants, they can be divided into five subgroups according to their amino acid sequence similarity: PIPs, plasma membrane intrinsic proteins; TIPs, tonoplast intrinsic proteins; NIPs, nodulin-26-like intrinsic proteins; SIPs, small and basic intrinsic proteins; and the uncharacterized intrinsic proteins (XIPs) (Chaumont *et al*. 2000; Sakurai *et al*. 2005; Park *et al*. 2010). AM fungi have been shown to transfer water from the soil to the root of the host plants (Ruth *et al*. 2011) and to regulate root hydraulic properties through the regulation of plant aquaporins (Ruiz-Lozano and Aroca 2010). The role of AM fungi in water transport and aquaporin function and regulation is poorly understood (Lehto and Zwiazek 2011). Fungal mycelia contain their own aquaporins, although little is known about their contribution to the water transport of mycorrhizal plants (Li *et al*.2013). Arbuscular mycorrhizal fungi will alter the aquaporin expression of the plants they are colonizing by increasing (Aroca *et al*. 2007), decreasing (Jahromi *et al*. 2008), or with no effect (Aroca *et al*. 2007; Jahromi *et al*. 2008) on them. These results support the idea that each aquaporin has its specific function under each environmental stress condition (Aroca *et al*. 2007) and that each plant will respond differently to each colonizing fungi. AM regulation of plant aquaporin genes under drought stress generally improves plant water status and drought tolerance (Aroca *et al*. 2007; Aroca and Li *et al*. 2013). In particular, the expression of genes encoding aquaporins has been demonstrated (Uehlein *et al*. 2007), and an aquaporin has been identified in AM fungal structures, both in the periarbuscular membrane and the extraradical mycelia (Aroca *et al*. 2009; Li *et al*. 2013). Both plant and fungal aquaporins are affected by stresses, including drought (Li *et al*. 2013). An earlier study in *Phaseolus vulgaris* inoculated with *G. intraradices* found the commonly observed positive AM-mediated effect on plant water content but also found different effects of AM plant responses to drought on the regulation of aquaporins (Aroca *et al*. 2007). The authors observed a lower expression of aquaporin genes in roots of mycorrhizal plants compared to nonmycorrhizal plants under drought conditions, suggesting that a mechanism of water conservation was employed by the AM plants.

5.6 Antioxidant enzymes-phytohormones and their mechanisms

Plants have developed various protective mechanisms to eliminate or reduce ROS, which are effective at different levels of stress-induced deterioration (Beak and Skinner, 2003).The enzymatic antioxidant system is one of the protective mechanisms including superoxide dismutase (SOD), which can be found in various cell compartments and it catalyses the disproportion of two O_2 radicals to H_2O_2 and O_2 (Scandalios, 1993). H_2O_2 is eliminated by various antioxidant enzymes such as catalases (CAT) (Scandalios, 1993) and peroxidases (POX) (Gara et al. 2003) which convert H_2O_2 to water. Other enzymes that are very important in the ROS scavenging system and function in the ascorbate-glutathione cycle are glutathione reductase (GR), monodehydro ascorbat reductase (MDHAR) and dehydroascorbate reductase (DHAR) (Candan and Tarhan 2003). Moreover, ROS are inevitable byproducts of normal cell metabolism (Martinz et al. 2001). But under normal conditions production and destruction of ROS is well regulated in cell metabolism (Mittler 2002). When a plant faces harsh conditions, ROS production will overcome scavenging systems and oxidative stress will burst. In these conditions, ROS attack vital biomolecules and disturb the cell metabolism and ultimately the cell causes its own death (Sakihama et al. 2002). These radicals are harmful at high concentrations causing oxidative damage to biomolecules, denaturation of proteins, and DNA mutations (Bowler et al. 1991). The amelioration of stress resistance by AM symbiosis is often related to the enhancement of antioxidant levels or activities in plants (Baslam and Goicoechea 2012). Ruiz- Sánchez et al. (2011) found that AM symbiosis ameliorated the response of plants to drought and salinity by improving photosynthetic performance but mainly through the accumulation of the antioxidant compound glutathione, which was concomitant with a reduction in oxidative damage to membrane lipids and to low cellular levels of hydrogen peroxide. In the same study, while glutathione levels increased, ascorbate levels decreased in mycorrhizal plants compared to nonmycorrhizal counterparts. This comprehensive study further supports the premise that mycorrhizal protection against drought and salinity-induced oxidative stress may be a crucial mechanism by which AM symbiosis increases the resistance of host plants to drought and salinity (Ruiz-Lozano 2003). Other potential ROS scavengers, flavonoids might also play a role in protecting mycorrhizal plants against oxidative damage: AM-mediated increases in the amounts of these compounds were sometimes found when plants were exposed to drought conditions (Abbaspour et al. 2012). AM symbiosis affected the allocation of carbon resources to different classes of isoprenoids such as the volatile nonessential isoprenoids (monoterpenes and sesquiterpenes) and the nonvolatile essential isoprenoids (abscisic acid (ABA), chlorophylls, and carotenoids (Asensio et al. 2012). By subjecting tomato plants to stressors such as drought and to an exogenous application of jasmonic acid, the AM symbiotic interaction in conditions where isoprenoids usually play a role in resistance to stress and in plant defense (Cazzonelli and Pogson 2010). These mechanisms that control the antioxidant capacity of AM plants are still not fully understood, but there have been some very interesting studies in AMF uncovering the genes that encode different proteins involved in the cellular defense against oxidative stress.

Hormonal regulation is achieved through a complex regulatory network that connects the different hormonal pathways enabling each hormone to assist or antagonize the others (Peleg and Blumwald 2011). Auxin, brassinosteroid, cytokinin (CK), and gibberellins (GA) are major developmental growth regulators, while abscisic acid (ABA), ethylene (ET), jasmonic acid (JA), salicylic acid (SA), and strigolactones (SL) are often implicated in stress responses. In general, ABA and JA frequently show similar biological effects (Aroca et al. 2013). It is known that low levels of ET and SA will promote root AM colonization (Herrera-Medina et al. 2003; Riedel et al. 2008) and that there is a relationship between the multipurpose signaling molecules of ethylene and the generation of ROS, but the interaction mechanism remains unclear (Wang et al. 2002).

Conclusion

To summarize, mycorrhizal plants employ various protective mechanisms to counteract drought and salinity stress. Considerable progress has been made in understanding the role of AM symbiosis in conferring drought and salinity resistance to plants, but different aspects still require attention for unraveling novel metabolites and hidden metabolic pathways. The effects of AMF in plant roots under these stresses are still not well understood although it is known that their presence will alter plant responses at the molecular, physiological, and biochemistry levels. The different responses found under salt and drought is usually related with the plant/host affinity and the different environmental circumstances of the study that would lead to differences on the rates of colonization at the root level. Exposure of plants inoculated with AMF to salinity and drought resulted in significant induction of antioxidative enzyme activities such as SOD, POX and CAT that could help the plants protect themselves from the oxidative effects of the ROS.

References

Abbaspour H, Saeid-Sar S, Afshari H, Abdel-Wahhab MA (2012) Tolerance of mycorrhiza infected Pistachio (*Pistacia vera* L.) seedlings to drought stress under glasshouse conditions. J. Plant Physiol. 169,704–709.

Adiku G, Renger M, Wessolek G, Facklam M, Hech-Bischoltz C (2001) Simulation of dry matter production and seed yield of common beans under varying soil water and salinity conditions. Agric. Water Manag. 47, 55–68.

Agre P, Sasaki S, Chrispeels MJ (1993) Aquaporins: a family of water channel proteins. Am. J. Physiol. 265, 461.

Aliasgharzadeh N, Rastin NS, Towfighi H, Alizadeh A (2001) Occurrence of arbuscular mycorrhizal fungi in saline soils of the Tabriz Plain of Iran in relation to some physical and chemical properties of soil. Mycorrhiza 11,119–122.

Al-Karaki GN, Al-Raddad A (1997) Effect of arbuscular mycorrhizal fungi and drought stress on growth and nutrient uptake of two wheat genotypes differing in drought resistance. Mycorrhiza 7, 83–88.

Al-Karaki GN, Clark RB (1998) Growth, mineral acquisition, and water use by mycorrhizal wheat grown under water stress. J. Plant Nutr. 21,263–276.

Al-Karaki GN (2000) Growth of mycorrhizal tomato and mineral acquisition under salt stress. Mycorrhiza 10,51–54.

Al-Karaki GN (2006) Nursery inoculation of tomato with arbuscular mycorrhizal fungi and subsequent performance under irrigation with saline water. Sci. Hortic. 109, 1–7.

Aroca R, Porcel R, Ruiz-Lozano JM (2007) How does arbuscular mycorrhizal symbiosis regulate root hydraulic properties and plasma membrane aquaporin in Phaseolus vulgaris under drought, cold or salinity stresses? New Phytol. 173,808–816.

Aroca R, Bago A, Sutka M, Paz JA, Cano C, Amodeo G (2009) Expression analysis of the first arbuscular mycorrhizal fungi aquaporin described reveals concerted gene expression between salt-stressed and nonstressed mycelium. Mol. Plant Microbe. Interact. 22, 1169–1178.

Aroca R, Porcel R, Ruiz-Lozano JM (2012) Regulation of root water uptake under abiotic stress conditions. J. Exp. Bot. 63, 43–57.

Aroca R, Ruiz-Lozano JM, Zamarren˜o AM, Paz JA, Garcý´a-Mina JM, Pozo MJ, Lo´pez-Ra´ez JA (2013) Arbuscular mycorrhizal symbiosis influences strigolactone production under salinity and alleviates salt stress in lettuce plants. J. Plant Physiol. 170, 47–55.

Asensio D, Rapparini F, Peñuelas J (2012) AMF root colonization increases the production of essential isoprenoids vs nonessential isoprenoids especially under drought stress conditions or after jasmonic acid application. Phytochem. 77, 149–161.

Asrar AA, Abdel-Fattah GM, Elhindi KM (2012) Improving growth, flower yield, and water relations of snapdragon (Antirhinum majus L.) plants grown under well-watered and waterstress conditions using arbuscular mycorrhizal fungi. Photosynthetica. 50,305–316.

Atkinson D (1994) Impact of mycorrhizal colonization on root architecture, root longevity and the formation of growth regulators. In: Impact of Arbuscular Mycorrhizas on Sustainable Agriculture and Natural Ecosystem (Ed) Gianinazzi S and Schuepp H, Springer-Birkhäuser, Basel, pp 89-99.

Augé, RM (2001) Water relations, drought and vesicular-arbuscular mycorrhizal symbiosis. Mycorrhiza. 11, 3–42.

Augé RM, Moore JL (2005) Arbuscular mycorrhizal symbiosis and plant drought resistance. In: Mehrotra VS (ed) Mycorrhiza: role and applications. Allied Publishers Limited, New Delhi, pp 136–157.

Bagyaraj DJ, and Padmavathi T Ravindra (1993) Mycorrhiza. In: Organics in Soil Health and Crop Production (Ed) Thampan PK, Peekay Tree Crops Development Foundation, Cochin, pp 185-200.

Ba´rzana G, Aroca R, Paz JA, Chaumont F, Martinez-Ballesta M, Carvajal M, Ruiz-Lozano JM (2012) Arbuscular mycorrhizal symbiosis increases relative apoplastic water flow in roots of the host plant under both well-watered and drought stress conditions. Ann. Bot. 109, 1009–1017.

Baslam M, Goicoechea N (2012) Water deficit improved the capacity of arbuscular mycorrhizal fungi (AMF) for inducing the accumulation of antioxidant compounds in lettuce leaves. Mycorrhiza 22, 347–359.

Beak KH, Skinner DZ (2003) Alteration of antioxidant enzyme gene expression during cold acclimation of near-isogenic wheat lines, Plant Sci. 165, 1221-1227.

Blumwald E, Aharon GS and Apse MP (2000) Sodium transport in plant cells. Biochemica et Biophysica Acta. 1465, 140-151.

Bohnert HJ, Nelson DE, Jensen RG (1995) Adaptations to environmental stresses. Plant Cell, 7, 1099-1011.

Bohnert HJ, Jensen RG (1996) Strategies for engineering water-stress tolerance in plants. Trends Biotechnol. 14, 89-97.

Bonfante P, Genre A (2010) Mechanisms underlying beneficial plant–fungus interactions in mycorrhizal symbiosis. Nat. Commun. 1(48), 1-11.

Bowler C, Slooten L, Vandenbranden S, De Rycke R, Botterman J, Sybesma C, Van Montagu M, Inze D (1991) Manganese superoxide dismutase can reduce cellular damage mediated by oxygen radicals in transgenic plants. EMBO J. 10, 1723–1732.

Bray DE (1997) Plant responses to water deficit. Trends Plant Sci. 2, 48-54.

Candan N, Tarhan L (2003) The correlation between antioxidant enzyme activities and lipid peroxidation levels in *Mentha pulegium* organs grown in Ca^{2+}, Mg^{2+}, Cu^{2+}, Zn^{2+} and M^{n2+} stress conditions. Plant Sci. 163, 769-779.

Cazzonelli CI, Pogson BJ (2010) Source to sink: regulation of carotenoid biosynthesis in plants. Trends Plant Sci. 15,266–274.

Chaumont F, Barrieu F, Jung R, Chrispeels MJ (2000) Plasma membrane intrinsic proteins from maize cluster in two sequence subgroups with differential aquaporin activity. Plant Physiol. 122, 1025–1034.

Compant S, van der Heijden MG, Sessitsch A (2010) Climate change effects on beneficial plant–microorganism interactions. FEMS Microbiol. Ecol. 73,197–214.

Corratge´ C, Zimmermann S, Lambilliotte RRL, Plassard C, Marmeisse R, Thibaud JB, Lacombe B, Sentenac H (2007) Molecular and functional characterization of a Na^+-K^+ transporter from the Trk family in the ectomycorrhizal fungus Hebeloma cylindrosporum. J. Biol. Chem. 282, 26057–26066.

Doubková P, Vlasáková E, Sudová R (2013) Arbuscular mycorrhizal symbiosis alleviates drought stress imposed on Knautia arvensis plants in serpentine soil. Plant Soil. 370, 149-161.

Ellis JR, Larsen HJ, Boosalis MG (1985) Drought resistance of wheat plants inoculated with vesicular-arbuscular mycorrhizae. Plant Soil 86, 369–378.

Evelin H, Kapoor R, Giri B (2009) Arbuscular mycorrhizal fungi in alleviation of salt stress: a review. Ann. Bot. 104, 1263–1280.

Facelli E, Smith SE, Facelli JM, Christophersen HM, Andrew SF (2010) Underground friends or enemies: model plants help to unravel direct and indirect effects of arbuscular mycorrhizal fungi on plant competition. New Phytol. 185, 1050–61.

Farooq M, Wahid A, Kobayashi N, Fujita D, Basra SMA (2009) Plant drought stress: effects, mechanisms and management. Agron. Sustain. Dev. 29, 185–212.

Feng G, Zhang FS, Li Xl, Tian CY, Tang C, Rengel Z (2002) Improved tolerance of maize plants to salt stress by arbuscular mycorrhiza is related to higher accumulation of soluble sugars in roots. Mycorrhiza 12, 185–190.

Flowers TJ, Colmer TD (2008) Salinity tolerance in halophytes. New Phytol. 179, 945–963.

Gara, LD, Pinto MC, Tommasi F (2003) The antioxidant systems vis-á-vis reactive oxygen species during plant-pathogen interaction. Plant Physiol. Biochem. 41, 863-870.

Giri B, Kapoor R, Mukerji KG (2003) Influence of arbuscular mycorrhizal fungi and salinity on growth, biomass and mineral nutrition of Acacia auriculiformis. Biol. Fert. Soils 38, 170–175.

Giri B, Kapoor R, Mukerji KG (2007) Improved tolerance of Acacia nilotica to salt stress by arbuscular mycorrhiza, Glomus fasciculatum may be partly related to elevated K/Na ratios in root and shoot tissues. Microbial Ecol. 54, 753–760.

Göhre V, Paszkowski U (2006) Contribution of the arbuscular mycorrhizal symbiosis to heavy metal phytoremediation. Planta 223, 1115–22.

Goicoechea N, Szalai G, Antolín MC, Sánchez-Díaz M, Paldi E (1998) Influence of arbuscular mycorrhizae and Rhizobium on free polyamines and proline levels in water-stressed alfalfa. J. Plant Physiol. 153, 706–711.

Graham JH, Syvertsen JP (1989) Vesicular-arbuscular mycorrhizas increase chloride concentration in citrus seedlings. New Phytol. 113, 29–36.

Hammer EC, Nasr H, Pallon J, Olsson PA, Wallander H (2011) Elemental composition of arbuscular mycorrhizal fungi at high salinity. Mycorrhiza 21, 117–129.

Herrera-Medina MJ, Gagnon H, Piche' Y, Ocampo JA, Garcý'a-Garrido JM, Vierheilig H (2003) Root colonization by arbuscular mycorrhizal fungi is affected by the salicylic acid content of the plant. Plant Sci. 164, 993–998.

He XH, Critchley C, Bledsoe C (2003) Nitrogen transfer within and between plants through common mycorrhizal networks (CMNs). Crit. Rev. Plant Sci. 22, 531–567.

Jahromi F, Aroca R, Porcel R, Ruiz-Lozano JM (2008) Influence of salinity on the in vitro development of Glomus intraradices and on the in vivo physiological and molecular responses of mycorrhizal lettuce plants. Microbial Ecol. 55, 45–53.

Juniper S, Abbott LK (1993) Vesicular–arbuscular mycorrhizas and soil salinity. Mycorrhiza 4, 45–57.

Juniper S, Abbott LK (2006) Soil salinity delays germination and limits growth of hyphae from propagules of arbuscular mycorrhizal fungi. Mycorrhiza 16, 371–379.

Kiers ET, Duhamel M, Beesetty Y, Mensah JA, Franken O, Verbruggen E (2011) Reciprocal Rewards Stabilize Cooperation in the Mycorrhizal Symbiosis. Science 333, 880–882.

Kishor PB, Hong Z, Miao GH, Hu CA, Verma DPS (1995) Overexpression of Ä1-pyrroline-5-carboxylate synthetase increases proline production and confers osmotolerance in transgenic plants. Plant Physiol. 108, 1387–1394.

Kishor PKB, Sangam S, Amrutha RN, Laxmi PS, Naidu KR, Rao KRSS (2005) Regulation of proline biosynthesis, degradation, uptake and transport in higher plants: its implications in plant growth and abiotic stress tolerance. Curr. Sci. 88, 424–438.

Läuchli, A. and E. Epstein. (1990). Plant responses to saline and sodic conditions. In K.K. Tanji (ed).Agricultural salinity assessment and management. ASCE manuals and reports on engineering practice No., 71. pp 113–137 ASCE New York.

Lehto T, Zwiazek JJ (2011) Ectomycorrhizas and water relations of trees: a review. Mycorrhiza 21, 71–90.

Li T, Hu YJ, Hao ZP, Li H, Wang YS, Chen BD (2013) First cloning and characterization of two functional aquaporin genes from an arbuscular mycorrhizal fungus Glomus intraradices. New Phytol. 197, 617–630.

Li X, Zhou W, Cao Y (1994) Acquisition of phosphorus by VA-mycorrhizal hyhpaefrom the dense soil. Plant Nutr. Fertil. Sci. 1, 55-6.

Lindermann RG. (1994) Role of VAM in biocontrol. In: Pfleger FL, Linderman RG. eds. Mycorrhizae and plant health. St. Paul: American Phytopathological Society, 1–26.

Leyval C, Joner EJ, Del Val C, Haselwandter K (2002) Potential of arbuscular mycorrhizal fungi for bioremediation. In: Mycorrhizal Technology in Agriculture (Ed) Gianinazzi S, Schüepp H, Barea JM and Haselwandter K, Springer- Birkhäuser, Basel, pp 175–186.

Ludwig-Müller J (2010) Hormonal responses in host plants triggered by arbuscular mycorrhizal fungi. In: Arbuscular Mycorrhizas: Physiology and Function (Ed) Koltai H and Kapulnik Y, Springer, New York, pp 169–190.

Maathuis FJM, Verlin D, Smith FA, Sanders D, Ferneáßndez JA and Walker NA (1996) The physiological relevance of Na^+-coupled K^+-transport. Plant Physiol. 112, 1609-1616.

Maeshima M (2001) Tonoplast transporters: Organization and function. Annu. Rev. Plant. Physiol. Plant Mol. Biol. 52, 469-497.

Marschner H Dell B. (1994) Nutrient uptake in mycorrhizal symbiosis. Plant Soil 159, 89-102.

Marschner H (1995) Mineral Nutrition of Higher Plants. Second edition. Academic Press, London, pp. 388–390.

Maurel C (2007) Plant aquaporins: novel functions and regulation properties. FEBS Lett. 581, 2227–2236.

Maurel C, Verdoucq L, Luu D-T, Santoni V (2008) Plant aquaporins: membrane channels with multiple integrated functions. Annu. Rev. Plant Biol. 59, 595–624.

Miransari M, Bahrami HA, Rejali F, Malakouti MJ, Torabi H (2007) Using arbuscular mycorrhiza to reduce the stressful effects of soil compaction on corn (*Zea mays* L.) growth. Soil Biol. Biochem. 39, 2014–2026.

Miransari M (2010) Contribution of arbuscular mycorrhizal symbiosis to plant growth under different types of soil stress. Plant Biol. 1, 563–569.

Mittler R, (2002) Oxidative stress, antioxidants and stress tolerance, Trends Plant Sci. 7, 405-410.

Morsomme P and Boutry M (2000) The plant plasma membrane H(+)-ATPase: structure, function and regulation. Biochem. Biophys. Acta 1465, 1-16.

Munns R (2002) Comparative physiology of salt and water stress. Plant Cell Environ. 25, 239–250.

Neumann PM (2008) Coping mechanisms for crop plants in drought-prone environments. Ann Bot. 101, 901–907.

Olfas R, Eljakaoui Z, Pardo JM, Belver A (2009) The Na^+/H^+ exchanger SOS1 controls extrusion and distribution of Na^+ in tomato plants under salinity conditions. Plant Signal Behav. 4, 973–976.

Parniske M (2008) Arbuscular mycorrhiza: the mother of plant root endosymbioses. Nature Rev. Microbiol. 6, 763–75.

Parry MAJ, Flexas J, Medrano H (2005) Prospects for crop production under drought: research priorities and future directions. Ann. Appl. Biol. 147, 211–226

Peng S, Ismail AM (2004) Physiological basis of yield and environmental adaptation in rice. In: Physiology and Biotechnology Integration for Plant Breeding, (Ed) Blum A, Nguyen H, Marcel Dekker, New York, pp 83–140.

Padmavathi T and Ranjini R (2011) Effect of Arbuscular mycorrhizal fungi on the growth of *Ocimum sanctum* and glomalin a soil related protein. Res. J. Biotechnol. 6(4), 44-50.

Padmavathi T, Rashmi D and Swetha S (2015) Effect of Rhizophagus spp and plant growth promoting *Acinetobacter junii* on *Solanum lycopersicon* and *Capsicum annuum*. Braz. J. Bot. 38 (2), 273-280.

Peleg Z, Blumwald E (2011) Hormone balance and abiotic stress tolerance in crop plants. Curr. Opin. Plant Biol. 14, 290–295.

Porcel R, Ruiz-Lozano JM (2004) Arbuscular mycorhhizal influence on leaf water potential, solute accumulation and oxidative stress in soybean plants subjected to drought stress. J. Exp. Bot. 55, 1743–1750.

Reynolds MP, Mujeeb-Kazi A, Sawkins M (2005) Prospects for utilizing plant-adaptive mechanisms to improve wheat and other crops in drought and salinity-prone environments. Ann. Appl. Biol. 146, 239–259.

Riedel T, Groten K, Baldwin IT (2008) Symbiosis between Nicotiana attenuate and *Glomus intraradices*: ethylene plays a role, jasmonic acid does not. Plant Cell Environ. 31, 1203–1213.

Rillig MC, Mummey DL (2006) Mycorrhizas and soil structure. New Phytol. 171, 41–53.

Ruiz-lozano JM, Azcon LR, Gomez M (1995) Effect of Arbuscular-Mycorrhizal *Glomus* Species on Drought Tolerance: Physiological and Nutritional Plant Responses. Appl. Environ. Microbiol. pp 456–460.

Ruiz-Lozano, J.M. and Azcón, R. (2000) Symbiotic efficiency and infectivity of an autochthonous arbuscular mycorrhizal *Glomus sp.* from saline soils and *Glomus deserticola* under salinity. Mycorrhiza 10, 137-143.

Ruiz-Lozano JM (2003) Arbuscular mycorrhizal symbiosis and alleviation of osmotic stress. New perspectives for molecular studies. Mycorrhiza13, 309–317.

Ruiz-Lozano JM, Aroca R (2010) Modulation of aquaporin genes by the arbuscular mycorrhizal symbiosis in relation to osmotic stress tolerance. In: Symbiosis and Stress (Ed) Sechback J and Grube M, Springer, Berlin. pp 357-374.

Ruiz-Lozano JM, Porcel R, Azcón C, Aroca R (2012) Regulation by arbuscular mycorrhizae of the integrated physiological response to salinity in plants: new challenges in physiological and molecular studies. J. Exp. Bot. 63, 4033–44.

Ruiz-Sánchez M, Armada E, Muñoz Y, de Salamone IEG, Aroca R, Ruiz-Lozano JM *et al* (2011) *Azospirillum* and arbuscular mycorrhizal colonization enhanced rice growth and physiological traits under well-watered and drought conditions. J. Plant Physiol. 168, 1031–1037.

Ruth B, Khalvati M, Schmidhalter U (2011) Quantification of mycorrhizal water uptake via high resolution on-line water content sensors. Plant Soil 342, 459–46.

Sakihama Y, Cohen MF, Grace SC, Yamasaki H (2002) Plant phenolic antioxidant and prooxidant activities: Phenolics-induced oxidative damage mediated by metals in plants. Toxicology 177, 67-80.

Sakurai J, Ishikawa F, Yamaguchi TM, Maeshima M (2005) Identification of 33 rice aquaporin genes and analysis of their expression and function. Plant Cell Physiol. 46, 1568–1577.

Scandalios JG (1993) Oxygen stress and superoxide dismutase. Plant Physiol. 101, 712-726.

Schachtman DP, Goodger JQD (2008) Chemical root to shoot signaling under drought. Trends Plant Sci. 13, 281–287.

Samaras Y, Bressan RA, Csonka LN, Garcia-Rios M, Paino D'Urzo M, Rhodes D (1995) Proline accumulation during water deficit. In: Environment and Plant Metabolism. Flexibility and acclimation (Ed) Smirnoff N, Bios Scientific, Oxford. pp 161-187.

Sheng M, Tang M, Chen H, Yang BW, Zhang FF, Huang YH (2008) Influence of arbuscular mycorrhizae on photosynthesis and water status of maize plants under salt stress. Mycorrhiza 18, 287–296.

Sheng M, Tang M, Zhang F, Huang Y (2011) Influence of arbuscular mycorrhiza on organic solutes in maize leaves under salt stress. Mycorrhiza. 21, 423–430.

Smith SE, Read DJ (1995) Mycorrhizal symbiosis, Academic Press, New York.

Smith SE, Facelli E, Pope S, Smith FA (2010) Plant performance in stressful environments: interpreting new and established knowledge of the roles of arbuscular mycorrhizal. Plant Soil 326, 3–20.

Smith SE, Jakobsen I, Grønlund M, Smith FA (2011) Roles of arbuscular mycorrhizas in plant phosphorus nutrition: interactions between pathways of phosphorus uptake in arbuscular mycorrhizal roots have important implications for understating and manipulating plant phosphorus acquisition. Plant Physiol. 156, 1050–1057.

Smith SE, Smith FA(2011) Roles of arbuscular mycorrhizas in plant nutrition and growth: new paradigms from cellular to ecosystem scales. Annu. Rev. Plant Biol. 62, 227–50.

Subramanian KS, Charest C (1999) Acquisition of N by external hyphae of an arbuscular mycorrhizal fungus and its impact on physiological responses in maize under drought-stressed and well-watered conditions. Mycorrhiza 9, 69–75.

Tavakkoli E, Rengasamy P, McDonald GK (2010) High concentrations of Na^+ and Cl^- ions in soil solution have simultaneous detrimental effects on growth of faba bean under salinity stress. J. Exp. Bot. 61, 4449–4459.

Thomson BD, Clarkson DT, Brain P (1990) Kinetics of phosphorus uptake by the germ tubes of the vesicular arbuscular mycorrhizal fungus, Gigaspora margarita. New Phytol. 116, 647–653.

Tyerman SD, Skerrett IM (1999) Root ion channels and salinity. Sci Hortic. 78, 175–235.

Uehlein N, Fileschi K, Eckert M, Bienert G, Bertl A, Kaldenhoff R (2007) Arbuscular mycorrhizal symbiosis and plant aquaporin expression. Phytochemistry 68, 122–129.

Veiga RSL, Jansa J, Frossard E, Van der Heijden MGA (2011) Can arbuscular mycorrhizal fungi reduce the growth of agricultural weeds? PLoS ONE. 6, e27825.

Wang B, Qiu Y-L (2006) Phylogenetic distribution and evolution of mycorrhizas in land plants. Mycorrhiza 16, 299–363.

Wang KLC, Li H, Ecker JR (2002) Ethylene biosynthesis and signaling networks. Plant Cell 14, S131–S151.

Xu G, Magen H, Tarchitzky J, Kafkaki U (2000) Advances in chloride nutrition. Adv. Agron. 68, 96–150

Yamato M, Ikeda S, IwaseK (2008) Community of arbuscular mycorrhizal fungi in a coastal vegetation on Okinawa island and effect of the isolated fungi on growth of sorghum under salt-treated conditions. Mycorrhiza 18, 241–249.

Yang HM, Zhang JH, Zhang XY (2005) Regulation mechanisms of stomatal oscillation. J. Integr. Plant Biol. 47, 1159–1172.

Ying-Ning ZOU, Qiang-Sheng WU (2011) Sodium chloride stress induced changes in leaf osmotic adjustment of trifoliate orange (*Poncirus trifoliate*) seedlings inoculated with mycorrhizal fungi. Not. Bot. Horti Agrobo. 39(2), 64-69.

Yooyongwech S, Phaukinsang N, Cha-Um S, Supaibulwatana K (2013) Arbuscular mycorrhiza improved growth performance in Macadamia tetraphylla L. grown under water deficit stress involves soluble sugar and proline accumulation. Plant Growth Regul. 69, 285–293.

Zuccarini P and Okurowska P (2008) Effects of mycorrhizal colonization and fertilization on growth and photosynthesis of sweet basil under salt stress. J. Plant Nutr. 31, 497–513.

10

Effect of Arbuscular Mycorrhizal Fungi and Abiotic Stress on Growth and Productivity of Important Cash Crops

Shinde B.P.

Abstract

Abiotic stresses lead to the overproduction of reactive oxygen species (ROS) in plants which are highly reactive and toxic and cause damage to proteins, lipids, carbohydrates and DNA which ultimately results in oxidative stress. The arbuscular mycorrhizal fungi (AMF)-symbiosis can confer a greater degree of tolerance against drought stress in many cash crops. The significance of heat shock proteins (Hsps) and chaperones in abiotic stress responses in plants, and co-operation among their different classes and their interactions with other stress-induced components is well known. Soil salinity is a major abiotic stress adversely affecting plant growth and crop production worldwide. Phyto-mycoremediation utilizing AMF as plant inoculants is regarded as a promising strategy to heavy metal pollution remediation. Plants have stress specific adaptive responses as well as responses which protect the plants from more than one environmental stress. The potential of phytoremediation of contaminated soil can be enhanced by inoculating hyper-accumulator plants with mycorrhizal fungi most appropriate for the contaminated site.

Keywords: Abiotic stress, Arbuscular mycorrhizal fungi, Cash crops, Plant productivity

1. Introduction

Abiotic stress is defined as the negative impact of non-living factors on the living organisms in a specific environment. The non-living variable must influence the environment beyond its normal range of variation to adversely affect the population performance or individual physiology of the organism in a significant way. Abiotic stress factors or stressors are naturally occurring, often intangible, factors such as intense sunlight or wind that may cause harm to the plants and animals in the area affected, e.g., water stress, salt stress, wind stress, heat stress, drought stress, nutrient stress, etc. Stresses can be studied classifying into the following different kinds:

i) **Drought stress:** It occurs when the available water in the soil is reduced and atmospheric conditions cause continuous loss of water by transpiration or evaporation. Drought stress tolerance is seen in almost all plants but its extent varies from species to species and even within species. Drought stress is characterized by reduction of water content, diminished leaf water potential and turgor loss, closure, nutrient metabolism and growth promoters.

ii) **Water stress:** Water stress may result in the arrest of photosynthesis, disturbance of metabolism and finally the death of plant. Water stress inhibits cell enlargement more than cell division. It reduces plant growth by affecting various physiological and biochemical processes, such as photosynthesis, respiration, translocation, ion uptake, carbohydrates, nutrient metabolism and growth promoters.

iii) **Salt stress:** There are wild plants that thrive in the saline environments along the sea shore, in estuaries and saline deserts. These plants, called halophytes, have distinct physiological and anatomical adaptations to counter the dual hazards of water deficit and ion toxicity. Salinity can affect any process in the plant's life cycle, so that tolerance will involve a complex interplay of characters.

iv) **Heat stress:** Heat stress often is defined as where temperatures are

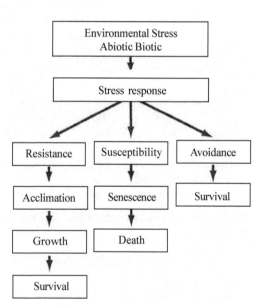

Fig. 1: Environmental stress and response by plants (*Source:* http:www.geo.arizona.edu/gallery/US/tuc_2.html)

hot enough for sufficient time that they cause irreversible damage to plant function or development. In addition, high temperatures can increase the rate of reproductive development, which shortens the time for photosynthesis to contribute to fruit or seed production.

v) **Wind stress:** Wind stress is visual stress incurred by wind causes damage to seedlings, breaking branches and even uprooting the whole plants. It helps to covert ground fires to crown fires.

vi) **Nutrient stress:** Nutrient stress is one of the major stresses causing dying back in plants. It is because of shortage of nutrients in soil such as nitrogen, phosphorus, iron, etc.

Abiotic stress responses are important for organisms such as plants because this type of organism cannot survive unless they are able to cope with environmental changes. The term 'abiotic stress' includes numerous stresses caused by complex environmental conditions, e.g. strong light, UV, high and low temperatures, freezing, drought, salinity, heavy metals and hypoxia (Hirayama et al. 2010). These stresses will increase in the near future because of global climate change, according to reports from the Intergovernmental Panel of Climate Change (http://www.ipcc.ch).

Abiotic stresses, especially salinity and drought, are the primary causes of crop loss worldwide. Plant adaptation to environmental stresses is dependent upon the activation of cascades of molecular networks involved in stress perception, signal transduction, and the expression of specific stress-related genes and metabolites (Vinocur et al. 2005). Studies on novel regulatory mechanisms involving use of small RNA molecules, chromatin modulation and genomic DNA modification have enabled us to recognize that plants have evolved complicated and sophisticated systems in response to complex abiotic stresses. (Hirayama et al. 2010)

Mycorrhizal associations vary widely in structure and function, but the most ubiquitous interaction is the arbuscular mycorrhizal (AM) symbiosis. This interaction forms between the roots of over 80% of all terrestrial plant species and Glomeromycete fungi. This ancient symbiosis confers benefits directly to the host plants growth and development through the acquisition of phosphate and other mineral nutrients from the soil by the fungus, while the fungus receives a carbon source from the host. The symbiosis may also enhance the plant's resistance to biotic and abiotic stresses. Additionally, AM fungi develop an extensive external hyphal network, which makes a significant contribution to the improvement of soil structure. Therefore, these fungi constitute an integral and important component of agricultural systems (Harrier et al. 2003). In this review, the effect of arbuscular mycorrhizal fungi (AMF) on the growth and production of important cash crops and its role in alleviating abiotic stress is discussed.

The term 'Cash crop" in Indian scenario is a crop that is cultivated for maximum profit gains. Cash crops in India form the strong base over which the Indian trade and commerce flourish both within and outside the country. The important cash crops are maize, wheat, cotton, barley, rice, sugarcane, oilseeds, pulses, vegetables etc. The cash crop varies from region to region due to the diverse climatic conditions in India.

2. Types of stress and crops

Gill *et al.* (2010) concluded that various abiotic stresses lead to the overproduction of reactive oxygen species (ROS) in plants which are highly reactive and toxic and cause damage to proteins, lipids, carbohydrates, DNA which ultimately results in oxidative stress. The antioxidant defense machinery protects plants against oxidative stress damages. Plants possess very efficient enzymatic [superoxide dismutase (SOD); catalase (CAT); ascorbate peroxidase (APX); glutathione reductase (GR); monodehydroascorbate reductase (MDHAR); dehydroascorbate reductase (DHAR); glutathione peroxidase (GPX); guaicol peroxidase (GOPX) and glutathione-*S*- transferase (GST)] and non-enzymatic [ascorbic acid (ASH); glutathione (GSH)] phenolic compounds, alkaloids, non-protein amino acids and α-tocopherols, antioxidant defense systems which work in concert to control the cascades of uncontrolled oxidation and protect plant cells from oxidative damage by scavenging of ROS. ROS also influence the expression of a number of genes and therefore control many processes like growth, cell cycle, programmed cell death (PCD), abiotic stress responses, pathogen defense, systemic signaling and development (Gill *et al.* 2010). Julkowska *et al.* (2015) reviewed early signaling events, such as phospholipid signaling, calcium ion (Ca^{2+}) responses, and reactive oxygen species (ROS) production, together with salt stress-induced abscisic acid (ABA) accumulation, which are brought into the context of long-term salt stress-specific responses and alteration of plant growth. Salt-induced quiescent and recovery growth phases rely on modification of cell cycle activity, cell expansion, and cell wall extensibility. The period of initial growth arrest varies among different organs, leading to altered plant morphology.

Abiotic stresses usually cause protein dysfunction. Maintaining proteins in their functional conformations and preventing the aggregation of non-native proteins are particularly important for cell survival under stress. Heat-shock proteins (Hsps)/chaperones are responsible for protein folding, assembly, translocation and degradation in many normal cellular processes, stabilize proteins and membranes, and can assist in protein refolding under stress conditions. They can play a crucial role in protecting plants against stress by re-establishing normal protein conformation and thus cellular homeostasis (Wang *et al.* 2004). Pozo *et al.* (2007) have studied relation of pathogen attack and AM symbiosis. As per their study, during mycorrhiza formation, modulation of plant defense responses occurs, potentially through cross-talk between salicylic acid and jasmonate dependent signaling pathways. This modulation may impact plant responses to potential enemies by priming the tissues for a more efficient activation of defense mechanisms.

Lopez-Raez *et al.* (2008) have done analytical experiments to identify hormonal status of plant (tomato) during AM symbiosis. Strigolactones are exuded into the soil, where they act as host detection signals for AMF, but also as germination stimulants for root parasitic plant seeds. Under phosphate limiting conditions, plants up-regulate the secretion of strigolactones into the rhizosphere to promote the formation of AM symbiosis. As per their study, AM symbiosis induces changes in transcriptional and hormonal profiles in tomato. Bompadre *et al.* (2014) studied the transplantation stress of Olive plant inoculated with AM fungi and concluded that the coinoculation improved olive plant growth and protected against oxidative stress. A combination of two AMF strains at the beginning of olive propagation produced vigorous plants successfully protected in field cultivation even with an additional cost at the beginning of plant growth.

Table 1: Diagrammatic representation of plant and symbiotic mycorrhizal strategies involved in acquisition of soil nutrient resources of different availabilities

Strategy / Resource	Plant strategies		Symbiotic strategies	
	Roots and root hairs	Exudates/ clusters	AM[a]	ECM[b] & ERM[c]
Soluble inorganic	All nutrients	P	P, Zn, (N)	P&N
Insoluble inorganic		P	(P)	P&N
Labile/soluble organic	All nutrients	(P)	(P)	P&N
Recalcitrant organic				P&N

Modified from Smith and Read (2008)

[a]Arbuscular Mycorrhiza, [b]Ectomycorrhiza, [c]Ericoid Mycorrhiza.

3. Drought stress

Wu *et al.* (2013) reviewed the AMF-induced tolerance to drought stress in citrus. They showed that AM-symbiosis can confer a greater degree of tolerance against drought stress in citrus, however, the underlying relevant mechanistic path-ways is still not fully understood. According to Porcel *et al.* (2004) soybean plants inoculated with AM fungi were protected against drought, as shown by their significantly higher shoot-biomass production. The AM symbiosis enhanced osmotic adjustment in roots, which could contribute to maintaining a water potential gradient favorable to the water entrance from soil into the roots. This enabled higher leaf water potential in AM plants during drought and kept the plants protected against oxidative stress, and these cumulative effects increased the plant tolerance to drought.

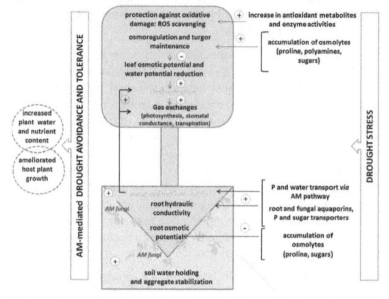

Fig. 2: Tolerance to drought stress due to AM fungi; *Source*: Rapparini *et al.* (2014)

The arid and contaminated soils are generally characterized by poor soil structure, low water holding capacity, lack of organic matter and nutrient deficiency. Therefore, in order to carry out successful reforestation, it is necessary to improve soil quality and the ability of the plant species to resist these harsh environments. In this respect, the application of organic amendments to the soil, prior to the inoculation of AM fungi, has been recommended (Medina *et al.* 2004). The beneficial effects of organic amendments include provision of plant nutrients, increased humus content and thereby increased water holding capacity, improved soil structure, and increased microbial activity (Caravaca *et al.* 2002).

Kohler *et al.* (2009) have studied positive effect of plant drought tolerance when AMF and PGPR are inoculated together. Azcon *et al.* (2013) stated that AMF induced improvements in plant root development and water uptake capacity, alteration of plant hormone balance, and protection against the oxidative stress generated by drought. Rapparini *et al.* (2014) suggested that AM plants withstand drought-induced oxidative stress by the increased production of antioxidant compounds that scavenge ROS and enhance the activities of antioxidant enzymes. AM fungal hyphae in the soil provide an efficient pathway for nutrient/ water uptake and transport, allowing a more efficient exploitation of the water and nutrient reservoirs in the soil where only fungal hyphae can grow, thereby bypassing the zones of water and nutrient depletion around the roots.

4. Temperature or Heat stress

Plants have evolved a variety of responses to elevated temperatures that minimize damage and ensure protection of cellular homeostasis. New information about the structure and function of heat stress proteins and molecular chaperones has become available. At the same time, transcriptome analysis of *Arabidopsis* has revealed the involvement of factors other than classical heat stress responsive genes in thermotolerance. Recent reports suggest that both plant hormones and reactive oxygen species also contribute to heat stress signaling. Additionally, an increasing number of mutants that have altered thermotolerance have extended our understanding of the complexity of the heat stress response in plants (Kotak *et al.* 2007). Wang *et al.* (2004) have summarized the significance of heat shock proteins (HSPs) and chaperones in abiotic stress responses in plants, and discussed the co-operation among their different classes and their interactions with other stress-induced components.

5. Salt stress

Soil salinity is a major abiotic stress adversely affecting plant growth and crop production worldwide. Increased salinization of arable land is expected to have destructive universal effects, resulting in 30% land loss within next 25 years and up to 50% by the middle of twenty-first century (Porcel *et al.* 2012; Kapoor *et al.* 2013 and Latef *et al.* 2014). Genetic analysis has defined the Salt Overly Sensitive (SOS) pathway, in which a salt stress induced calcium signal is probably sensed by the calcium binding protein SOS3 which then activates the protein kinase SOS2. The SOS3–SOS2 kinase complex regulates the expression and activity of ion transporters such as SOS1 to re establish cellular ionic homeostasis under salinity. The ICE1 (Inducer of CBF Expression 1)–CBF (C Repeat Binding Protein) pathway is critical for the regulation of the cold responsive transcriptome and acquired freezing tolerance, although at present the signaling events that activate the ICE1 transcription factor during cold stress are not known. Both ABA dependent and independent signaling pathways appear to be involved in osmotic stress tolerance. Components of mitogen activated protein kinase (MAPK) cascades may act as converging points of multiple abiotic as well as biotic stress signaling pathways (Chinnusamy *et al.* 2004).

Castor bean inoculated with fungi may enlarge the C pool of the coastal saline soil (Zhang *et al.* 2014). Yano-Melo *et al.* (2003) investigated the plant (banana) tolerance to soil salinity by inoculation with isolates of *Acaulospora scrobiculata* Trappe, *Glomus clarum* Nicolson & Schenck and *Glomus etunicatum* Becker & Gerdemann under glasshouse conditions. They found that the salt tolerance of banana as measured by leaf number and plant height increased considerably in presence of *Glomus* isolates. Inoculation with specific AMF therefore constitutes an alternative method to reduce banana plant stress caused by soil salinization. Kohler *et al.* (2009) investigated the influence of inoculation with a plant growth-promoting rhizobacterium, *Pseudomonas mendocina* Palleroni, alone or in combination with AMF, *Glomus intraradices* (Schenk & Smith) or *Glomus mosseae* (Nicol & Gerd.) Gerd. & Trappe, on antioxidant enzyme activities (catalase and total peroxidase), phosphatase activity, solute accumulation, growth and mineral nutrient uptake in leaves of *Lactuca sativa* L. cv. Tafalla affected by three different levels of salt stress. They concluded that the mycorrhizal inoculation treatments only were effective in increasing shoot biomass at the medium salinity level. Feng *et al.* (2002) have concluded after study with maize plants that their improved tolerance to salt stress is related to the higher accumulation of soluble sugars in roots.

6. Heavy metal or Nutrient stress

Among soil microorganisms, AMF are important plant symbionts living in association with the roots of most land plants (80%) and occurring also in heavy metal contaminated soils. In such soils, AMF are critical in the establishment and fitness of plants, affecting the physico-chemical characteristics of the soil and enhancing metal immobilization. For these reasons phyto-mycoremediation utilizing AMF as plant inoculants is regarded as a promising strategy to heavy metal pollution remediation (Turrini *et al.* 2012). The most commonly associated nutrient with mycorrhizal benefit is phosphorus (P) which is highly immobile in most of the soils and limiting plant growth and reproduction. The main areas where the benefits of introducing inoculum of AMF into a plant growth system will increase the growth of plant where they are lacking indigenous inoculum of AMF, the effect is subsequently to increase early growth and nutrient uptake by phosphate. This uptake and transfer of mineral elements is done by the AMF once established in agro-systems (Dodd, 2000). Sudhakara *et al.* (2002) have studied the influence of aluminium on mineral nutrition of ectomycorrhizal fungi and concluded that the mycelia biomass had decreased. Rai *et al.* (2001) studied the distribution of metals in edible parts of aquatic plants like *Trapa* and *Ipomoea*. *Pseudomonas* sp. and fungal spp. were studied to remove heavy metals from mixed industrial effluents by Patel *et al.* (2005).

The stress-severity and optimal resource allocation hypotheses predict mutualistic symbiotic benefits to increase with the degree of metabolic imbalance and environmental stress. Aghili *et al.* (2014) have gathered evidence for direct involvement of AMF as they have studied association with wheat and found positive correlations between Zn uptake from soil and frequency of fungal symbiotic nutrient exchange organelles, as well as the quantitative abundance of AMF of the genera *Funneliformis* and *Rhizophagus*. AMF have repeatedly been demonstrated to alleviate heavy metal stress of plants (Hildebrandt *et al.* 2007). They have studied the colonization of plants by AMF in heavy metal soils, the depositions of heavy metals in plant and fungal structures and the potential to use AMF-plant combinations in phytoremediation.

AMF can contribute to plant growth, particularly in disturbed or heavy metal contaminated sites, by increasing plant access to relatively immobile minerals such as P, improving soil texture by binding soil particles into stable aggregates that resist wind and water erosion, and by binding heavy metals into roots that restricts their translocation into shoot tissues. Furthermore, the fungi can accelerate the revegetation of severely degraded lands such as coal mines or waste sites containing high levels of heavy metals (Gaur *et al.* 2004). Mycorrhizal associations increase the absorptive surface area of the plant due to extramatrical fungal hyphae exploring rhizospheres beyond the root-hair zone, which in turn enhances water and mineral uptake. Alori *et al.* (2012) investigated the potential of indigenous AMF (*Scutellospora reticulata* and *Glomus pansihalos*) in Southern Guinea Savanna ecological zone of Nigeria to enhance phytoremediation of soils contaminated with Aluminium (Al) and Manganese (Mn) and concluded that *Scutellospora reticulata* and *Glomus pansihalos* have the potential for use in phytoremediation of soils polluted with Aluminium and Manganese. Turnau *et al.* (2006) carried out research within the last 15 years on the role of mycorrhizal fungi in phytoremediation of Zinc wastes located in southern Poland and showed that plants conventionally introduced in such places disappear relatively soon, while those appearing during natural succession are better adapted to harsh conditions. Khan *et al.* (2006) reviewed the role of phytochelators in making the heavy metals bio-available to the plant and their symbionts in enhancing the uptake of bio-available heavy metals. Bafeel (2008) showed that inoculation of the host plants with AMF can protect them from the potential toxicity caused by increased uptake of Pb. They also concluded that arbuscular mycorrhizae have the potential in phytoremediation of the heavy metal contaminated soils, particularly in the presence of legumes. Rufyikiria *et al.* (2004) performed experiments in which subterranean clover inoculated with the AMF *Glomus intraradices* was grown on soil containing six levels of ^{238}U in the range 0–87 mg kg^{-1}. They concluded that AMF increased the shoot dry matter and P concentration in roots and shoots, while in most cases; it decreased the Ca, Mg and K concentrations in plants. The AMF influenced U concentration in plants only in the treatment receiving 87 mg U kg^{-1} soil.

According to Carrascoa *et al*. (2006) reduced plant uptake of toxic metals, particularly lead, could be involved in the beneficial effects of AMF on plant development in Mediterranean salt marshes contaminated with mining wastes. Wang *et al*. (2007) performed pot culture experiment to study heavy metal (HM) phytoaccumulation from soil contaminated with Cu, Zn, Pb, and Cd by maize (*Zea mays L.*) inoculated with AMF and showed that consortia of AMF could benefit HM phytoextraction and therefore, show potential in the phytoremediation of HM-contaminated soils. Azcóna *et al*. (2009) studied the significance of treated agrowaste residue and autochthonous inoculates (AMF and *Bacillus cereus*) on bacterial community structure and phytoextraction to remediate soils contaminated with heavy metals and showed that the microbial inoculants and amendment used can favour plant growth and the phytoextraction process and concomitantly modify bacterial community in the rhizosphere. Wua *et al*. (2005) evaluated the effects of four biofertilizers containing AMF (*Glomus mosseae* or *Glomus intraradices*) with or without N-fixer (*Azotobacter chroococcum*), P solubilizer (*Bacillus megaterium*) and K solubilizer (*Bacillus mucilaginous*) on soil properties and the growth of *Zea mays*. They showed that the application of biofertilizer containing mycorrhizal fungus and three species of bacteria can significantly increase the growth of *Zea mays*.

According to Karimi *et al*. (2011) AMF associations are integral, functioning parts of plant roots and are widely recognized as enhancing plant growth on severely disturbed sites, including those contaminated with HM. The authors have made a detailed review highlighting the interaction between HM-contaminated soils and AMF. Singh *et al*. (2013) analysed the diversity of AMF in wheat agro-climatic regions of India and found that the species present in the agricultural fields, which are constantly subjected to human intervention, are restricted subset of those which would occur naturally in the region and even the cultural practices probably could exert a strong selective pressures on AMF communities. Ricalde (2002) reviewed the phenomenon of dispersal, distribution and establishment of AMF. They also identified the various factors which were responsible for affecting the establishment of AMF (Figure 3).

Brito *et al*. (2014) investigated whether intact extraradical mycelium (ERM) is more effective than other forms of propagule from indigenous AMF in providing protection against stress to a host plant. The response of wheat (*Triticum aestivum L.*) to Mn toxicity was studied in a two-phase greenhouse experiment. They concluded that AMF colonization starts from an intact ERM and develops faster than from others source of inoculum and this greatly enhances the role of AMF in protecting against Mn toxicity of sensitive plants like wheat. According to Reddy *et al.* (2005) AMF association is found in 80% of the plant species and most plant families are mycorrhizal except Cruciferae, Chenopodiaceae, Caryophyllaceae and Cyperaceae. A detailed review was made related to the molecular investigations carried out during the last two decades all over the world and also project areas for future research in India. Gamalero *et al.* (2009) reported that both plant growth promoting bacteria (PGPB) and AMF can be used to facilitate the process of phytoremediation and the growth of plants in metal-contaminated soils.

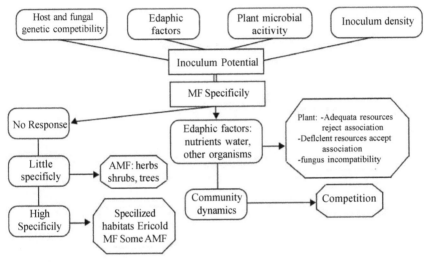

Fig. 3: Factors affecting the establishment of mycorrhizal fungi

7. Stresses in combinations

Mittler (2006) revealed that the response of plants to a combination of two different abiotic stresses is unique and cannot be directly extrapolated from the response of plants to each of the different stresses applied individually. Knight *et al.* (2001) have discussed the 'cross-talk' between different signaling pathways and question whether there are any truly specific abiotic stress signaling responses. Plants have stress specific adaptive responses as well as responses which protect the plants from more than one environmental stress. There are multiple stress perception and signaling pathways, some of which are specific, but others may cross talk at various steps (Chinnusamy *et al.* 2004). Fujita *et al.* (2006) reviewed that some studies have revealed several molecules, including transcription factors and kinases, as promising candidates for common players that are involved in crosstalk between stress signaling pathways. Emerging evidence suggests that hormone signaling pathways regulated by abscisic acid, salicylic acid, jasmonic acid and ethylene, as well as ROS signaling pathways, play key roles in the crosstalk between biotic and abiotic stress signaling. Identification of the genes involved in the production of various antioxidants and enzymes controlling the synthesis of various osmoregulators will provide further insights into the molecular basis of the mechanism (Evelins *et al.* 2009). Isolation of the indigenous and presumably stress-adapted AM fungi can be a potential biotechnological tool for inoculation of plants for successful restoration of degraded ecosystems (Mathur *et al.* 2007). The appropriate management of ecosystem services rendered by AM will impact on natural resource conservation and utilization with an obvious net gain for human society (Gianinazzi *et al.* 2010).

Conclusion

Our limited knowledge of stress-associated metabolism remains a major gap in our understanding; therefore, comprehensive profiling of stress-associated metabolites is most relevant to the successful molecular breeding of stress-tolerant crop plants. Plants exhibit a variety of responses to abiotic stresses that enable them to tolerate and survive adverse conditions. As we learn more about the signaling pathways leading to these responses, it is becoming clear that they constitute a network that is interconnected at many levels. Forward and reverse genetic analysis in combination with expression profiling will continue to uncover many signaling components, and biochemical characterization of the signaling complexes will be required to determine specificity and cross talk in abiotic stress signaling pathways. Studying stress-induced changes in growth dynamics can be used for screening to discover novel genes contributing to salt stress tolerance in model species and crops.

Tolerance to a combination of different stress conditions, particularly those that mimic the field environment, should be the focus of future research programs aimed at developing transgenic crops and plants with enhanced tolerance to naturally occurring environmental conditions. The role of salt overly sensitive (SOS) genes with respect to mycorrhizal application needs to be uncovered. The potential of phytoremediation of contaminated soil can be enhanced by inoculating hyper-accumulator plants with mycorrhizal fungi most appropriate for the contaminated site. The potential role of arbuscular mycorrhizal fungi (AMF) in phytoremediation of heavy metal contaminated soils and water is becoming evident although there is need to completely understand the ecological complexities of the plant-microbe-soil interactions and their better exploitation in remediation strategies. The effect of AMF and abiotic stress on growth and productivity was reviewed and analyzed. Using different mechanisms, the plant by itself or in association with mycorrhizal fungi can tolerate or survive the stress. However, in the presence of the fungi, plant ability to resist the stress increases as a result of morphological and physiological changes. Production of different solutes, plant hormones, antioxidant products, extensive network of the mycorrhizal plant roots, and enhanced nutrient uptake are all among the processes that make the plant to survive under stress.

References

Aghili F, Jansa J, Khoshgoftarmanesh AH, Afyuni M, Schulin R, Frossard E, Gamper HA (2014) Wheat plants invest more in mycorrhizae and receive more benefits from them under adverse than favorable soil conditions. Appl. Soil Ecol., 84, 93-111.

Alori E, Fawole O (2012) Phytoremediation of Soils Contaminated with Aluminium and Manganese by Two Arbuscular Mycorrhizal Fungi. J. Agric. Sci. 4(8), 246-252.

Azcón R, Medina A, Aroca R, Ruiz-Lozano JM (2013) Abiotic stress remediation by the arbuscular mycorrhizal symbiosis and rhizosphere bacteria/yeast interactions. In: Molecular Microbial Ecology of the Rhizosphere (Ed) de Bruijn FJ, Wiley, Hoboken, pp 991-1002.

Azcóna R, Medinab A, Roldánc A, Biród B, Vivase A (2009) Significance of treated agrowaste residue and autochthonous inoculates (Arbuscular mycorrhizal fungi and *Bacillus cereus*) on bacterial community structure and phytoextraction to remediate soils contaminated with heavy metals. Chemosphere 75(3), 327–334.

Bafeel S O (2008) Contribution of Mycorrhizae in Phytoremediation of Lead Contaminated Soils by *Eucalyptus rostrata* Plants. World Appl. Sci. J. 5(4), 490-498.

Bompadre MJ, Pérgola M, Fernández Bidondo L, Colombo RP, Silvani VA, Pardo AG, Godeas AM (2014) Evaluation of Arbuscular Mycorrhizal Fungi Capacity to Alleviate Abiotic Stress of Olive (*Olea europaea* L.) Plants at Different Transplant Conditions. Sci. World J. DOI:10.1155/2014/378950.

Brito I, Carvalho M, Alho L, Goss MJ (2014) Managing arbuscular mycorrhizal fungi for bioprotection: Mn toxicity. Soil Biol. Biochem. 68, 78-84.

Caravaca, F, Barea JM, Figueroa D, Roldán A (2002) Assessing the effectiveness of mycorrhizal inoculation and soil compost addition for enhancing reforestation with *Olea europaea* subsp. *sylvestris* through changes in soil biological and physical parameters. Appl. Soil Ecol. 20, 107-118.

Carrascoa L, Caravacaa F, Álvarez-Rogelb J, Roldána A (2006) Microbial processes in the rhizosphere soil of a heavy metals-contaminated Mediterranean salt marsh: A facilitating role of AM fungi. Chemosphere 64 (1), 104–111.

Chinnusamy V, Schumaker K, Zhu JK (2004) Molecular genetic perspectives on cross talk and specificity in abiotic stress signaling in plants. J. Expt. Bot. 55(395), 225-236.

Dodd JC (2000) The role of arbuscular mycorrhizal fungi in agro-and natural ecosystems. Outlook Agric., 29 (1), 55-55.

Evelins H, Kapoor R, Giri B (2009) Arbuscular mycorrhizal fungi in alleviation of salt stress: a review. Ann. Bot. 104(7), 1263-1280.

Feng G, Zhang F, Li X, Tian C, Tang C, Rengel Z (2002) Improved tolerance of maize plants to salt stress by arbuscular mycorrhiza is related to higher accumulation of soluble sugars in roots. Mycorrhiza 12(4), 185-190.

Fujita M, Fujita Y, Noutoshi Y, Takahashi F, Narusaka Y, Yamaguchi-Shinozaki K, Shinozaki K (2006) Crosstalk between abiotic and biotic stress responses: a current view from the points of convergence in the stress signaling networks. Curr. Opin. Plant Biol. 9(4), 436-442.

Gamalero E, Lingua G, Berta G, Glick, BR (2009) Beneficial role of plant growth promoting bacteria and arbuscular mycorrhizal fungi on plant responses to heavy metal stress. Can. J. Microbiol. 55(5), 501-514.

Gaur A and Adholeya A (2004) Prospects of arbuscular mycorrhizal fungi in phytoremediation of heavy metal contaminated soils Curr. Sci., 86(4), 528-534.

Gianinazzi S, Gollotte A, Binet MN, Van Tuinen D, Redecker D, Wipf D (2010) Agroecology: the key role of arbuscular mycorrhizas in ecosystem services. Mycorrhiza, 20(8), 519-530.

Gill SS, Tuteja N (2010) Reactive oxygen species and antioxidant machinery in abiotic stress tolerance in crop plants. Plant Physiol. Biochem. 48(12), 909-930.

Harrier LA, Watson, CA (2003) The role of arbuscular mycorrhizal fungi in sustainable cropping systems. Adv. Agron. 79, 185-225.

Hildebrandt U, Regvar M, Bothe H (2007) Arbuscular mycorrhiza and heavy metal tolerance. Phytochemistry 68(1), 139-146.

Hirayama T, Shinozaki K (2010) Research on plant abiotic stress responses in the post genome era: Past, present and future. Plant J. 61(6), 1041-1052.

Julkowska MM, Testerink, C (2015) Tuning plant signaling and growth to survive salt. Trends Plant Sci., 20(9), 586-594.

Kapoor R, Evelin H, Mathur P, Giri B (2013) Arbuscular Mycorrhiza: Approaches for abiotic stress tolerance in crop plants for sustainable agriculture, In: Plant Acclimation to Environmental Stress (Ed) Tuteja N and Gill SS, Springer, New York, pp 359-401.

Karimi Akbar, Habib Khodaverdiloo, Mozhgan Sepehri and Mirhassan Rasouli Sadaghiani (2011) Arbuscular mycorrhizal fungi and heavy metal contaminated soils. Afr. J. Microbiol. Res. 5(13), 1571-1576.

Khan AG (2006) Mycorrhizoremediation - an enhanced form of phytoremediation. J. Zhejiang Univ. – Sci. B 7(7), 503-514.

Knight H, Knight, MR (2001) Abiotic stress signaling pathways: specificity and cross-talk. Trends Plant Sci. 6(6), 262-267.

Kohler J, Hernández JA, Caravaca F, Roldán A (2009) Induction of antioxidant enzymes is involved in the greater effectiveness of a PGPR versus AM fungi with respect to increasing the tolerance of lettuce to severe salt stress. Environ. Expt. Bot. 65(2), 245-252.

Kotak S, Larkindale J, Lee U, von Koskull-Döring P, Vierling E, and Scharf KD (2007) Complexity of the heat stress response in plants. Curr. Opin. Plant Biol. 10(3), 310-316.

Latef AAHA, Miransari M (2014) The Role of Arbuscular Mycorrhizal Fungi in Alleviation of Salt Stress. In: Use of Microbes for the Alleviation of Soil Stress (Ed) Miransari M, Springer, New York, pp 23-38.

López-Ráez JA, Charnikhova T, Gómez-Roldán V, Matusova R, Kohlen W, De Vos R, Verstappen F, Puech-Pages V, Bécard G, Mulder P, et al. (2008) Tomato strigolactones are derived from carotenoids and their biosynthesis is promoted by phosphate starvation. New Phytol. 178, 863–874.

Mathur N, Joginder S, Sachendra B, Afshan Q and Anil V (2007) Arbuscular Mycorrhizal Fungi: A Potential Tool for Phytoremediation. J. Plant Sci. 2, 127-140.

Medina, A, M Vassilev, MM Alguacil, A Roldan, R Azcon (2004) Increased plant growth, nutrient uptake, and soil enzymatic activities in a desertified mediterranean soil amended with treated residues and inoculated with native mycorrhizal fungi and plant growth promoting yeast. Soil Sci. 169, 260-270.

Mittler R (2006) Abiotic stress, the field environment and stress combination. Trends Plant Sci. 11(1), 15-19.

Patel KP, Patel KC, Solanki SJ, Patel NN (2005) Optimization of conditions for effective removal of heavy metals by Pseudomonas sp and fungi sp to purify mixed industrial effluents. Pollut. Res. 24(4), 763-766.

Porcel R, Aroca R, Ruiz-Lozano, JM (2012) Salinity stress alleviation using arbuscular mycorrhizal fungi, A review. Agron. Sustain. Dev. 32(1), 181-200.

Porcel R, Ruiz-Lozano, JM (2004) Arbuscular mycorrhizal influence on leaf water potential, solute accumulation, and oxidative stress in soybean plants subjected to drought stress. J. Expt. Bot. 55(403), 1743-1750.

Pozo MJ, Azcón-Aguilar C (2007) Unraveling mycorrhiza-induced resistance. Curr. Opin. Plant Biol. 10(4), 393-398.

Rai UN, Sinha S (2001) Distribution of metals in aquatic edible plants. Environ. Monit. Assess. 70 (3), 241-252.

Rapparini, F, Peñuelas, J (2014) Mycorrhizal fungi to alleviate drought stress on plant growth In: Use of Microbes for the Alleviation of Soil Stresses (Ed) Miransari M, Springer, New York, pp 21-42.

Reddy SR, Pindi PK, Reddy SM (2005) Molecular methods for research on arbuscular mycorrhizal fungi in India: Problems and Prospects. Curr. Sci. 89(10), 25.

Ricalde SLC (2002) Dispersal, Distribution and Establishment of Arbuscular Mycorrhizal Fungi: A Review. Bol. Soc. Bot. México, 71, 33-44.

Rufyikiria G, Huysmansa L, Wannijna J, Heesa MV, Leyvalb C and Jakobsenc I (2004) Arbuscular mycorrhizal fungi can decrease the uptake of uranium by subterranean clover grown at high levels of uranium in soil. Environ. Pollut. 130(3), 427–436.

Singh R and Adholeya A (2013) Diversity of AM (Arbuscular mycorrhizal) Fungi in Wheat Agro-climatic Regions of India. Virol. Mycol. 2(2),1-9.

Smith SE, Read DJ (2008). Mycorrhizal Symbiosis. 3rd edition, Academic Press, London.

Sudhakara RM, Babita K, Bay G, Ramamurthy V (2002) Influence of aluminium on mineral nutrition of the ectomycorrhizal fungi *Pisolithus* sp and *Canthsrellus cibarius.* Water Air Soil Pollut. 135, 55-64.

Turnau K, Orlowska E, Ryszka P, Zubek S, Anielska T, Gawronski S, Jurkiewicz A (2006) Role of Mycorrhizal Fungi in Phytoremediation and Toxicity Monitoring of Heavy Metal Rich Industrial wastes in Southern Poland. Soil Water Pollut. Monit. Prot. Remediat. 3–23.

Turrini A, Stefano B, Emanuele A, Manuela G (2012) Phyto-mycoremediation: Morphological and molecular characterization of arbuscular mycorrhizal fungi from a heavy metal polluted ash dump downtown. Venice Environ. Eng. Manage. J. 11(3), pages S47.

Vinocur B, Altman A (2005) Recent advances in engineering plant tolerance to abiotic stress: achievements and limitations. Curr. Opin. Biotechnol. 16(2), 123-132.

Wang W, Vinocur B, Shoseyov O, Altman A (2004) Role of plant heat-shock proteins and molecular chaperones in the abiotic stress response. Trends Plant Sci. 9(5), 244-252.

Wang FY, Lina XU, Yina R (2007) Effect of Arbuscular Mycorrhizal Fungal Inoculation on Heavy Metal Accumulation of Maize Grown in a Naturally Contaminated Soil. International J. Phytoremediat., 9, 345-353.

Wu QS, Srivastava AK and Zou YN (2013) AMF-induced tolerance to drought stress in citrus: A review, Sci. Hortic. 164,77–87.

Wua SC, Caob ZH, Lib ZG, Cheunga KC, Wonga MH (2005) Effects of biofertilizer containing N-fixer, P and K solubilizers and AM fungi on maize growth: a greenhouse trial Geoderma, 125, (1–2), 155–166.

Yano-Melo AM, Saggin OJ, Maia LC (2003) Tolerance of mycorrhized banana (*Musa* sp cv Pacovan) plantlets to saline stress. Agric. Ecosyst. Environ. 95(1), 343-348.

Zhang HS, Zai XM, Wu XH, Qin P, Zhang WM (2014) An ecological technology of coastal saline soil amelioration Ecol. Eng., 67, 80-88.

11

Arbuscular Mycorrhizal Fungi: Role in Alleviating Salt Stress in Crop Plants

Muthukumar T., Bagyaraj D.J. and Ashwin R.

Abstract

Soil salinization is an important factor limiting crop production, mainly in the arid and semiarid regions of the world. Nevertheless, plants naturally have evolved several physiological, biochemical and molecular mechanisms to establish and thrive in salt stressed soils. In addition to breeding crops for salt tolerance, recent emphasis is on exploring the possible role of plant-microbe interactions in amelioration of salt stress in plants. It is now well established that arbuscular mycorrhizal (AM) symbiosis alleviates the effect of salt stress in plants. Further, available evidence does suggest that AM symbiosis regulates many of the plant's salt stress alleviation mechanisms. However, the molecular basis for many of these mechanisms is yet to be ascertained. In this chapter, we summarize the current knowledge on the role of AM symbiosis in alleviation of salt stress in crop plants, which now appears to be a promising strategy to improve crop yield in salt stressed soils.

Keywords: AM fungi, Antioxidant, Crop yield, Mechanisms of salt tolerance, Nutrient uptake

1. Introduction

Plants growing in cultivated systems are affected by various abiotic and biotic stresses, which negatively influence crop growth and yield. Of the different abiotic stresses, soil salinity is a major challenge to sustainable agriculture. Soil salinity, whether primary or secondary in nature can damage soil structure and fertility (Rengasamy 2006). Though salt-affected soils occur naturally in arid and semiarid regions, agriculture in these regions depends mostly on the ground water resources due to the scarcity of the surface fresh water resources. Irrigated agriculture contributes to well over 30% of the global agricultural production in spite of the fact that only about 17% of the world's agricultural soils is under irrigation (Hillel 2000; Pitman and Läuchli 2002). During irrigation, the dissolved mineral ions in ground water accumulate in the upper part of the soil profile. Thus, secondary salinization of irrigated lands is of major concern for global food production. Various estimates indicate that around 20 to 50% of the irrigated lands are salt-affected (Flowers 1999; Ghassemi *et al.* 1995). The coincidence of irrigation and salinization threatens the sustainability of high agricultural productivity (Flowers and Yeo 1995). In addition, the possible incursion of seawater can lead to tidal intrusion of saline water into rivers and aquifers in coastal areas (Flowers 1999). Some of the impact of soil salinity includes reduction in soil quality, low agricultural production, reduced economic returns, and high reclamation and management costs (Manchanda and Garg 2008). Although the response of crops to salinity varies with species, plant metabolism is affected resulting in reduced plant growth and yield (Fig. 1). Several strategies like phase farming, alley farming, intercropping, and precision farming are used to ameliorate yield reduction and sustainable management of the salt affected lands. However, implementation of these is often limited by cost and good-quality water resources (Manchanda and Garg 2008).

Plants naturally are associated with a wide range of soil microorganisms that influence their growth and health. Some of these microorganisms are known to improve plant's performance under adverse environmental conditions. Arbuscular mycorrhizal (AM) fungi belonging to the phylum Glomeromycota is an important group of fungi as they associate with more than 80% of land plants, including most crop species (Smith and Read 2008; Bagyaraj 2011; Bagyaraj 2014). The influence of AM symbiosis on plant growth depends mainly on the ability of the fungi to take-up and transfer limiting soil nutrients, especially phosphorus (P) to plant roots in exchange for plant photosynthates (Desai *et al.* 2016). In addition, AM fungal symbiosis is known to improve plant's tolerance to various stresses. The role of AM fungi in imparting salinity tolerance has been extensively reviewed in the recent past (Bothe 2012; Porcel *et al.* 2012; Ruiz-Lozano *et al.* 2012). In this chapter, we discuss the current knowledge on the impact of salinity on AM fungi and role of AM symbiosis on modulation of host response to salinity.

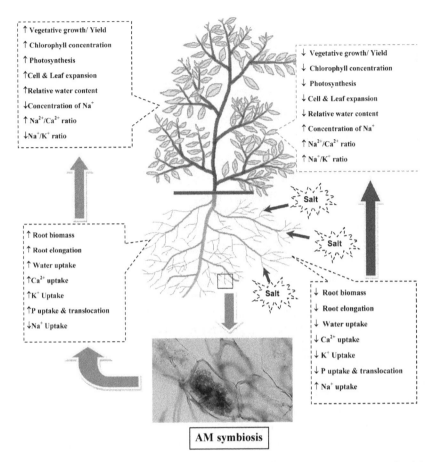

AM symbiosis

Fig. 1: Influence of salt stress on plants and the role of arbuscular mycorrhizal (AM) fungi in amelioration of salt stress. Arrows indicate increase (↑) or decrease (↓) of the parameter concerned

2. Distribution of AM fungi in saline soils

The occurrence of AM fungi in soils of different salinity levels has been reported by several researchers (Kumar and Ghose 2008; Wilde *et al.* 2009). Despite the common occurrence of AM fungi, plants in saline habitats like the mangroves often tend to be non-mycorrhizal (Brundrett 2009). Spore populations of AM fungi can range from low or absent (e.g., Barrow *et al.* 1997) to high as in non saline soils (Alisgharzadeh *et al.* 2001). Wilde *et al.* (2009) reported AM fungal spore populations of 36-951/ 10 g soil from salt marshes at the Dutch Island of Terschelling and Schreyahn, Northern Germany. Similarly, García and Mendoza (2007) reported AM fungal spore densities ranging from 73-89 per g of soil during different seasons of the year in a saline-sodic soil of Argentina. The early report on AM fungal symbiosis in halophytes dates back to 1928 (Mason 1928) and since then many researchers have reported this symbiosis in halophytes (García and Mendoza 2007; Kumar and Ghose 2008; Wilde *et al.* 2009). In spite of the presence of AM symbiosis in halophytes, increasing soil salinity reducing AM fungal colonization has been reported in several crop plants (Talaat and Shawky 2013, Kadian *et al.* 2013, Estrada *et al.* 2013b).

3. Salinity and mycorrhiza formation

It has been well established that the presence of host roots is not essential for the germination of AM fungal spores. Further, earlier studies have also shown that soil factors especially salinity could negatively influence AM fungal spore germination and subsequent development of the germinating hyphae (Campagnac and Khasa 2014). As the germination of AM fungal spores consist of different phases like hydration, activation, germ tube emergence and hyphal growth (Tommerup 1984), it is always interesting to identify the stage of spore germination influenced by salinity. In a recent study, Campagnac and Khasa (2014) have shown that in *Rhizophagus irregularis* (=*Glomus intraradices*), the germ tube emergence and hyphal growth were the phases most affected by salinity. However, other morphological features like the lateral branching of the hyphae or the hyphal diameter was not affected. A similar response was also observed in *Scutellospora calospora* by Juniper and Abbott (1993); nevertheless, the hyphal morphology of *Gigapsora margarita* was modified under saline conditions. An *in vitro* root compartmental study investigating the influence of salinity and hyphal production in two isolates of *R. irregularis* indicated that 75 and 150 mM NaCl negatively affected the hyphal length produced. In contrast, by eight weeks of culture, the same amount of NaCl stimulated sporulation of *R. irregularis* (Estrada *et al.* 2013c). Increased sporulation can be regarded as a stress perception of the fungus as spores are a form of resistant propagules to tide over adverse environmental conditions. In the same study, Estrada *et al.* (2013c) also showed that branched absorbing structures (BAS) were more negatively influenced by salt concentrations during early stages than at later stages of culture. Though it has been well resolved that species of AM fungi differ in their sensitivity to salinity, information on the response of isolates of an AM fungus to salinity is limited. Recently, Campagnac and Khasa (2014) screened three isolates of *R. irregularis* originating from saline soils, and two isolates from non-saline soils for their ability to tolerate salinity. The results of this experiment clearly showed that the isolates originating from saline soils were better adapted to salinity compared to those from non-saline soils. This supports studies where AM fungal isolates originating from saline soils promoted plant growth better than those originating from non-saline soils (Estrada *et al.* 2013a,b). Contrarily, studies also indicate that AM fungal isolates originating from non-saline soils could promote plant growth better or similar to those originating from saline soils (Hadad *et al.* 2012; Tian *et al.* 2004).

4. Influence of AM fungi on plant growth and yield

4.1 Plant growth

Soil salinity results in stunted growth in plants due to inhibition of cell elongation (Nieman 1965). In spite of their sensitivity to salinity, roots are less affected than shoots (Munns and Termaat 1986). The beneficial effects of AM fungi on plant growth under saline conditions have been demonstrated in tomato (Al-Karaki 2006), maize (Feng *et al.* 2000), wheat (Talaat and Shawky 2014b), egg plant, Sudan grass (Mohammad and Mittra 2013), and pepper (Abdel Latef and Chaoxing 2011; Cekic *et al.* 2012). Zarei and Paymaneha (2014) determined the effects of AM fungi [*Claroideoglomus etunicatum* (=*Glomus etunicatum*) and *R. irregularis*] and salt concentrations (2-8 dS m^{-1} as NaCl) on growth of rough lemon rootstocks. Increasing salinity decreased all measured characteristics of plants after 4.5 months of growth. However, shoot and root dry weights were significantly higher in AM seedlings than non-mycorrhizal seedlings at all salinity levels. A study on the interaction between AM fungus, *Funneliformis mosseae* (=*Glomus mosseae*), and salinity stress on plant growth, in chickpea indicated that salt stress resulted in a noticeable decline in shoot and root dry matter accumulation and shoot-to-root ratios (Garg and Chandel 2011a). The plants inoculated with *F. mosseae* accumulated more shoot and root biomasses than non-mycorrhizal plants under salt stress. However, the shoot-to-root ratios of AM plants were almost similar or slightly lower compared to nonmycorrhizal plants (Garg and Chandel 2011a). In a pot culture, Abdel-Fattah and Asrar (2012) examined the effects of three different arbuscular mycorrhizal fungi [*F. mosseae*, *Glomus deserticola* and *Racocetra gregaria* (=*Gigaspora gregaria*)] on growth of wheat in saline soil. Under saline condition, AM inoculation significantly increased growth responses like plant height, number of leaves, leaf area, and shoots and root dry weights. The root by shoot ratios of AM plants was significantly lower than non-AM plants (Abdel-Fattah and Asrar 2012). A similar response of wheat to *F. mosseae* inoculation has also been reported by Sheng *et al.* (2008). Estrada *et al.* (2013a) investigated the influence of native AM fungi (*Glomus* spp.) isolated from a saline soil on maize plants. The results of the study indicated that plants inoculated with two out of the three native AM fungi had the highest shoot dry biomass at all salinity levels. Pepper plants inoculated with *Rhizophagus clarus* (=*Glomus clarum*) and grown at high salinity levels had higher plant biomass compared to non- AM plants in spite of a decline in AM colonization with increasing salinity (Kaya *et al.* 2009).

4.2 Crop yield

Like plant growth, salinity is known to negatively influence crop yield. Although, information on AM mediated amelioration of crop yield is limited, available evidence indicate that yield parameters of AM plants are always higher compared to non-mycorrhizal plants grown at similar salinity levels. Huang et al. (2013) examined the role of F. mosseae on growth and yield of tomato under saline conditions. Colonization of AM fungus alleviated the decrease in tomato yield by 5-30%, juice extracts ratio by 4-6% and total soluble solids of tomato fruits by 4-6%. In a study, chickpea (Cicer arietinum) colonized by F. mosseae or Acaulospora laevis in a saline soil (4 dSm^{-1}) had 45-58% more pods and 138-184% higher pod mass compared to non-mycorrhizal plants (Kadian et al. 2013). Further, chickpea plants colonized by both the fungi had 37-49% more pods which were 56-110% heavier compared to those colonized by an individual fungus. Daei et al. (2009) and Talaat and Shawky (2012) investigated the efficacy of a mixture of AM fungi on growth and yield of wheat cultivars under salt stress in a greenhouse over two seasons. Mycorrhizal colonization protected wheat plants against the detrimental effect of salinity and increased all the yield parameters like spike number, grain number, weight of spikes, grain yield and grain weight. In addition, grains of AM plants had higher protein content than non-AM plants suggesting an improvement in seed quality. Cucumber plants colonized by R. irregularis had significantly higher fruit production under alkaline conditions (10 mM NaHCO$_3$ and 0.5 gl^{-1} CaCO$_3$) compared to non-mycorrhizal plants (Rabie 2005). However, there was no significant variation in fruit quality.

4.3 Salinity and root architecture

Studies on the influence of AM fungi on plant root morphology under salt stress are limited. Nevertheless, these limited studies do suggest that AM fungi could modify plant's root architecture thereby enhancing the efficiency of the root systems under saline conditions. Mycorrhizal citrus seedlings raised under 100 mM NaCl had significantly higher root length; root projected area and root surface area than non-mycorrhizal seedlings (Wu et al. 2010). However, root diameter was neither influenced by salinity nor by AM inoculation. Contrarily, in a greenhouse experiment, maize plants inoculated with F. mosseae and grown in five concentrations of NaCl (0, 0.5, 1.0, 1.5, and 2.0 g/kg dry substrate) had lower specific root length, lower percentage of root length in the 0 – 0.2 mm diameter class, and higher percentage of root length in both the 0.2 – 0.4 mm and 0.4 – 0.6 mm diameter classes irrespective of NaCl concentrations (Sheng et al. 2009). This shows that mycorrhizal plants have the tendency to shift towards to a coarser root system. An increased efficiency of the root system in response to mycorrhization has often been attributed to the greater salt tolerance of mycorrhizal plants. Modifications in the proportion of endogenous plant hormones like gibberellins and cytokinins (Berta *et al.* 1993) as well as increased levels of polysaccharides in root hairs (Zangaro *et al.* 2005) has often been cited as the cause for changes in root architecture in AM plants.

4.4 Nutrient uptake

Soil salinity reduces the availability of major nutrients like N, P and K because of their precipitation, changes in nutrient metabolism or competition for binding sites (Evelin *et al.* 2009). Increased nutrient acquisition by the host plant is a well-known response to AM fungal association (Smith and Read 2008). The improved growth and productivity of AM plants in saline soils was primarily attributed to AM-mediated increase in the uptake of limiting nutrients, especially P, as soil salinity increases the mycorrhizal dependency of plants (Beltrano *et al.* 2013; Garg and Chandel 2011a, b; Manchanda and Garg 2011; Promita and Mohan 2013). Irrigation of mung bean (*Vigna radiata*) with different concentrations of sea water (0-30%) has shown to increase root colonization by *R. clarus* especially the root length with arbuscules (Rabie 2005). This indicates an increase in mycorrhizal dependence as formation of arbuscules is regulated by the nutrient demand of the host (Smith and Read 2008). Several studies have shown increased uptake and accumulation of both macro and micronutrients in plant tissues (see Smith and Read 2008 and references therein). This increase in nutrient uptake by AM plants can range from one to several folds depending on crop species, AM fungus and the salinity levels. However, AM symbiosis has been shown to increase as well as decrease uptake of some mineral nutrients like Cu, Zn, Na^+ and Cl^- depending on plant species and salinity levels (Evelin *et al.* 2009). The increased nutrient acquisition by AM plants is facilitated by the extensive extraradical hyphal network of the fungus which explores greater volume of soil than non AM plants (Smith and Read 2008). Matamoros *et al.* (1999) estimated that up to 80% of the plant's P demand is satisfied by the soil hyphal network. Further, the activities of acid and alkaline phosphatase activities in roots were reported to be higher for mycorrhizal than non-mycorrhizal mung bean under increasing salinity (Rabie 2005).

Although, the exact mechanism through which AM fungi increase the uptake of N under salt stress is not known, available evidence does indicate that AM plants accumulate more N in their tissues than their non-AM counterparts. In addition, AM association under salt stress has been shown to improve nodulation and nitrogen fixation in legumes (Garg and Chandel 2011a, b; Manchanda and Garg 2011). Nitrate reductase (NR), the first enzyme in the NO_3^- assimilation pathway is a substrate induced plant enzyme (Campbell 1988). Therefore, its level drops sharply under adverse environmental conditions due to the low influx of NO_3^- from soil into plant roots. Higher NR activity was correlated with increased tolerance of AM plants to stress (Ruiz-Lozano and Azcón 1996). Talaat and Shawky (2012) showed that NR activity was higher in mycorrhizal wheat plants exposed to salt stress than non-mycorrhizal plants.

Mycorrhizal association can improve plant's K^+ uptake under salt stress resulting in a higher K^+:Na^+ ratio (Kadian et al. 2013). This increased K^+ uptake under salt stress by mycorrhizal plants may modify the negative influence of Na^+ on plant growth and metabolism (Evelin et al. 2009). In addition to facilitating nutrient uptake, AM fungi can also improve the nutrient use efficiency of colonized plants under salt stress conditions resulting in greater biomass per unit of nutrient taken (Rabie 2005).

5. Mechanisms of salinity tolerance

5.1 Biochemical changes

Salt-induced changes in plants are primarily due to water deficit rather than specific effects of salts (Javid et al. 2011). The soil water potential becomes more negative with the accumulation of salt in the soil. To counteract this effect, and avoid dehydration, plants reduce their water potential by osmoregulation or osmotic adjustment to maintain a continuous flow of water from the soil to roots. Osmoregulation is achieved through active accumulation of organic ions like K^+ and Cl⁻ or solutes such as proline, glycine, betaine, souble sugars, pinitol and mannitol (Flowers and Colmer 2008; Ruiz-Luzano et al. 2012). This osmotic adjustment enables plants to maintain cell turgor and its related processes like stomatal opening, cell expansion and photosynthesis (Fig. 1).

5.1.1 Proline

Proline is one of the most common osmolyte studied in various plant stresses including salt stress. This compound is known to maintain the osmotic stability of the cells and protect cells against the detrimental effects of salinity through stabilizing sub-cellular structures, buffering cellular redox potential and scavenging free radicals (Chen and Dickman 2005). Changes in proline concentrations in response to AM colonization under salinity is often contradictory. Many studies have reported higher concentrations of proline in AM plants than in non-AM plants under saline conditions (Dodd and Pérez-Alfocea 2012; Garg and Baher 2013). Contrarily, low concentrations of proline have also been reported in AM than in non-AM plants under soil salinity (Jahromi et al. 2008; Sheng et al. 2011). These variations among studies were attributed to the different plant and AM fungal species involved, including genotypes of varied levels of saline tolerance (Ruiz-Lozano *et al*. 2012). Nevertheless, proline accumulation in plants appears to be more influenced by salinity than by AM symbiosis. In Pearl millet, proline accumulation in shoots of *Rhizophagus fasciculatus* (=*Glomus fasciculatum*) colonized plants was 16-232% lower than the non-AM plants 90 days after inoculation in different levels of salinity. In contrast, proline content of *R. fasciculatus* colonized roots were 14-76% higher for the same day and levels of salinity (Borde *et al*. 2011). It has been suggested that roots maintain higher concentration of proline to facilitate the osmotic balance between water-absorbing root cells and the substrate for water absorption under salt stress (Evelin *et al*. 2009).

5.1.2 Betaines

Betaines are quaternary ammonium compounds that accumulate in plants under salt stress. Wheat cultivars colonized by a mixture of *Glomus* spp., accumulated more glycinebetaine than non-mycorrhizal ones at varying levels of soil salinity (0.1-9.4 dSm^{-1}) (Talaat and Shawky 2014a). Higher concentrations of glycinebetaine under salt stress (EC= 4-8 dSm^{-1}) has also been reported in root nodules of pigeonpea when the plants were mycorrhizal (Manchanda and Garg 2011).

5.1.3 Sugar concentrations

An increase in the concentrations of soluble sugars has also been reported in mycorrhizal plants under salt stress. Talaat and Sahwky (2011) found a positive correlation between total soluble-sugar concentration and mycorrhization of wheat plants. A similar response was also reported in pigeonpea colonized by *F. mosseae* (Garg and Chandel 2011b). Though the accumulation of total soluble-sugar concentration in mycorrhizal plants could likely be due to the sink effect of the fungus (Feng *et al*. 2002), it could protect the plant against the detrimental effect of salinity.

5.1.4 Phytohormones

a) **Strigolactones:** Strigolactones are apocarotenoid phytohormones produced in the roots of both monocot and dicot plants (Xie *et al*. 2010). They are also found in root exudates and are known to regulate the above- and below-ground plant processes, including AM symbiosis (Kohlen *et al*. 2012). Initially, strigolactones were identified as signal molecules involved in the host detection of root parasitic plants of the family Orobanchaceae (López-Ráez *et al*. 2011). Concurrently, it was also found that these molecules could as well favour AM symbiosis. As nutrient stress induces strigolactone production in plants, it was presumed as a mediator of plant response to stressful environmental conditions (García-Garrido *et al.* 2009). Further, the increased strigolactone production is also known to promote AM fungal development and symbiosis establishment in different plant species (López-Ráez *et al.* 2008). Recent studies have also shown that strigolactones could be important in the establishment of AM symbiosis, in addition to its stress protection function as it was found to be essential to complete arbuscule formation as well as to sustain colonization in plant roots (see López-Ráez *et al.* 2011). In a recent study, Aroca *et al.* (2013) showed that salinity could negatively influence strigolactone production in non-AM lettuce plants. However, strigolactone production increased up to five folds under saline conditions when the lettuce plants were colonized by the AM fungus *R. irregularis*. This increase in strigolactone production in AM lettuce plants under saline conditions occurred in spite of a reduction in AM fungal colonization (Aroca *et al.* 2013).

b)   **Abscisic acid and Jasmonic acid:** *Abscisic* acid (ABA) and jasmonic acid (JA) are the two endogenous phytohormones that are known to increase with increasing salinity (Javid *et al.* 2011). Increased concentrations of ABA in response to salinity has been reported in plant species like mustard (He and Cramer 1996), common bean (Cabot *et al.* 2009), maize (Cramer and Quarrie 2002) and rice (Kang *et al.* 2005). Aroca *et al.* (2013) showed that concentrations of ABA increased with increasing levels of NaCl in lettuce colonized by *R. irregularis*. However, the concentrations in AM-plants were either lower or were almost similar compared to non-AM plants. There is reliable evidence to suggest that ABA is a signaling hormone whose production is regulated by environmental factors and the physiological status of the plants (Javid *et al.* 2011). For example, ABA concentrations in roots of lettuce plants colonized by *R. irregularis* was almost four folds higher under non-saline conditions, whereas, it was lower than non-AM plants in 50 and 100 mM NaCl (Jahromi *et al.* 2008). In addition, a coincidence in ABA pattern in plant tissues with those of ABA biosynthesis gene *LsNCED2*, indicates the production of ABA in response to soil salinity (Aroca *et al.* 2013). The low concentration of ABA in AM colonized plants compared to non-AM plants at similar levels of salinity suggest that AM symbiosis could reduce the deleterious effect of salt stress to a certain extent.

Jasmonic acid is an important cellular regulator actively involved in various plants' developmental processes and senescence (Wasternack and Hause 2002). This plant hormone also activates plant defense mechanisms under various biotic and abiotic stresses including salinity (Cheong and Choi

2003). Increased concentrations of jasmonic acid in plant tissues in response to salt stress has been reported in crops like tomato (Pedranzani *et al.* 2003), barley (Kramell *et al.* 2000), and rice (Kang *et al.* 2005). Although, the exact mechanism underlying jasmonic acid mediated amelioration of salt stress is yet to be investigated, available evidences do indicate increased protein synthesis, reduced concentrations of Na^+ in plant tissues and an enhancement in the uptake of major ions in response to jasmonc acid application to salt stressed plants (Javid *et al.* 2011). Application of jasmonic acid has been shown to favour (Landgraf *et al.* 2012) as well as inhibit (Ludwig-Muller *et al.* 2002) AM development. However, the effect seems to depend on the AM fungal species involved, concentration, timing and nutritional conditions (Gutjahr and Paszkowski 2009). In spite of contradictory results, most of the studies do suggest that jasmonate signaling is important for the development of AM symbiosis (Gutjahr and Paszkowski 2009). For example, JA-deficient spr2 tomato mutants developed very less AM colonization, but AM colonization was restored by methyl jasmonate application (Tejeda-Sartorius *et al.* 2008). As the events involved in early stages of AM development may be quite different from those at later stages, there is a possibility of a varied role of jasmonic acid at different stages of the symbiosis (Foo *et al.* 2013). Nevertheless, the direct role of AM fungi in modulating the synthesis of endogenous jasmonic acid in plants under salt is yet to be ascertained.

5.1.5 Ionic homeostatis

Increased uptake of Na^+ in plants under saline conditions increases reactive oxygen species (ROS) accumulation and has a deleterious effect on certain enzymes and photosynthesis (Mahajan and Tuteja 2005). However, salt tolerant plants have evolved strategies like the prevention of Na^+ entry into roots, sequestration into vacuoles and transport to and allocation within the leaf (Ruiz-Lozano *et al.* 2012). Many effects of Na^+ on plant metabolism arise from its ability to compete with K^+ ions for binding sites. Studies have shown that AM colonization enhances the uptake of K^+ and Ca^{2+} in saline soils and prevent Na^+ transport to shoot tissues thereby preventing Na^+ induced salt stress (Talaat and Shawky 2011). This enables AM plants to maintain a higher K^+:Na^+ ratio, thereby neutralizing the negative effects of Na^+ on plant metabolism. Further, Hammer *et al.* (2011) showed that AM fungi can selectively take up K^+ and Ca^{2+} ions from the soil solution and avoid the uptake of Na^+. In addition, AM symbiosis has also been shown to possess a buffering effect whereby Na^+ is maintained within accepted limits in plant tissues under high salt concentrations (Evelin *et al.* 2009; Hammer *et al.* 2011). This fact is supported by a recent observation where root colonization of fenugreek plants by *R. irregularis* limited the uptake of Na^+ at increasing soil salinity (Evelin *et al.* 2012). Gene expression in ion homeostasis in mycorrhizal plants under salt stress is limited to a study by Ouziad *et al.* (2006) where AM fungi failed to alter the expression of *LeNHX1* and *LeNHX2* genes in tomato. Ruiz-Lozano *et al.* (2012) suggested that a better understanding is essential on the possible regulation of genes encoding for known ion transporters involved in the response of AM symbiosis to salinity. Further research on the role of AM fungi on the uptake of Na^+ and K^+ and their distribution and compartmentation within plant cells under saline stress would help to unravel the mechanisms involved in the improved salinity tolerance of AM plants.

5.1.6 Antioxidants

Salt stress results in a large number of metabolic changes in plants resulting in the dramatic elevation of production of ROS (Mittler 2002). This enhanced production of ROS results in the ROS-associated injuries in plants (Miller *et al.* 2010). The accumulation of ROS at any point depends on the balance between ROS production and ROS scavenging (Miller *et al.* 2010). Plants have evolved both non-enzymatic as well as enzymatic antioxidant systems to effectively scavenge the excess ROS present within cells (Sharma *et al.* 2012). The nonenzymatic antioxidants include ascorbate (AsA), carotenoids, glutathione (GSH), phenolics and tocopherols. The enzymatic antioxidants include catalase (CAT), ascorbate peroxidase (APX), superoxide dismutase (SOD), guaiacol peroxidase (GPX), monodehydroascorbate reductase (MDHAR), dehydroascorbate reductase (DHAR), and glutathione reductase (GR) (Noctor and Foyer 1998). On the other side, ROS also acts as signal transduction molecules in metabolic imbalances during stress in plants. This activates acclimation and defense mechanisms, which would in turn counteract stress-associated oxidative stress (Sharma *et al.* 2012). An investigation on the accumulation of ROS in *R. irregularis* colonized roots of clover, maize and tobacco indicated the involvement of both AM colonized root cortical cells and the fungal partner in ROS production under stress. A small level of ROS was produced in the extraradical hyphae and spores of the AM fungus in response to stress (Fester and Hause 2005). Intracellular accumulation of H_2O_2 in AM plants was observed in the cytoplasm adjacent to intact and collapsing fungal structures. In contrast, intercellular H_2O_2 was located on the surface of fungal hyphae. The accumulation of ROS in AM roots appears to be similar to those of senescing legume root nodules (Fester and Hause 2005).

Though the involvement of the antioxidant defense system in plant's response to salt stress is well known, information is scarce regarding the role of AM symbiosis on antioxidant defense under saline conditions. Studies involving various hosts and AM fungal species at different levels of soil salinity have shown that AM symbiosis modulate the production of non-enzymatic and enzymatic antioxidants (Table 1).

5.1.7 Glomalin production

Glomalin is a glycoprotein produced by AM fungi, which is known to play an important role in soil aggregation (Rillig 2004). Glomalin levels in the soil and *in vitro* cultures are often negatively related with hyphal length (Lovelock *et al*. 2004), suggesting that its production might be induced by a stress response. Further, it is believed that glomalin could be involved in minimizing possible cytosolic damages of Na^+ induced protein misfolding (Maathuis and Amtmann 1999). Hammer and Rillig (2011) found that NaCl stress strongly increased glomalin production in hyphae of *R. irregularis* under *in vitro* conditions. However, the response of glomalin production to salt stress was found to be substantially different from those induced by the low-water potential manifested by the addition of glycerol. This suggests that some physiochemical property of NaCl like the direct toxicity of ions might be involved in the induction of glomalin production (Hammer and Rillig 2011).

5.2 Physiological modifications

5.2.1 Improved water status

Soil salinity could induce physiological drought decreasing plant growth and yield. A study examining the influence of AM fungus (*F. mosseae*) on water status of maize indicated that AM symbiosis significantly improved the water status of the plants (Sheng *et al*. 2008). The improved water status in turn could indirectly enhance the capacity of gas exchange and photosynthesis. The enhanced water status of mycorrhizal plants under salt stress could be due to the direct uptake and transport of water by AM fungal hyphae (Jiang and Huang 2003), changes in root architecture and root activities (Berta *et al*. 1993).

5.2.2 Aquaporins

Increasing soil salinity decreases the water potential of the soil preventing the inflow of water into roots from the soil. It is now well established that AM fungi can take up water from the soil and transfer it into plant roots under limiting conditions. Aquaporins are small transmembrane proteins (21 to 34 kD) commonly present in plants, facilitating the transport of certain small molecules in addition to water across biological membranes (Maurel *et al*. 2008). The capacity of water permeation mediated by the aquaporin water channels is

Table 1: Influence of arbuscular mycorrhizal (AM) symbiosis on antioxidant system of salt stressed plants

Salinity levels	Crop species	AM fungus	Plant organ	Non-Enzymes*		Enzymes*						Reference
				AsA	GSH	CAT	APX	SOD	POX	MDA	GR	
0-1.0% NaCl	Lycopersicon esculentum	Funneliformis mosseae	Root			I	I	I	I	I	I	He et al. (2007)
0-100 μmol m⁻²S⁻¹	Zea mays	Rhizophagus irregularis, Claroideoglomus etunicatum, Septoglomus constrictum	Shoot			I		I				Estrada et al. (2013b)
			Root			I		I				
4-8 dSm⁻¹	Cajanus cajan	F. mosseae	Nodules			I		I	I			Manchanda and Garg (2011)
2.2-12 dSm⁻¹	L. esculentum	F. mosseae	Leaves			I	I		I	I		Abdel Latef and Chaoxing (2011)
0-100 mM Nacl	Capsicum annum	F. mosseae	Shoot			I		I	I	D		Abdel Latef and Chaoxing (2014)
			Root			I		I	I	D		
0.9-7.1 dSm⁻¹	L. esculentum	F. mosseae	Leaves	I		I		I	I	I		Huang et al. (2010)
			Root	I		I		I	I	I		
4.7, 9.4 dSm⁻¹	Triticum aestivum	Mixture of Glomus spp.	Leaves	I	I	I		I	I		I	Talaat and Shawky (2014a)
6,12 dSm⁻¹	Glycine max	F. mosseae	Nodulated roots			I		I	I			Younesi et al. (2013)
2, 4 g NaCl/Kg 100-300 m M NaCl	Lactuca sativa Pennisetum glaucum	R. irregularis Rhizophagus fasciculatus	Shoot Root Shoot			I I		I	I I I			Kohler et al. (2009) Borde et al. (2011)
3.13-9.38 dSm⁻¹	Triticum aestivum	mixture of Glomus spp.	Leaves			I		I	I			Talaat and Shawky (2011)

The non enzymatic antioxidants include ascorbate (AsA) and glutathione (GSH). The enzymatic antioxidants include catalase (CAT), ascorbate peroxidase (APX), superoxide dismutase (SOD), peroxidase (POX), Malnoaldehyde (MDA), glutathione reductase (GR)
*I, D indicate increase and decrease respectively.

10–100 times higher than that by diffusion (Agre *et al.* 2002). Therefore, aquaporins are beneficial to the osmoregulation of plants under various hydric stresses (Maurel *et al.* 2008). In addition, aquaporins also mediate osmoregulation through their involvement in intracellular Ca^{2+} signaling in response to osmotic stresses (Benfenati *et al.* 2011). In plants, seven types of aquaporins have been reported so far. These include: plasma membrane intrinsic proteins (PIPs), tonoplast intrinsic proteins (TIPs), NOD26-like intrinsic proteins (NIPs), small basic intrinsic proteins (SIPs), GlpF-like intrinsic proteins (GIPs), hybrid intrinsic proteins (HIPs) and X-intrinsic proteins (XIPs) (Danielson and Johanson 2008). Studies have shown that AM fungi may up-regulate or down-regulate plant aquaporin genes in roots to increase root hydraulic conductivity (Aroca *et al.* 2008; Ruiz-Lozano *et al.* 2009). Mycorrhization of *Lotus japonicus* with *Gigaspora margarita* has been shown to upregulate two putative aquaporin genes, *LjNIP1* and *LjXIP1* which demonstrate the presence of mycorrhiza-inducible aquaporins in plants (Giovannetti *et al.* 2012). Further, two functional aquaporin genes, namely *GintAQPF1* and *GintAQPF2*, from *R. irregularis* have been cloned recently (Li *et al.* 2013). The expression of these genes was upregulated in the extraradical mycelium of the fungus under water stress conditions. These clearly suggest that AM fungi not only transports more water to the host plant by regulating their own aquaporin activities, but may also regulate the expression of plant aquaporin genes to improve plant water relations (Li *et al.* 2013).

5.2.3 Photosynthesis

Photosynthesis is one of the primary processes to be affected by salinity resulting in reduced crop growth and productivity (Munns *et al.* 2006). The effects of salt stress on photosynthesis can be primary or secondary in nature. The primary effect includes decreased CO_2 availability caused by diffusion limitations through alterations in the stomatal functioning and the mesophyll (Flexas *et al.* 2007) or changes in photosynthetic metabolism itself (Lawlor and Cornic 2002). The secondary effect can arise from the oxidative stress and can seriously affect leaf photosynthetic machinery (Ort 2001). Photosynthetic response to salinity stress is highly complex (Chaves *et al.* 2009). Further, the reduction in photosynthetic rate induced by salt stress may result in an over reduction of the reaction centers PSII. The balance excess energy if not dissipated by the plant may damage the photosynthetic machinery (Ruiz-Lozano *et al.* 2012). Few studies have examined the role of AM symbiosis on photosynthesis under salt stress. Generally, an improvement in photosynthetic efficiency has often been reported in AM plants growing under salt stress (Ruiz-Lozano *et al.* 2012). The main effect of AM symbiosis in imparting salinity tolerance appears to be the result of enhanced photosynthetic capacity through elevated gas exchange and the efficiency of photochemistry and non-photochemistry of PSII. Further, regulation in energy bifurcation between photochemical and non-photochemical events also has been shown to play an important role in salinity tolerance of mycorrhizal plants (Ruiz-Lozano et al. 2012). Under salt stress, AM plants have been shown to possess higher chlorophyll content and better water status than their non-mycorrhizal counter parts (Abdel Latef and Chaoxing 2011; Beltrano et al. 2013). In tomato, AM colonization has been shown to improve the net assimilation rates by increasing the stomatal conductance and maintenance of photosynthetic capacity by protecting the photochemical process from salinity (Hajiboland et al. 2010). Nevertheless, as photosynthetic efficiency in plant depends on the activities of enzymes like Rubisco (ribulose 1,5-biphosphate carboxylase/ oxygenase), it is important to understand the influence of salinity on these enzymes as suggested by Ruiz-Lozano et al. (2012).

5.2.4 Reduction in electrolytic leakage

Cell membranes play an important role in regulating the uptake and transport of ions. Cell membranes are the initial sites of stress injury and damages caused by environmental stresses. This induces changes in membrane potentials and permeability reducing the membrane's ability to take up and retain solutes (Khatkar and Kuhad 2000). Salt stress induces severe electrolytic leakage. Differences exist at membrane levels between saline sensitive and saline tolerant cultivars (Khatkar and Kuhad 2000). But, mycorrhizal plants are generally less affected than non-mycorrhizal plants (Ibrahim et al. 2011). Colonization of two wheat cultivars by a mixture of *Glomus* spp., was shown to reduce electrolytic leakage by 7.8-18% at salinity levels ranging from 0.1-9.4 dSm^{-1} (Talaat and Shawky 2014a).

Conclusion and future perspectives

Increasing crop productivity under salt-stress is a major challenge. Such attempts require a clear understanding of the plant and other related mechanisms involved in response to salt stress. Crop adaptation to salinity involves the synthesis of osmoprotectants and triggering of antioxidant response to control cellular damage. Further, changes in water uptake and distribution, and maintenance of photosynthesis to near normal levels enable glycophytes to thrive under saline conditions. Currently, there is enough evidence to believe that AM symbiosis influence and regulate many of these mechanisms. Nevertheless, many physiological aspects and molecular basis for AM based regulation of these processes are unknown. For some of the aspects (e.g., ionic homeostatis) studies are limited and results of these studies are rather inconclusive. Most studies on the amelioration of salinity tolerance have involved a limited number of crops and AM fungal species. Further, changes in signal transduction between salt-stressed plants and AM fungi are virtually not known. Though studies have identified certain AM fungal isolates that could establish a functional symbiosis under saline conditions, the basis for this salinity tolerance is yet to be explored. An understanding of this would enable transfer of salinity tolerance to other AM fungal taxa. This will enable to realize the goal of scaling up crop yield in saline soils.

References

Abdel Latef AAH, Chaoxing H (2011) Effect of arbuscular mycorrhizal fungi on growth, mineral nutrition, antioxidant enzymes activity and fruit yield of tomato grown under salinity stress. Sci. Hortic. 127, 228–233.

Abdel Latef AA, Chaoxing H (2014) Does the inoculation with *Glomus mosseae* improve salt tolerance in pepper plants? J. Plant Growth Regulat. 33, 644-653.

Abdel-Fattah GM, Asrar A-WA (2012) Arbuscular mycorrhizal fungal application to improve growth and tolerance of wheat (*Triticum aestivum* L.) plants grown in saline soil. Acta Physiol. Plant. 34, 267–277.

Agre P, King LS, Yasui M, Guggino WB, Ottersen OP, Fujiyoshi Y, Engel A, Nielsen S (2002) Aquaporin water channels – from atomic structure to clinical medicine. J. Physiol. 542, 316.

Aliasgharzadeh N, Saleh Rastin N, Towfighi H, Alizadeh A (2001) Occurrence of arbuscular mycorrhizal fungi in saline soils of the Tabriz Plain of Iran in relation to some physical and chemical properties of soil. Mycorrhiza 11, 119–122.

Al-Karaki GN (2006) Nursery inoculation of tomato with arbuscular mycorrhizal fungi and subsequent performance under irrigation with saline water. Sci. Hortic. 109, 1–7.

Aroca R, Vernieri P, Ruiz-Lozano JM (2008) Mycorrhizal and non-mycorrhizal *Lactuca sativa* plants exhibit contrasting responses to exogenous ABA during drought stress and recovery. J. Exp. Biol. 59, 2029– 2041.

Aroca R, Ruiz-Lozanoa JM, Zamarrênob ÁM, Paza JA, García-Minab JM, Pozoa MJ, López-Ráeza JA (2013) Arbuscular mycorrhizal symbiosis influences strigolactone production under salinity and alleviates salt stress in lettuce plants. J. Plant Physiol. 170, 47–55.

Bagyaraj DJ (2011) Microbial Biotechnology for Sustainable Agriculture, Horticulture and Forestry. New India Publishing Agency, New Delhi.

Bagyaraj DJ (2014) Ecology of arbuscular mycorrhizal fungi. In, Microbial Diversity and Biotechnology in Food Security (Ed) Kharwar RN, Upadhyay R, Dubey N, Raghuwanshi R, Springer India, New Delhi, pp 133–148.

Barrow JR, Havstad KM, McCaslin BD (1997) Fungal root endophytes in four wing saltbrush, *Atriplex canescens*, on arid rangeland of southwestern USA. Arid. Soil Res. Rehab. 11, 177–185.

Beltrano J, Ruscitti M, Arango MC, Ronco M (2013) Effects of arbuscular mycorrhiza inoculation on plant growth, biological and physiological parameters and mineral nutrition in pepper grown under different salinity and P levels. J. Soil Sci. Plant. Nutr. 13, 123–141.

Benfenati V, Caprini M, Dovizio M, Mylonakou MN, Ferroni S, Ottersen OP, Amiry-Moghaddam M (2011) An aquaporin-4/transient receptor potential vanilloid 4 (AQP4/TRPV4) complex is essential for cell-volume control in astrocytes. Proc. Nat. Acad. Sci., USA 108, 2563–2568.

Berta G, Fusconi A, Trotta A (1993) VA mycorrhizal infection and the morphology and function of root systems. Environ. Exp. Bot. 33, 159–173.

Borde M, Dudhane M, Jite P (2011) Growth photosynthetic activity and antioxidant responses of mycorrhizal and non-mycorrhizal bajra (*Pennisetum glaucum*) crop under salinity stress condition. Crop. Prot. 30, 265–271.

Bothe H (2012) Arbuscular mycorrhiza and salt tolerance of plants. Symbiosis 58, 7–16.

Brundrett MC (2009) Mycorrhizal associations and other means of nutrition of vascular plants, understanding the global diversity of host plants by resolving conflicting information and developing reliable means of diagnosis. Plant Soil 320, 37–77.

Cabot C, Sibole JV, Barcelo J, Poschenrieder C (2009) Abscisic acid decreases leaf Na+exclusion in salt treated *Phaseolus vulgaris* L. J. Plant Growth Regul. 28, 187–192.

Campagnac E, Khasa DP (2014) Relationship between genetic variability in *Rhizophagus irregularis* and tolerance to saline conditions. Mycorrhiza 24,121–129.

Campbell WH (1988) Nitrate reductase and its role in nitrate assimilation in plants. Physiol. Plant 74, 214–219.

Cekic FO, Unyayar S, Ortas I (2012) Effects of arbuscular mycorrhizal inoculation on biochemical parameters in *Capsicum annuum* grown under long term salt stress. Turk. J. Bot. 36, 63–72.

Chaves MM, Flexas J, Pinheiro C (2009) Photosynthesis under drought and salt stress, regulation mechanisms from whole plant to cell. Ann. Bot. 103, 551–560.

Chen C, Dickman MB (2005) Proline suppresses apoptosis in the fungal pathogen *Colletotrichum trifolii*. Proc. Nat. Acad. Sci., USA 102, 3459–3464.

Cheong JJ, Choi YD (2003) Methyl jasmonate as a vital substance in plants. Trends Genet. 19, 409–413.

Cramer GR, Quarrie SA (2002) Abscisic acid is correlated with the leaf growth inhibition of four genotypes of maize differing in their response to salinity. Funct. Plant Biol. 29, 111–115.

Daei G, Ardekani MR, Rejali F, Teimuri S, Miransari M (2009) Alleviation of salinity stress on wheat yield, yield components, and nutrient uptake using arbuscular mycorrhizal fungi under field conditions. J. Plant Physiol. 166, 617–625.

Danielson J, Johanson U (2008) Unexpected complexity of the aquaporin gene family in the moss *Physcomitrella patens*. BMC Plant Biol. 8, 45.

Desai S, Praveen Kumar G, Amalraj LD, Bagyaraj DJ, Ashwin R (2016) Exploiting PGPR and AMF biodiversity for plant health management. In, Microbial Inoculants in Sustainable Agricultural Productivity, Vol. 1, Research Perspectives (Ed) Singh DP, Singh HB, Ratna Prabha), Springer India, pp 145–160.

Dodd IC, Pérez-Alfocea F (2012) Microbial amelioration of crop salinity stress. J. Exp. Bot. 63, 3415–3428.

Estrada B, Aroca R, Azcón-Aguilar C, Barea JM, Ruiz-Lozano JM (2013a) Importance of native arbuscular mycorrhizal inoculation in the halophyte *Asteriscus maritimus* for successful establishment and growth under saline conditions. Plant Soil 370, 175–185.

Estrada B, Aroca R, Barea JM, Ruiz-Lozano JM (2013b) Native arbuscular mycorrhizal fungi isolated from a saline habitat improved maize antioxidant systems and plant tolerance to salinity. Plant Sci. 201–202 , 42–51.

Estrada B, Aroca R, Maathuis FJM, Barea JM, Ruiz-Lozano J (2013c) Arbuscular mycorrhizal fungi native from a Mediterranean saline area enhance maize tolerance to salinity through improved ion homeostasis. Plant Cell Environ. 36, 1771–1782.

Evelin H, Giri B, Kapoor R (2012) Contribution of *Glomus intraradices* inoculation to nutrient acquisition and mitigation of ionic imbalance in NaCl-stressed *Trigonella foenum-graecum*. Mycorrhiza 22, 203-217.

Evelin H, Kapoor R, Giri B (2009) Arbuscular mycorrhizal fungi in alleviation of salt stress, a review. Ann. Bot. 104,1263–1280.

Feng G, Zhang FS, Li XL, Tian CY, Tang C, Rengel Z (2002) Improved tolerance of maize plants to salt stress by arbuscular mycorrhiza is related to higher accumulation of soluble sugars in roots. Mycorrhiza 12,185–190 .

Feng G, Li X, Zhang F, Li S (2000) Effect of AM fungi on water and nutrition status of corn plants under salt stress. Chin. J. Appl. Ecol. 11, 595–598.

Fester T, Hause G (2005) Accumulation of reactive oxygen species in arbuscular mycorrhizal roots. Mycorrhiza 15, 373–379.

Flexas J, Diaz-Espejo A, Galmés J, Kaldenhoff R, Medrano H, Ribas-Carbo M (2007) Rapid variations of mesophyll conductance in response to changes in CO_2 concentration around leaves. Plant Cell Environ. 30,1284–1298.

Flowers TJ, Colmer TD (2008) Salinity tolerance in halophytes. New Phytol 179, 945–963.

Flowers TJ (1999) Salinisation and horticultural production. Sci. Hortic. 78, 1–4.

Flowers TJ, Yeo AR (1995) Breeding for salinity resistance in crop plants, Where next? Aust. J. Plant Physiol. 22, 875–884.

Foo E, Ross JJ, Jones WT, Reid JB (2013) Plant hormones in arbuscular mycorrhizal symbioses, an emerging role for gibberellins. Ann. Bot. 111, 769–779.

García HV, Mendoza RE (2007) Arbuscular mycorrhizal fungi and plant symbiosis in a saline-sodic soil. Mycorrhiza 17, 167–174.

García-Garrido JM, Lendzemo V, Castellanos-Morales V, Steinkellner S, Vierheilig H (2009) Strigolactones, signals for parasitic plants and arbuscular mycorrhizal fungi. Mycorrhiza 19, 449–59.

Garg M, Chandel S (2011a) Effect of mycorrhizal inoculation on growth, nitrogen fixation, and nutrient uptake in Cicer arietinum (L.) under salt stress. Turk. J. Agric. For. 35, 205–214.

Garg M, Chandel S (2011b)The effects of salinity on nitrogen fixation and trehalose metabolism in mycorrhizal Cajanus cajan (L.) Millsp., plants. J. Plant Growth Regul. 30, 490–503.

Garg N, Baher N (2013) Role of arbuscular mycorrhizal symbiosis in proline biosynthesis and metabolism of Cicer arietinum L. (chickpea) genotypes under salt stress. J. Plant Growth Regul. 32, 767–778

Ghassemi F, Jakeman AJ, Nix HA (1995) Salinisation of Land and Water Resources. University of New South Wales Press Ltd, Canberra, Australia.

Giovannetti M, Balestrini R, Volpe V, Guether M, Straub D, Costa A, Ludewig U, Bonfante P (2012) Two putative-aquaporin genes are differentially expressed during arbuscular mycorrhizal symbiosis in Lotus japonicas. BMC Plant Biol. 12, 186.

Gutjahr C, Paszkowski U (2009) Weights in the balance, jasmonic acid and salicylic acid signaling in root–biotroph interactions. Mol. Plant Microbe Interact. 22, 763–772.

Hadad MA, Al-Hashmi HS, Mirghani SM (2012) Tomato (Lycopersicon esculentum Mill.) growth in response to salinity and inoculation with native and introduced strains of mycorrhizal fungi. Int. Res. J. Agric. Sci. Soil Sci. 2, 228–233.

Hajiboland R, Aliasgharzadeh A, Laiegh SF, Poschenrieder C (2010) Colonization with arbuscular mycorrhizal fungi improves salinity tolerance of tomato (Solanum lycopersicum L.) plants. Plant Soil 331, 313–327.

Hammer E, Rillig MC (2011) The influence of different stresses on glomalin levels in an arbuscular mycorrhizal fungus - salinity increases glomalin content. PLoS ONE 6 (12), e28426.

Hammer EC, Nasr H, Pallon J, Olsson PA, Wallander H (2011) Elemental composition of arbuscular mycorrhizal fungi at high salinity. Mycorrhiza 21,117–129.

He T, Cramer GR (1996) Abscisic acid concentrations are correlated with leaf area reductions in two salt-stressed rapid cycling Brassica species. Plant Soil 179, 25–33.

He ZQ, He CX, Zhang ZB, Zou ZR, Wang HS (2007) Changes in antioxidative enzymes and cell membrane osmosis in tomato colonized by arbuscular mycorrhizae under NaCl stress. Colloids Surfaces B. 59, 128–133.

Hillel D (2000) Salinity Management for Sustainable Irrigation. The World Bank, Washington DC.

Huang J-C, Lai W-A, Singh S, Hameed A, Young C-C (2013) Response of mycorrhizal hybrid tomato cultivars under saline stress. J. Soil Sci. Plant Nutr. 13, 469–484.

Huang Z, He C-X, He ZQ, Zou Z-R, Zhang Z-B (2010) The effects of arbuscular mycorrhizal fungi on reactive oxyradical scavenging system of tomato under salt tolerance. Agric. Sci. China 9, 1150–1159.

Ibrahim AH, Abdel-Fattah GM, Emam FM, Abd El-Aziz MH, Shokr AE (2011) Arbuscular mycorrhizal fungi and spermine alleviate the adverse effect of salinity stress on electrolytic leakage and productivity of wheat plants. Phyton 51, 261–276.

Jahromi F, Aroca R, Porcel R, Ruiz-Lozano JM (2008) Influence of salinity on the *in vitro* development of *Glomus intraradices* and on the *in vivo* physiological and molecular responses of mycorrhizal lettuce plants. Microbial Ecol 55, 45–53.

Javid MG, Sorooshzadeh A, Moradi F, Sanavy SAMM, Allahdadi I (2011) The role of phytohormones in alleviating salt stress in crop plants. AJCS 5, 726–734.

Jiang XY, Huang Y (2003) Mechanism of contribution of mycorrhizal fungi to plant saline-alkali tolerance. Ecol. Environ. 12, 353–356.

Juniper S, Abbott LK (1993) Vesicular-arbuscular mycorrhizas and soil salinity. Mycorrhiza 4, 45–57.

Kadian N, Yadav K, Badda N, Aggarwal A (2013) AM fungi ameliorates growth, yield and nutrient uptake in *Cicer arietinum* L., under salt stress. Russian Agric. Sci. 39, 321–329.

Kang DJ, Seo YJ, Lee JD, Ishii R, Kim KU, Shin DH, Park SK, Jang SW, Lee IJ (2005) Jasmonic acid differentially affects growth, ion uptake and abscisic acid concentration in salt-tolerant and salt-sensitive rice cultivars. J. Agron. Crop Sci. 191, 273–282.

Kaya C, Ashraf M, Sonmez O, Aydemir S, Tuna AL, Cullu MA (2009) The infuence of arbuscular mycorrhizal colonisation on key growth parameters and fruit yield of pepper plants grown at high salinity. Sci. Hortic. 121, 1–6.

Khatkar D, Kuhad MS (2000) Short-term salinity induced changes in two wheat cultivars at different growth stages. Biol. Plant. 43, 629–632.

Kohlen W, Charnikhova T, Lammers M, Pollina T, Tóth P, Haider I, Pozo MJ, de Maagd RA, Ruyter-Spira C, Bouwmeester HJ, López-Ráez JA (2012) The tomato CAROTENOID CLEAVAGE DIOXYGENASE8 (SlCCD8) regulates rhizosphere signaling, plant architecture and affects reproductive development through strigolactone biosynthesis. New Phytol. 196, 535–47.

Kohler J, Hernández JA, Caravaca F, Roldán A (2009) Induction of antioxidant enzymes is involved in the greater effectiveness of a PGPR versus AM fungi with respect to increasing the tolerance of lettuce to severe salt stress. Environ. Exp. Bot. 65, 245–252.

Kramell R, Miersch O, Atzorn R, Parthier B, Wasternack C (2000) Octadecanoid-derived alteration of gene expression and the 'oxylipin signature' in stressed barley leaves. Implications for different signaling pathways. Plant Physiol. 123, 177–187.

Kumar T, Ghose M (2008) Status of arbuscular mycorrhizal fungi (AMF) in the Sunderbans of India in relation to tidal inundation and chemical properties of soil. Wetlands Ecol. Manage. 16, 471–483.

Landgraf R, Schaarschmidt S, Hause B (2012) Repeated leaf wounding alters the colonization of *Medicago truncatula* roots by beneficial and pathogenic microorganisms. Plant Cell Environ. 35, 1344–1357.

Lawlor DW, Cornic G (2002) Photosynthetic carbon assimilation and associated metabolism in relation to water deficits in higher plants. Plant Cell Environ. 25, 275–294.

Li T, Hu Y-J, Hao Z-P, Li H, Wang Y-S, Chen B-D (2013) First cloning and characterization of two functional aquaporin genes from an arbuscular mycorrhizal fungus *Glomus intraradices*. New Phytol. 197, 617–630.

López-Ráez JA, Charnikhova T, Gómez-Roldán V, Matusova R, Kohlen W, De Vos R, Verstappen F, Puech-Pages V, Bécard G, Mulder P, Bouwmeester H (2008) Tomato strigolactones are derived from carotenoids and their biosynthesis is promoted by phosphate starvation. New Phytol. 178, 863–74.

López-Ráez JA, Pozo MJ, García-Garrido JM (2011) Strigolactones, a cry for help in the rhizosphere. Botany 89,513–22.

Lovelock CE, Wright SF, Clark DA, Ruess RW (2004) Soil stocks of glomalin produced by arbuscular mycorrhizal fungi across a tropical rain forest landscape. J. Ecol. 92, 278–287.

Ludwig-Muller J, Benett RN, Garcý´a-Garrido JM, Piche´ Y, Vierheilig H (2002) Reduced arbuscular mycorrhizal root colonization in *Tropaeolum majus* and *Carica papaya* after jasmonic acid application cannot be attributed to increased glucosinolate levels. J. Plant Physiol. 159, 517–523.

Maathuis FJM, Amtmann A (1999) K^+ nutrition and Na^+ toxicity, The basis of cellular K+/Na+ ratios. Ann. Bot. 84, 123–133.

Mahajan S, Tuteja N (2005) Cold, salinity and drought stresses, An overview. Arch Biochem Biophy 444, 139–158.

Manchanda G, Garg N (2008) Salinity and its effects on the functional biology of legumes. Acta Physiol. Plant. 30, 595–618.

Manchanda G, Garg N (2011) Alleviation of salt-induced ionic, osmotic and oxidative stresses in *Cajanus cajan* nodules by AM inoculation. Plant Biosyst. 145, 88–97.

Mason E (1928) Note on the presence of mycorrhizae in the roots of saltmarsh plants. New Phytol. 27, 193–195.

Matamoros MA, Baird LM, Escuredo PR, Dalton DA, Minchin FR, Iturbe-Ormaetxe I, Rubio Maria C, Moran JF, Gordon AJ, Becana M (1999) Stress-induced legume root nodule senescence, physiological, biochemical and structural alterations. Plant Physiol. 121, 97–111.

Maurel C, Verdoucq L, Luu DT, Santoni V (2008) Plant aquaporins, membrane channels with multiple integrated functions. Annu. Rev. Plant Biol. 59, 595–624.

Miller GAD, Suzuki N, Ciftci-Yilmaz S, Mittler R (2010) Reactive oxygen species homeostasis and signalling during drought and salinity stresses. Plant Cell Environ. 33, 453–467.

Mittler R (2002) Oxidative stress, antioxidants and stress tolerance. Trends Plant Sci. 7, 405–410.

Mohammad A, Mittra B (2013) Effects of inoculation with stress adapted arbuscular mycorrhizal fungus *Glomus deserticola* on growth of *Solanum melongena* L. and *Sorghum sudanese* Staph., seedlings under salinity and heavy metal stress conditions. Arch. Agron. Soil Sci. 59, 173–183.

Munns R, James RA, Läuchli A (2006) Approaches to increasing the salt tolerance of wheat and other cereals. J. Exp. Bot. 57, 1025–1043.

Munns R, Termaat A (1986) Whole plant responses to salinity. Aust. J. Plant Physiol. 13, 143–160.

Nieman RH (1965) Expression of bean leaves and its suppression by salinity. Plant Physiol. 40, 156–161.

Noctor G, Foyer CH (1998) Ascorbate and glutathione, keeping active oxygen under control. Annu. Rev. Plant Biol. 49, 249–279.

Ort DR (2001) When there is too much light. Plant Physiol. 125, 29–32.

Ouziad F, Wilde P, Schmelzer E, Hildebrandt U, Bothe H (2006) Analysis of expression of aquaporins and Na+/H+ transporters in tomato colonized by arbuscular mycorrhizal fungi and affected by salt stress. Environ. Exp. Bot. 57, 177–186.

Pedranzani H, Racagni G, Alemano S, Miersch O, Ramirez I, Pena-Cortes H, Taleisnik E, Machado-Domenech E, Abdala G (2003) Salt tolerant tomato plants show increased levels of jasmonic acid. Plant Growth Regul. 41, 149–158.

Pitman MG, Läuchli A (2002) Global impact of salinity and agricultural ecosystems. In, Läuchli A, Lüttge U (eds) Salinity, Environment - Plants – Molecules. Kluwer Academic Publishers, The Netherlands, pp 3–20.

Porcel R, Aroca R, Ruiz-Lozano JM (2012) Salinity stress alleviation using arbuscular mycorrhizal fungi. A review. Agron. Sustain. Dev. 32, 181–200.

Promita D, Mohan K (2013) Growth response and dependency of *Arachis hypogaea* L. on two AM fungi under salinity stress. Indian J. Agric. Biochem. 26, 18–24.

Rabie GH (2005) Influence of VA-mycorrhizal fungi and kinetin on the response of mung bean plants to irrigation with seawater. Mycorrhiza 15, 225–230.

Rengasamy P (2006) World salinization with emphasis on Australia. J. Exp. Bot. 57, 1017–1023.

Rillig MC (2004) Arbuscular mycorrhizae, glomalin, and soil aggregation. Can. J. Soil Sci. 84, 355–363.

Ruiz-Lozano JM, Azcón R (1996) Mycorrhizal colonization and drought stress exposition as factors affecting nitrate reductase activity in lettuce plants. Agric. Ecosyst. Environ. 60, 175–181

Ruiz-Lozano JM, del Mar Alguacil M, Barzana G, Vernieri P, Aroca R (2009) Exogenous ABA accentuates the differences in root hydraulic properties between mycorrhizal and non mycorrhizal maize plants through regulation of PIP aquaporins. Plant Mol. Biol. 70, 565–579.

Ruiz-Lozano JM, Porcel R, Azcón C, Aroca R (2012) Regulation by arbuscular mycorrhizae of the integrated physiological response to salinity in plants, new challenges in physiological and molecular studies. J. Exp. Bot. 63, 4033–4044.

Sharma P, Jha AB, Dubey RS, Pessarakli M (2012) Reactive oxygen species, oxidative damage, and antioxidative defense mechanism in plants under stressful conditions. J. Bot. DOI,10.1155/2012/217037

Sheng M, Tang M, Chen H, Yang B, Zhang F, Huang Y (2008) Influence of arbuscular mycorrhizae on photosynthesis and water status of maize plants under salt stress. Mycorrhiza 18, 287–296.

Sheng M, Tang M, Chen H, Yang B, Zhang F, Huang Y (2009) Influence of arbuscular mycorrhizae on the root system of maize plants under salt stress. Can. J. Microbiol. 55, 879–886.

Sheng M, Tang M, Zhang FF, Huang YH (2011) Influence of arbuscular mycorrhiza on organic solutes in maize leaves under salt stress. Mycorrhiza 21, 423–430.

Smith SE, Read DJ (2008) Mycorrhizal symbiosis. Academic Press Inc, San Diego.

Talaat NB, Shawky BT (2011) Influence of arbuscular mycorrhizae on yield, nutrients, organic solutes, and antioxidant enzymes of two wheat cultivars under salt stress. J. Plant Nutr. Soil Sci. 174, 283–291.

Talaat NB, Shawky BT (2012) Influence of arbuscular mycorrhizae on root colonization, growth and productivity of two wheat cultivars under salt stress. Arch. Agron. Soil Sci. 58, 85–100.

Talaat NB, Shawky BT (2013) Modulation of nutrient acquisition and polyamine pool in salt-stressed wheat (*Triticum aestivum* L.) plants inoculated with arbuscular mycorrhizal fungi. Acta. Physiol. Plant 35, 2601–2610.

Talaat NB, Shawky BT (2014a) Modulation of the ROS-scavenging system in salt-stressed wheat plants inoculated with arbuscular mycorrhizal fungi. J. Plant Nutr. Soil Sci. 177, 199–207.

Talaat NB, Shawky BT (2014b) Protective effects of arbuscular mycorrhizal fungi on wheat (*Triticum aestivum* L.) plants exposed to salinity. Environ. Expt. Bot. 98, 20–31.

Tejeda-Sartorius M, de la Vega OM, Delano-Frier JP (2008) Jasmonic acid influences mycorrhizal colonization in tomato plants by modifying the expression of genes involved in carbohydrate partitioning. Physiol. Plant 133, 339–353.

Tian CY, Feng G, Li XL, Zhang FS (2004) Different effects of arbuscular mycorrhizal fungal isolates from saline or non-saline on salinity tolerance of plants. Appl. Soil Ecol. 26, 143–148.

Tommerup I (1984) Effect of soil water potential on spore germination by vesicular–arbuscular mycorrhizal fungi. Trans. Br. Mycol. Soc. 83, 193–202.

Wasternack C, Hause B (2002) Jasmonates and octadecanoids, signals in plant stress responses and development. Prog. Nucleic Acid Res. Mol. Biol. 72, 165–221.

Wilde P, Manal A, Stodden M, Sieverding E, Hildebrandt U, Bothe H (2009) Biodiversity of arbuscular mycorrhizal fungi in roots and soils of two salt marshes. Environ. Microbiol. 11, 1548–1561.

Wu QS, Zou YN, He XH (2010) Contributions of arbuscular mycorrhizal fungi to growth, photosynthesis, root morphology and ionic balance of citrus seedlings under salt stress. Acta Physiol. Plant. 32, 297–304.

Younesi O, Moradi A, Namdari A (2013) Influence of arbuscular mycorrhiza on osmotic adjustment compounds and antioxidant enzyme activity in nodules of salt-stressed soybean (*Glycine max*). Acta Agriculturae Slovenica 101, 219–230.

Xie X(1), Yoneyama K, Yoneyama K (2010) The strigolactone story. Ann. Rev. Phytopathol. 48, 93–117.

Zangaro W, Nishidate FR, Camargo FRS, Romagnoli GG, Vandressen J (2005) Relationships among arbuscular mycorrhizas, root morphology and seedlings growth of tropical native woody species in southern Brazil. J. Trop. Ecol. 21, 529–540.

Zarei M, Paymaneha Z (2014) Effect of salinity and arbuscular mycorrhizal fungi on growth and some physiological parameters of *Citrus jambheri*. Arch. Agron. Soil Sci. 60, 993–1004.

12

Contribution of AMF in the Remediation of Drought Stress in Soybean Plants

Abhishek Bharti, Shivani Garg, Anil Prakash and Mahaveer P. Sharma

Abstract

Amongst various oilseed crops grown across the world, soybean (*Glycine max* L. Merrill) is globally important as a high source of protein and oil for human consumption and is also being used as potential feed for animals. Soybean is grown under rain-fed conditions in the tropics and subtropics in the marginal lands with less amount of external application of chemical fertilizers. The frequent climate variations have created a number of biotic and abiotic stresses, drought being a major abiotic stress which adversely affects the soybean productivity. Thus, to enhance the productivity of soybean, besides managing the nutrients, stress management is of utmost importance. There is a great opportunity of application of microbes especially arbuscular mycorrhizal fungi (AMF) where its application alone or in combination with plant growth promoting rhizobacteria (PGPR) can help in nutrient mobilization and remediation of plant stresses. AMF are a ubiquitous group of soil fungi known to colonize roots of plants belonging to more than ninety per cent of plant families. In this chapter, detailed account of work done on the mycorrhizal symbiosis and its role in alleviation of drought stress has been presented. We have also described mechanisms underlying drought tolerance in AM-colonized soybean plants. The information on understanding interactions between AMF and PGPR and agronomic practices is also provided which would help in the abiotic stress tolerance of soybean plants. In addition, a brief account on the production methods of AMF and availability of AMF inocula is also provided.

Keywords: Commercial AM inocula, Drought, Mycorrhizal fungi, Soybean

1. Introduction

Soybean [(*Glycine max* (L.) Merril] containing about 40-42% protein and 18-22% oil is emerging as one of the fast growing oilseed crop in the world. In India, Malwa plateau of Central India is the hub for soybean cultivation, the total production during 2014 was about 11.64 mt from 10.02 m hectare (Anonymous 2014-15). Although, the spread of soybean in different parts of the country resulted into parallel growth of oil industry but for the past few years soybean is facing climatic challenges where the soybean yields are declining. When compared to other countries the productivity of soybean per unit area is very low in India and productivity stagnated due to recurrence of drought, the low nutrient use efficiency of crop, nutrient deficiency in soil and other biotic and abiotic stresses. According to FAO's report, 2013; drought has become more frequent and intense worldwide and is a major factor causing alterations in plant morphology, physiology, nutrient uptake and metabolism, ultimately affecting the plant's growth and development adversely (Evelin *et al.* 2009). Various studies have shown that drought stress leads to decrease in soybean yield up to 50% (Sadeghipour and Abbasi, 2012).

Numerous mitigation strategies can be used to cope with such impacts. However, most of them being based are either a long-term in nature like breeding drought tolerance varieties or based on use of chemical fertilizers which are currently becoming inaccessible and costly to farmers. Hence there is a need to develop eco-friendly and low cost biological methods for the management of abiotic stress. Microorganisms such as AMF could be successfully exploited for the purpose, as they possess unique properties such as tolerance to extremities, ubiquity, genetic diversity and interactions with soil and crop plants. Besides influencing the physico-chemical properties of rhizospheric soil through production of exopolysaccharides and formation of biofilm, microorganisms can also influence higher plants' response to abiotic stresses like drought through different mechanisms like induction of osmo-protectants, heat shock proteins etc. in plant cells. Use of these microorganisms can help crop plants to cope with drought stress. They also provide excellent models for understanding the stress tolerance, adaptation and response mechanisms that can be subsequently engineered into crop plants to cope with stresses (Grover *et al.* 2010).

Among all, AMF are cosmopolitan in all soil types and form symbiotic association with the roots of over 90 % terrestrial plant species. They have coevolved with plants over 400 million years ago. They improve adaptation capability of plants to different types of stress like drought. The soil surrounding the plant roots, i.e., rhizosphere, is an ecological niche which is integral to many biochemical reactions and is deeply influenced by the root exudates. Here, microorganisms have multiple roles such as plant growth promotion, degradation of natural or synthetic compounds. Some of them can act as pathogens as well. PGPR are also known to elicit systemic tolerance to abiotic stresses in plants, apart from the plants' own survival mechanisms such as drought avoidance (through reduced transpiration losses following stomatal closure or water storage in plant tissues) and drought tolerance (through accumulation and translocation of assimilates, osmotic adjustments or maintenance of cell wall elasticity). The plants' adaptations to drought stress can cause changes in belowground C input through higher root production and turnover which in turn influences the functional structure and activity of the microbial community in the rhizosphere (Grayston *et al.* 1998). Several ecophysiological studies have demonstrated that AMF symbiosis is a key component in helping plants to cope with water stress, as demonstrated in a number of host plants. Mycorrhizal colonization improves drought resistance of plants as a consequence of enhancing nutritional status, especially P and water status which in turn enhances plant growth and productivity. Further, plants develop their own survival strategies as well to increase tolerance against drought (Bohnert *et al.* 2006). In this chapter, we describe how AMF contribute in combating drought stress, particularly in soybean plant. Mycorrhizal symbiosis can potentially improve water uptake by plants, mainly through improved nutrient uptake which can be attributed to the enhanced absorbing surface area provided by AMF hyphae in the soil together with the fungal ability to take up water from soil having low water potential. Bohnert *et al.* 2005 have studied signaling interactions in the AMF-soybean symbiosis that is expected to regulate the response of mycorrhizal soybean to drought stress.

2. Role of AMF in enhanced drought stress alleviation in soybean

Drought stress affects many physiological plant processes which in turn induce premature senescence in root nodules of legumes including soybean, thereby decreasing their ability to fix atmospheric nitrogen. Various symptoms of drought stress include reduction in leaf size, extension in stem length and root proliferation. These cause disturbance in plant water relations leading to reduced water-use efficiency (WUE). Symbiosis between AMF and most plants helps nutrient and mineral uptake, and increase plant stress tolerance. Studies have established that association of roots of numerous plant species with AMF increased uptake of various minerals from the soil, including water, and also help host plant's ability to grow under conditions of drought stress (Barea *et al.* 2005). However, in the last two decades, comparatively fewer researchers have taken up studies on symbiosis between AMF and soybean (Table1).

Table 1: Influence of AMF on drought tolerance of soybean (past two decades)

Inoculation	Condition/Stress	Reference
AMF-*Glomus mosseae, Glomus etunicatum* with and without *B. japonicum*	Different moisture stress /field capacity levels in pots	Aliasgharzad *et al.* 2006
Septoglomus constrictum, Glomus sp., and *Glomus aggregatum* & mixture	Water Stress /Drought stress	Grumberg *et al.* 2015
Glomus intraradices	*Environmental chamber	Porcel *et al.* 2003
Glomus mosseae, Glomus intraradices	*Environmental chamber	Ruiz-Lozano *et al.* 2001
G. mosseae, G. intraradices and *Piriformospora indica*	Water stress maintained by stopped water spraying /Drought stress	Rathod *et al.* 2011
Glomus intraradices	Water stress maintained by stopped water spraying /Drought stress	Liu *et al.* 2015
Glomus intraradices	*Environmental chamber	Porcel R & Ruiz-Lozano JM, 2004
G. deserticola, G. etunicatum, G. intraradices, G. fasciculatum, G. mosseae, G. caledonium, G. occultum	**Environmental chamber	Ruiz-Lozano *et al.* 1995
Glomus intraradices	Leaf osmotic potential variation	Auge *et al.* 2001

*Environmental Chamber conditions: 70–80% RH, day/night temperatures of 25/15°C, and a photoperiod of 16 h at a photosynthetic photon flux density (PPFD) of 460–500 μmol m^{-2}s^{-1}

**Environmental Chamber conditions: 70–80% RH, day/night temperatures of 25/15°C, and a photoperiod of 14 h. The photosynthetic photon flux density was 500 μmol m^{-2}s^{-1}

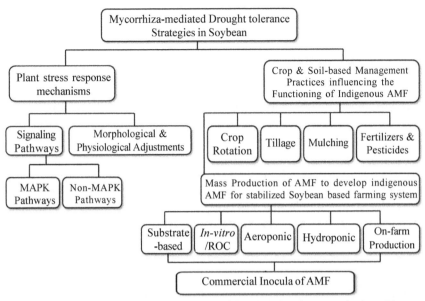

Fig. 1: Mycorrhiza-mediated approaches in providing drought tolerance to Soybean Plants

In soil, water scarcity is tightly linked to low nutrient availability and so various hypotheses have been formulated to explain the underlying plant nutrition mechanisms. Improved nutrient uptake by AMF is a fundamental mechanism that can alleviate the adverse effects of water stress on plant growth (Rapparini and Penuelas, 2014*)*. One of the most common explanations for the improved nutrient status in mycorrhizal plants is the enhanced surface area for absorption that is provided by AMF hyphae (Ruiz-Lozano *et al.* 2003).

3. Mechanisms of drought tolerance

Plants usually interact with soil microorganisms that make it more efficient in coping with environmental stress such as drought. (Azcon *et al.* 2013). The AMF mediated response in soil causes many physiological and biochemical changes in water availability owing to a number of mechanisms that enable drought resistance in plants. Communication between AMF and host plant starts when a spore comes into contact with a host plant root, since the cell walls are the outermost boundary of the root; they are the first interface for interactions with soil-borne microbes. Based on this notion, it has earlier been postulated that interactions of the cell walls between plants and microbes may play an important role in symbiotic interactions. This relation is important for both, plants and AMF. After infection, hyphae penetrate deep into the parenchyma cortex. Here, fungal development culminates in the differentiation of intracellular haustoria, known as arbuscules. These fungal structures, which establish a large surface of contact with the plant protoplast, are attributed to a key role in reciprocal nutrient exchange between the plant cells and the AMF symbionts. Outer layer of plant root's cell wall participates in many interactions with the soil environment. Evidences suggest that plant itself plays a central role in softening and remodeling of the cell wall during symbioses.

In higher plants, drought stimuli are presumably perceived by osmosensors (that are yet to be identified) and then transduced down the signaling pathways, which activate downstream drought responsive genes to display tolerance effects. The tolerance involves not only the activities of protein receptors, kinases, transcription factors, and effectors but also the production of metabolites as messengers for transducing the signals. Drought tolerance is of multigenic nature, involving complex molecular mechanisms and genetic networks. The signaling pathway of drought stress is similar to those of osmotic stresses that have been reviewed in detail by Ahuja *et al.* (2010).

3.1. MAPK pathway

Mitogen activated protein kinase (MAPK) cascades play important roles in the stress response in both plants and microorganisms. The symbiosis established between AMF and plants can enhance plant drought tolerance, which might be closely related to the fungal MAPK response and the molecular dialogue between fungal and soybean MAPK cascades. Many studies support the role of protein kinases in stress signaling (Bartels *et al.* 2010). In plants, the drought responsive signal transduction of the MAPK family (MAPK, MAPKK, MEKK, MAPKKK, MKK) as well as the MAPK phosphatases (MKP) family have been relatively well-studied in *A. thaliana* and rice but remained under-explored in soybean, although a PA-responsive MAPK has been identified in soybean (Lee S *et al.* 2001). In soybean, two-component histidine kinases (GmHK07, GmHK08, GmHK09, GmHK14, GmHK15, GmHK16 and GmHK17) and receptor-like protein kinases (GmCLV1A, GmCLV1B, GmRLK1, GmRLK2, GmRLK3 and GmRLK4) have been identified as candidates of osmosensors. However, direct evidence for their functions to perceive stress signals in soybean is still missing. Drought stress upregulates the level of MAPK transcripts in mycorrhiza-colonized soybean roots. Hence, there might exist a molecular dialogue between the two symbionts to regulate the mycorrhizal soybean drought-stress response. Meanwhile, the changes in hydrogen peroxide, soluble sugar, and proline levels in mycorrhizal soybean as well as in the accelerated exchange of carbon and nitrogen in the symbionts were contributable to drought adaptation of the host plants. Thus, it can be preliminarily inferred that the interactions of MAPK signals on both sides, symbiotic fungus and plant, might regulate the response of symbiosis and, thus, improve the resistance of mycorrhizal soybean to drought stress. (Zhilei *et al*. 2014*)*. Extensive study has demonstrated AM-mediated plants are resistant to drought conditions, but the underlying mechanisms have not yet been clearly elucidated.

3.2. Non-MAPK signaling pathways

On the other hand, some non-MAPK type protein kinases found in soybean may be related to drought responses. ABA-mediated signaling pathway: Abscisic acid (ABA) helps plants' response to abiotic stresses by regulating their physiology (Zhang *et al*. 2006). Its biosynthesis, accumulation, and catabolism are all crucial for the transduction of ABA-mediated signals. It is synthesized in various cell types including root cells, parenchyma cells, and mesophyll cells. Under drought stress, ABA is transported to guard cells to control stomatal aperture (Wilkinson S *et al*. 2010). ABA reaching the target tissues and cells will be recognized and the signals will be transduced down the ABA signalosome (Umezawa *et al*. 2010), including ABA receptors (PYR/PYL/RCAR), negative regulators (e.g. group A protein phosphatases 2C) and positive regulators (e.g. SnRK-type kinases). Components of this system have been discovered in soybean. For example, GsAPK is a SnRK-type kinase from wild soybean that is up-regulated by drought stress in both leaves and roots, but down-regulated by ABA treatment in roots (Yang *et al*. 2012). Accumulation of ABA in response to drought is associated with changes in Ca^{2+} (second messenger) and reactive oxygen species (ROS) levels. Drought-induced changes in cytosolic Ca^{2+} level activate the signaling pathways for downstream stress responses (Xiong *et al*. 2002). Various types of Ca^{2+}-binding proteins: CaMs (calmodulins), CMLs (CaM-like proteins), CDPKs (Ca^{2+}-dependent protein kinases), and CBLs (calcineurin B-like proteins) (Falco *et al*. 2010) act as Ca^{+2} sensors.

Expression of the soybean CaM (GmCaM4) in transgenic *A. thaliana* activated a R2R3 type MYB transcription factor which in turn up-regulated several drought-responsive genes, including *P5CS* (encoding a proline anabolic enzyme) (Yoo *et al*. 2005). While the application of Ca^{2+} affects the nodulation of soybean (Bell *et al*.1989), the gene encoding a soybean CaM binding protein was found to be differentially expressed in soybean nodules under drought stress (Clement *et al*. 2008). In isolated soybean symbiosome membrane, a CDPK was demonstrated to phosphorylate an aquaporin called nodulin 26 during drought conditions and hence enhance the water permeability of the membrane. (Guenther *et al*. 2003), Besides Ca^{2+}, phosphatidic acid (PA) and the intermediates of inositol metabolism are also second messengers for signal transduction (Xue *et al*. 2009). However, there are only very limited evidence supporting the involvement of phospholipid signaling in drought stress response of soybean.

Cellular ROS gets accumulated during drought stress that triggers the generation of hydrogen peroxide, a signaling molecule that will activate ROS scavenging mechanisms (Cruz *et al.* 2008). In soybean, exogenous application of hydrogen sulphide alleviates symptoms of drought stress, probably *via* triggering an antioxidant signaling mechanism (Zhang *et al.* 2010).The ubiquitin-mediated protein degradation pathway: this is also an integral part of the signal transduction network (Zhou *et al.* 2010). This pathway directs the degradation of target proteins by the 26S proteasome and is responsive to drought stress. Two ubiquitin genes and one gene encoding ubiquitin conjugating enzyme were identified as differentially expressed genes in nodulated soybean under drought stress (Clement *et al.* 2008). Over expression of the ubiquitin ligase gene *GmUBC2* enhances drought tolerance in *A. thaliana*, *via* up-regulating the expression of genes encoding ion transporters (AtNHX1 and AtCLCa), a proline biosynthetic enzyme (AtP5CS), and a copper chaperone (AtCCS) (Zhou *et al.* 2010).

3.3. Morphological and physiological adjustments of soybean under drought stress

Various morphological traits indicate drought stress response in soybean (Liu *et al.* 2005). For instance, root distribution, which is measured in terms of horizontal and vertical root length or dry matter in soil of different depths (Benjamin and Nielsen, 2006). Liu *et al*, 2005 have also reported a positive correlation between drought tolerance and dry root weight/ plant weight; total root length/ plant weight, and root volume/ plant weight. Further, the root to shoot ratio increases under water deficit conditions (Wu and Cosgrove, 2000), indicating a higher sensitivity of shoot than roots towards drought conditions. This happens due to differential changes in cell wall composition involving thickening of shoot cell wall and expansion of root cell wall by catalytic enzymes and stiffening agents. However, there are only limited reports on related studies in soybean. The study on GmRD22 from soybean suggested a relationship between osmotic stress and cell wall metabolism. GmRD22 is a BURP-domain containing protein localized in the apoplast, which may play a role in stress tolerance by regulating lignin content of cell wall under stress, presumably through interacting with peroxidases of the cell wall (Wang *et al.* 2012). Changes in leaf morphology may also play a role in drought tolerance. Some cultivars take advantage from the maintenance of leaf area which provides a possible benefit for the growth of soybean plant after the stress is relieved (Manavalan *et al.* 2009). Under stress, drought tolerant soybean cultivars exhibited a larger leaf area when compared with less tolerant cultivars; this can be attributed to a reduction in stomatal conductance and photosynthetic rate in the tolerant cultivar (Stolf *et al.* 2010).

3.4 Physiological adjustments

Plants undergo physiological adjustments at various levels to cope up with drought stress. For instance, it is important for the soybean leaves to adjust stomatal conductance to prevent excessive water loss. Stolf *et al.* 2010, reported a higher degree of reduction in stomatal conductance as compared to drought sensitive cultivar BR16, 30 days after water stress. Maintenance of cell turgidity is another crucial adjustment. To maintain cell turgidity under stress, osmotic adjustment is a common mechanism which involves active accumulation of solutes in cells (Manavalan *et al.* 2009). Drought tolerant soybean cultivar PI 416937 has been found to maintain a lower solute and higher water potential as compared to sensitive cultivar Forrest, which in turn resulted in a higher seed weight and yield than Forrest under drought (Sloane *et al.* 1990). In soybean, drought stress up-regulates the expression of the *P5CS* gene which encodes the enzymeÄ1-pyrroline-5-carboxylate synthase, a key enzyme in proline biosynthesis (Porcel R *et al.* 2004). Knocking down this gene hampered survival under drought stress (Ronde *et al.* 2000). However, the involvement of proline accumulation in drought stress adjustment in soybean awaits further confirmation. Cellular biochemical adjustments under drought stress involve scavenging of ROS. Under normal situation, ROS including singlet oxygen, superoxide radical, hydrogen peroxide, and hydroxyl radical are continuously synthesized and eliminated in plant cells as "by-products" of photosynthesis, photorespiration, and respiration in chloroplast and mitochondria (Foyer *et al.* 2003). Under drought stress, ROS accumulates when its production outweighs removal (Agarwal *et al.* 2005). The over-produced ROS attacks cellular components including nucleic acids, protein, and lipid and eventually leading to cell death (Mittler 2002). ROS scavenging enzymatic activities of superoxide dismutase, catalase, and glutathione peroxidase have been found to increase in 5 soybean germplasms under drought stress (Masoumi *et al.* 2010). Ectopic expression of the *GmPAP3* gene (encoding a mitochondria localized purple acid phosphatase) from soybean significantly reduces ROS accumulation and thereby alleviates osmotic stress (Li *et al.* 2008). Drought stress can also cause misfolding of ER proteins resulting in unfolded protein response (Liu *et al.* 2010). A number of genes have been identified as potential candidates for integration of ER stress signaling responses by global expression-profiling analyses on soybean leaves (Irsigler *et al.* 2007). Moreover, over expression of soybean BiP (binding protein); an ER-resident molecular chaperone enhanced drought tolerance in soybean (Valente *et al.* 2009).

4. Interaction of AMF with other plant growth promoting rhizobacteria (PGPR)

Plant Growth Promoting Rhizobacteria (PGPR) have important role in improving soil fertility. Soil surrounding the plant roots, i.e., rhizosphere is an ecological niche which is integral to many biochemical reactions and is deeply influenced by the root exudates. Here, PGPR play a critical role in plant growth promotion or act on pathogens inhibiting plant growth and also take part in microbial degradation of natural or synthetic compounds. PGPR are also known to elicit systemic tolerance to abiotic stresses such as drought stress. Several ecophysiological studies have demonstrated that AMF symbiosis is a key component in helping plants to cope with water stress and in increasing drought resistance. Investigations on a number of host plants and fungal species have shown that mycorrhizal colonization improves drought resistance of the marigold plants as a consequence of enhancing nutritional status, especially P and water status which in turn enhances plant growth and productivity. Further, plants develop their own survival strategies as well to increase tolerance against drought (Bohnert et al. 2006).

AMF, a key member of the rhizosphere, form a symbiotic relationship with the host plant and cast a positive effect on its growth and nutrient uptake. Proliferation of AMF symbioses with the host plant can positively be influenced by PGPR through various mechanisms such as increased spore germination and hyphal permeability in plant roots. Although there are evidences that combined interactions between AMF and PGPR can promote plant growth, better understanding of the interactions between AMF and other microorganisms is necessary for maintaining soil fertility and enhancing crop production (Ramasamy et al. 2011). Commonly reported plant growth promotion mechanisms by bacteria are the morphological and physiological changes of the root system, increased roots and root hairs that facilitate more nutrients and water absorption. Higher water and nutrient uptake by roots cause improved water status of plants (Wu and Cosgrove, 2000). PGPR inoculation may help to improve crop resistance against abiotic stress conditions. Exopolysaccharides produced by *Pseudomonas mendocina* binds to soil cat ions including Na and reduce the Na available for plant uptake. Glycoprotein (glomalin) produced by AMF can act as an insoluble glue to stabilize soil aggregates. Drought is a major limitation for crop production in rain-fed ecosystems Synergistic effect of co-inoculated bacteria and AMF help in restoring plant growth under drought conditions (Marulanda et al. 2009).The use of indigenous drought tolerant *G. intraradices* strain along with native bacterium reduced 42% water requirement for the production of *Retama sphaerocarpa*. AMF have 500% increased shoot fresh weight (SFW) compared to uninoculated control plants. Interestingly, these AMF plants co-inoculated with *Azospirillum* showed a further increase of 12% in SFW. In drought stressed conditions, combined inoculation of AMF and *Azospirillum* increased SFW by 103% compared to the uninoculated control. Similar results were also observed by Franzini *et al*, 2010. Combined inoculation of ACC deaminase positive *Psudomonas putida* and *Gigaspora rosea* showed increased plant growth and improved root architecture. The results also showed that ACC deaminase producing PGPR strain along with AMF can improve the survivability of plants under stressed conditions. PGPR are associated with plant roots and affect plant productivity and immunity; however, recent work by several groups show that PGPR also induce systemic tolerance to drought. (Yang *et al*. 2009). Specific combinations of autochthonous or allochthonous inoculants also contribute to plant drought tolerance by changing proline and antioxidative activities. However, non-inoculated plants have low relative water and nutrients contents; shoot proline accumulation and glutathione reductase activity, but the highest superoxide dismutase activity, stomatal conductance and electrolyte leakage. Microbial activities irrespective of the microbial origin seem to be coordinately functioning in the plant as an adaptive response to modulated water stress tolerance and minimizing the stress damage (Ortiz *et al*. 2015).

5. Agronomic practices favouring AMF for alleviating drought

Various management strategies have been proposed to cope up with drought stress. Agronomic practices include integrated cropping systems mainly involving crop and soil management practices. In order to increase our ability to optimize management of AMF in field situations, there is a need for more basic information on the seasonal variation in the development of AMF in different crop species and how this is influenced by agricultural practices. The development of a diverse AMF population which can adapt to management and environmental changes is likely to be a key factor in improving the sustainability of low input and organic cropping systems. The results of such studies show that the interactions between plants and AMF are complex and their expression for response is strongly dependent on the environmental factors. Therefore there is a prerequisite to understand the environmental and cultural factors for effective management of AMF.

Crop rotation modifies the population of soil microflora, including AMF by changing the availability of soil nutrients and beneficially reducing the deleterious organisms for the benefit of crop health. Agricultural management thus influences both the presence of AMF and their activity. Organic agriculture is used, here, as a model for low input agriculture systems. Around 60 years ago Sir Albert Howard, a founder of the organic movement suggested that, the presence of an effective AMF symbiosis is essential to plant health. Studies relating to the effects of different composts on spore populations have revealed that both fungal species and the nature of the organic material are likely to be related to the concentration and availability of nutrients. Within crop rotations, colonized roots and hyphae are an important source of inocula for the cropping sequence (Boswell et al. 1998). There are lots of results for the so-called 'rotation effect' in addition to the straight nutritional beneficial effects of N-fixation by legumes in rotations. Cook, 1986 has attributed the 'rotation effect' to improved root health, in which AMF may have an important role. Azcon et al (1996) have addressed the beneficial effects of AMF in reducing disease susceptibility. As a result of finding large changes in mycorrhizal communities associated with different rotations under the same soil and climatic conditions, Hendrix et al (1995) suggested a role for AMF in the rotation effect. Another promising effect of combining rotations with low input systems is weed control. There is evidence that tillage can restrict P nutrition early in the season by disrupting the extraradical hyphae. Tillage may cause a shift in AMF and host plant communities (Douds et al. 1995). A study on various cropping systems in soybean resulted in significantly higher AMF spore count (Sharma et al. 2012). Inclusion of maize in the rotation irrespective of tillage systems showed comparatively higher phosphatase activities.

Traditional mulching involves covering of the field with straw. The mulch can trap moisture and hence retain soil water. The degrading organic mulch also adds humus to the soil and improves the water holding capacity of the soil that will help AMF to grow very efficiently. It will also help plants to grow under stressed environment. In China, plastic mulch has been widely used on soybean inter planted with maize, potato or cucumber. For example, a study conducted in Shouyang County of the Shanxi Province, China suggested that mulching cultivation with hole-sowing or row-sowing techniques can increase soybean yield up to 23.4% and 50.6%, respectively (Guo et al. 2007).

6. Methods of producing AMF and its availability

AMF has been found to be effective as it focuses on developing a healthy soil and enables plants to become better attuned to the environment through an improved nutrient absorption. Therefore its availability has becomes utmost importance. Since AMF are obligate symbionts, require host to complete its life cycle, therefore production under laboratory conditions is still a complex and difficult to practice by most of growers. There are various methods currently being followed to practice.

6.1 On-farm production

Traditionally, the on farm method of inoculum production has been developed in several ways in a number of countries. The on-farm technology is cost-effective and can be easily transferred to farmers (Douds et al. 2006; Sharma and Sharma, 2008). In this system, AMF production is done by farmers on their own property, under natural conditions using indigenous or introduced AMF isolates (Douds et al. 2008, 2012). Indigenous AMF species seem to be more efficient in some situations as they are locally adapted to the soil conditions (Sreenivassa et al. 1992). Commercial AMF produced using these systems are available in several countries; however, the costs associated with the technology of AMF production, including establishment of single cultures of AMF species, shipping and handling, and development of the carrier substrate are to be borne by farmers and nursery owners (Douds et al. 2006).

6.2 Substrate based

Another promising method of production of AMF inoculum uses sand and vermiculite irrigated with nutrient solution. Large-scale multiplication of AMF for field applications is generally carried out in substrate-based, substrate-free, and in vitro systems (Ijdo et al. 2011). Organic amendments added to the substrate can stimulate sporulation of AMF and replace the nutrient solution. The production of spores varies among the tested AMF and according to the organic source added to the substrate. The vermicompost promotes higher sporulation of certain AMF such as *Acaulospora longula* in relation to other AMF species and substrates (Ijdo et al. 2011).

6.3 Aeroponic production of AMF

Production of AMF by aeroponic system enables the production of cleaner spores and facilitates uniform nutrition of colonized plants. This system allows for efficient production of AMF free of any physical substrate. The colonized root material can be sheared resulting in inocula with high propagule densities. Aeroponically produced inocula shows tremendous impact on plant growth and health and can become a key aspect for sustainable agriculture (Singh et al. 2012). Multiple techniques have been developed in the past for the mass production of AMF. In the present time, in vitro cultivation methods such as hydroponic system and root organ culture have been used for the mass production of AMF. These methods not only maintain the quality of AMF propagule but they can also be developed as cost effective methods for the mass propagation of AMF.

6.4 Hydroponic production of AMF

AMF inoculum can be produced hydroponically where by roots of plants supported on a solid medium or structure are submerged in a reservoir of a nutrient solution such as dilute Hoagland's solution or Hewitt's solution with low phosphorus concentration. (Mosse et al. 1984) Full-strength Hewitt's solution (Thompson et al. 1986) consists of (mg/L) Ca 160, K 156, N 114 (NO_3 50–100%). S 112 or 240, P 41, Mg 36, Na 246 or 62, Cl 284, Fe 2.8, Mn 0.55, B 0.33, Zn 0.065, Cu 0.015, Mo 0.015, Co 0.015. The roots of the plant grow through the band of support structure or medium into a nutrient reservoir. Air is continuously bubbled throughout the solution. The nutrient solution is changed at regular intervals. In a submerged sand system, it is necessary to change the medium at an interval of 3–4 days (Thompson et al. 1986) Distilled or deionized water is added to the reservoir as needed. Nine to ten weeks after transplanting, plant tops are cut and roots recovered from the reservoir, processed as needed, and either used immediately or stored for use at a later time. Alternatively, AMF roots can be produced by growing suitable nurse plants in a sand matrix submerged in a nutrient solution conducive for AMF development (Thompson et al. 1986). Fine roots are sampled and examined for AMF colonization. For instance, Macdonald et al (1981) have described a compact autoclavable hydroponic culture system for the production of axenic AMF formed between *Trifolium parviflorum* and *Glomus caledonius*.

6.5 *In-vitro* production of AMF or ROC (Root Organ Culture)

In-vitro culture of AMF was achieved for the first time in the early 1960s (Mosse, 1962). Since then, various pioneering steps were aimed at axenic culturing of AMF, continuous cultures of vigorous ROCs (Ri T-DNA-transformed) have been obtained through transformation of roots by the soil bacterium *A. rhizogenes* or mass scale production of AMF was achieved by root organ culture in small containers (Tiwari and Adholeya, 2003). These are methods that help mass production of roots in a very short span of time. The ROC is an attractive mass multiplication method for providing viable, rapid and pure inocula. Different production systems have been derived from the basic ROC in petri plates. For example, Gadkar et al. (2006) further developed a system where a petri plate containing a ROC was used to initiate fungal propagation in a separate compartment filled with sterile expanded clay balls. Recently monoxenic culture of *G. intraradices* with Ri T-DNA transformed roots in two-compartment petri dishes was also achieved. (Sharma et al. 2015). A derived plant *in vitro* production system has also achieved and a patent filed (Declerck et al. 2009) where each pre-inoculated *in vitro* produced plant (Voets et al. 2009) is individually introduced into a sterile growth tube. A nutrient solution circulates in that closed system flowing on the mycorrhizal roots. A list of commercial formulations of AM inocula produced by either of the above methods is given in Table 2. The most common method being bulk inocula containing a mixture of spores, colonized roots, hyphae and substrate from the pot containers, usually grown in sterilized soil or soil less media. For substrate-host based method, in addition to use of selected AMF, there is a need to optimize the best substrate and its optimal forms and time of production to obtain maximum sporulation and production of AMF.

Table 2: List of commercial sources/products/suppliers of AM inoculants

Company	Country
Symplanta GmbH & Co. Kg	Munich, Germany
Ag Bio Inc., Westminster	Colorado, USA
Accelerator Horticultural Products	Ohio, USA
Bio-Organics Supply,	Camarillo California, USA
Becker –Underwood, Ames	Iowa, USA
Bio Scientific, Inc., Avondale	Arizona, USA
Eco Life Corporation,	Moorpark California, USA
First Fruits Sarasota,	Florida, USA
J.H. Biotech, Inc., Ventura,	California, USA
First Fruits	Sarasota, Florida, USA
J.H. Biotech, Inc.,	Ventura, California, USA
Mikro-Tek Inc.,	Ontario, Canada
Mycorrhizal Applications, Grants Pass	Oregon, USA
Bio Grow TM	North America
Plant health Care, Inc.,	Pennsylvania, USA
Mycor TM VAM Mini Plug TM	North America
Premier Horticulture, Red Hill	Pennsylvania, USA
Premier Tech	Quebec, Canada
Reforestation Technologies, Salinas	California, USA
Roots Inc., Independence	Montana, USA
T & J Enterprises, Spokane	Washington, USA
TIPCO, Inc., Knoxvile	Tennesse, USA
Tree of Life Nursery, San Juan Capistrano	California, USA
Tree Pro, West Lafayette	Indiana, USA
Biological Crop Protection Ltd	Kent, UK
Bio-organics	Medillin, Columbia
Biorize	Dijon, France
Central Glass Co., Chemicals Section	Tokyo, Japan
Global Horticare	Lelystad, The Netherlands
Idemitsu Kosan Co.,	Sodegaura, Chile

(Contd.)

MicroBio, Ltd	Royston, Hertz, UK
N-Viron Sdn Bhd	Malaysia
PlantWorks Ltd., Sittingbourne	UK
Triton Umweltschutz GmbH	Bitterfeld, Germany
KCP Sugar and Industries Corporation Ltd	Andra Pradesh, India
Cadila Pharmaceuticals Ltd	Ahmedabad, India
Symbiotic Sciences Pvt Ltd	Gurgaon, Haryana, India
Symbiom (ViaTerra LLC, Jacksonville)	FL, USA

Source: Modified from Sharma and Adholeya (2008)

Conclusion

Soybean is nutritionally and economically important crop. Its growth is affected by various environmental changes such as climatic variations, biotic and abiotic stresses like drought stress. Due to drought stress, soybean production is very much affected worldwide. In this chapter, we have explained effects of AM-mediated drought stresses on soybean and the underlying drought responsive mechanisms. The studies pertaining to morphological, physiological, and molecular changes occurring during stress conditions have led to accumulation of considerable information regarding possible methods of overcoming these stresses. Mycorrhizal plants employ various protective mechanisms to counteract drought stress. The accumulated physiological, biochemical, and molecular data based on classical approaches will benefit from the various 'omic' techniques and their combinations. An in-depth investigation using the advanced methodologies could help to elucidate the mechanisms of drought avoidance and/or tolerance induced by AM symbiosis and to discriminate the drought-induced processes of the protective mechanisms regulated by AM symbiosis. Soybean productivity, under drought, can be improved by integrating all technologies and knowledge involved. Furthermore, some key efforts are needed to identify efficient strains of PGPR and AMF, which can enhance tolerance against biotic or abiotic stresses. Many studies have shown large amounts of hyphal biomass and higher indigenous AMF in crop rotations but the combined application of AMF and PGPRare yet to be streamlined. Most importantly AM inoculum production method particularly ROC systems needs critical evaluation and quality checks by third party/stake holders. The methods for checking the quality assurances need to be standardized. Finally, potential commercial formulations need to be subjected to regulatory requirements and quality checks.

Acknowledgements

The authors would like to thank the Director, ICAR-Indian Institute of Soybean Research, Indore for providing the necessary facilities. Funding from ICAR-AMAAS network subproject to MPS and SRF to AB and SG is also gratefully acknowledged.

References

Agarwal S, Sairam R, Srivastava G, Meena R (2005) Changes in antioxidant enzymes activity and oxidative stress by abscisic acid and salicylic acid in wheat genotypes. Biol. Plant. 49, 541-550.

Aliasgharzad N, Neyshabouri MR, Salimi G (2006) Effects of arbuscular mycorrhizal fungi and *Bradyrhizobium japonicum* on drought stress of soybean. Biology 61, 324-328.

Anonymous, (2014-15). Director's Report and Summary Tables of Experiments 2014-15, All India Coordinated Research Project on Soybean, ICAR-Directorate of Soybean Research, Indore India pp 329.

Ahuja I, Vos D, Rich CH, Bones AM, Hall RD (2010) Plant molecular stress responses face climate change. Trends Plant Sci. 15(12), 664-674.

Auge RM, Kubikova E, Moore JL (2001) Foliar dehydration tolerance of mycorrhizal cowpea, soybean and bush bean. New Phytol. 151, 535-541.

Azcon R, Aguilar C, Barea JM, (1996) Arbuscular mycorrhizas and biological control of soil borne plant pathogens. Mycorrhiza 6, 457-464.

Azcon R, Medina A, Aroca R, Ruiz-Lozano JM (2013) Abiotic stress remediation by the arbuscular mycorrhizal symbiosis and rhizosphere bacteria/yeast interaction. In: Molecular Microbial Ecology of the Rhizosphere (Ed) France J and Bruijn D, John Wiley and Sons, pp 991-1002.

Barea JM, Pozo MJ, Azcon R, Azcon-Aguilar C (2005) Microbial cooperation in the rhizosphere. J. Exp. Biol. 56, 1761-1778.

Bartels S, Besteiro MAG, Lang D, Ulm R (2010) Emerging functions for plant MAP kinase phosphatases. Trends Plant Sci. 15(6), 322-329.

Bell RW, Edwards DG, Asher CJ (1989) External calcium requirements for growth and nodulation of six tropical food legumes grown in flowing solution. Aust. J. Agric. Res. 40(1), 85 - 96.

Benjamin JG, Nielsen DC (2006) Water deficit effects on root distribution of soybean field pea and chickpea. Field Crop Res. 97(2-3), 248-253.

Bohnert HJ, Gong QQ, Li PH, Ma SS (2006) Unraveling abiotic stress tolerance mechanisms getting genomics going. Curr. Opin. Plant Biol. 9(2), 180-188.

Boswell EP, Koide RT, Shumway DL, Addy H D (1998) Winter wheat cover cropping, VA mycorrhizal fungi and maize growth and yield. Agric. Ecosyst. Environ. 67, 55-65.

Clement M, Lambert A, Herouart D, Boncompagni E (2008) Identification of new up-regulated genes under drought stress in soybean nodules. Gene 426(1-2), 15-22.

Cook RJ (1986) Interrelationships between plant health and the sustainability of agriculture. Am. J. Alternative Agr. I, 19-25.

Cruz CMH (2008) Drought stress and reactive oxygen species: Production, scavenging and signaling. Plant Signal Behav. 3(3), 156.

Declerck S, IJdo M, Fernandez K, Voets L, de la Providencia I (2009) Method and system for in vitro mass production of arbuscular mycorrhizal fungi. WO/2009/09022

Douds DD, Galvez L, Janke RR, Wagoner P (1995) Effect of tillage and farming system upon populations and distribution of vesicular-arbuscular mycorrhizal fungi. Agric. Ecosyst. Environ. 52, 111-118.

Douds DDJ, Lee J, Rogers L, Lohman ME, Pinzon N, Ganser S (2012) Utilization of inoculum of AMF produced on-farm for the production of *Capsicum annuum*: a summary of seven years of field trials on a conventional vegetable farm. Biol. Agric. Hortic. 28, 129-14.

Douds DDJ, Nagahashi G, Pfeffer PE, Kayser WM, Reider C (2006) On farm production of AM fungus inoculums in mixtures of compost and vermiculite. Bioresour. Technol. 97, 809-818.

Douds DDJ, Nagahashi G, Reider C, Hepperly PR (2008) Choosing a mixture ratio for the on farm production of AM fungus inoculum in mixtures of compost and vermiculite. Compost Sci. Util. 16, 52-60.

Evelin H, Kapoor R, Giri B (2009) Arbuscular mycorrhizal fungi in alleviation of salt stress. Ann. Bot. 104, 1263-80.

Falco DT, Bender K, Snedden W (2010) Breaking the code: Ca2+ sensors in plant signalling. Biochem. J. 425, 27-40.

FAO (2013) FAOSTAT. Rome, Italy: FAO publications. Available: http://faostat3.fao.org/home/index.html.

Foyer CH, Noctor G (2003) Redox sensing and signalling associated with reactive oxygen in chloroplasts, peroxisomes and mitochondria. Physiol. Plantarum. 119 (3), 355-364.

Franzini VI, Azcon R, Mendes FL, Aroca R (2010) Interactions between *Glomus* species and *Rhizobium* strains affect the nutritional physiology of drought-stressed legume hosts. J. Plant Physiol. 167(8), 614-9.

Gadkar V, Driver JD, Rillig MC (2006) A novel in vitro cultivation system to produce and isolate soluble factors released from hyphae of arbuscular mycorrhizal fungi. Biotechnol. Lett. 28, 1071–1076.

Grayston SJ, Campbell CD, Lutze JL, Gifford RM (1998) Impact of elevated CO_2 on the metabolic diversity of microbial communities in N-limited grass swards. Plant Soil 203, 289-300.

Grover M, Ali Sk Z, Sandhya V, Rasul A, Venkateswarlu B (2010) Role of microorganisms in adaptation of agriculture crops to abiotic stresses. World J. Microbiol. Biotechnol. 27, 1231-1240.

Grumberg BC, Urcelay C, Shroeder MA, Vargas-Gil S, Luna CM (2015) The role of inoculum identity in drought stress mitigation by arbuscular mycorrhizal fungi in soybean. Biol. Fertil. Soils 51, 1-10.

Guenther JF, Chanmanivone N, Galetovic MP, Wallace IS, Cobb JA, Roberts DM (2003) Phosphorylation of soybean nodulin 26 on serine 262 enhances water permeability and is regulated developmentally and by osmotic signals. Plant Cell 15(4), 981-991.

Guo ZL, Sun CQ, Liang N (2007) Impacts of plastic mulching on water saving and yield increasing of dry land spring soybean and its density effect. Chinese J. Eco Agric. 15(1), 205-206.

Hendrix J W, Guo B Z, An ZQ (1995) Divergence of mycorrhizal fungal communities in crop production systems. Plant Soil 170, 131-140.

Ijdo M, Cranenbrouck S, Declerck S (2011) Methods for large-scale production of AMF past, present, and future. Mycorrhiza 21, 1-16.

Irsigler A, Costa M, Zhang P, Reis P, Dewey R. Boston R, Fontes E (2007) Expression profiling on soybean leaves reveals integration of ER-and osmotic-stress pathways. BMC Genomics 8(1), 431.

Lee S, Hirt H, Lee Y (2001) Phosphatidic acid activates a wound activated MAPK in *Glycine max*. Plant J. 26(5), 479-486.

Li WYF, Shao G, Lam HM (2008) Ectopic expression of GmPAP3 alleviates oxidative damage caused by salinity and osmotic stresses. New Phytol. 178(1), 80-91.

Liu JX, Howell SH (2010) bZIP28 and NF-Y transcription factors are activated by ER stress and assemble into a transcriptional complex to regulate stress response genes in Arabidopsis. Plant Cell 22(3), 782-796.

Liu Y, Gai JY, Lu HN, Wang YJ, Chen SY (2005) Identification of drought tolerant germplasm and inheritance and QTL mapping of related root traits in soybean (*Glycine max* (L.) Merr.). Acta Genet. Sin. 32(8), 855-863.

Liu Z, Li Y, Ma L, Wei H, Zhang J, He X, Tian C (2015) Coordinated regulation of arbuscular mycorrhizal fungi and soybean mapk pathway genes improved mycorrhizal soybean drought tolerance. Mol. Plant-Microbe Interact. 28, 408-419.

Macdonald RM (1981) Routine production of axenic vesicular-arbuscular mycorrhizas. New Phytol. 89, 87-93.

Manavalan LP, Guttikonda SK, Tran LSP, Nguyen HT (2009) Physiological and molecular approaches to improve drought resistance in soybean. Plant Cell Physiol. 50(7), 1260-1276.

Marulanda A, Barea JM, Azcon R (2009) Stimulation of plant growth and drought tolerance by native microorganisms (AMF and bacteria) from dry environments: mechanisms related to bacterial effectiveness. J. Plant Growth Regul. 28, 115-124.

Masoumi H, Masoumi M, Darvish F, Daneshian J, Nourmohammadi G, Habibi D (2010) Change in several antioxidant enzymes activity and seed yield by water deficit stress in soybean (Glycine max L.) cultivars. Not. Bot. Horti Agrobo. 38(3), 86-94.

Mittler R (2002) Oxidative stress, antioxidants and stress tolerance. Trends Plant Sci. 7(9), 405-410.

Mosse B (1962) The establishment of AMF under aseptic conditions. J. Gen. Microbiol. 27, 509-520.

Mosse B, Thompson JP (1984) Vesicular–arbuscular endomycorrhizal inoculum production. I. Exploratory experiments with beans (*Phaseolus vulgaris*) in nutrient flow culture. Can. J. Bot. 62, 1523-1530.

Ortiz N, Armada E, Duque E, Roldan A, Azcona R (2015) Contribution of arbuscular mycorrhizal fungi and/or bacteria to enhancing plant drought tolerance under natural soil conditions: Effectiveness of autochthonous or allochthonous strains. J. Plant Physiol. 174, 87-96.

Porcel R, Ruiz-Lozano JM (2004) Arbuscular mycorrhizal influence on leaf water potential, solute accumulation, and oxidative stress in soybean plants subjected to drought stress. J. Exp. Bot. 55 (403), 1743-1750.

Porcel R, Azcón R, Ruiz-Lozano JM (2004) Evaluation of the role of genes encoding for Ä-pyrroline-5-carboxylate synthetase (P5CS) during drought stress in arbuscular mycorrhizal and plants. Physiol. Mol. Plant Pathol. 65(4), 211-221.

Porcel R, Barea JM, Ruiz-Lozano JM (2003) Antioxidant activities in mycorrhizal soybean plants under drought stress and their possible relationship to the process of nodule senescence. New Phytol. 157, 135-143.

Ramasamy K, Joe MM, Kim K, Lee S, Shagol C, Rangasamy A, Chung J, Islam MR, Tongmin S (2011) Synergistic effects of arbuscular mycorrhizal fungi and plant growth promoting rhizobacteria for sustainable agricultural production. Korean J. Soil Sci. Fert. 44(4), 637-649.

Rapparini F and Penuelas J (2014) Mycorrhizal fungi to alleviate drought stress on plant growth In: Use of Microbes for the Alleviation of Soil Stresses (Ed) Miransari M, Springer Science+Business Media, New York, pp 21-42.

Rathod DP, Brestic M, Shao HB (2011) Chlorophyll a fluorescence determines the drought resistance capabilities in two varieties of mycorrhized and non-mycorrhized Glycine max Linn. Afr. J. Microbiol. Res. 5, 4197-4206.

Ronde JA, Spreeth MH, Cress WA (2000) Effect of antisense L-Ä1-pyrroline-5-carboxylate reductase transgenic soybean plants subjected to osmotic and drought stress. Plant Growth Regul. 32(1), 13-26.

Ruiz-Lozano JM, Azcon R, Gomez M (1995) Effects of arbuscular-mycorrhizal glomus species on drought tolerance: physiological and nutritional plant responses. Appl. Environ. Microbiol. 61, 456-460.

Ruiz-Lozano JM, Collados C, Barea JM, Azcón R (2001) Arbuscular mycorrhizal symbiosis can alleviate drought-induced nodule senescence in soybean plants. New Phytol. 151, 493-502.

Ruiz-Lozano JM (2003) Arbuscular mycorrhizal symbiosis and alleviation of osmotic stress. New perspectives for molecular studies. Mycorrhiza.13, 309-317.

Sadeghipour O, Abbasi S (2012) Soybean response to drought and seed inoculation. World Appl. Sci. J. 17(1), 55-60.

Sharma MP and Adholeya A (2008) Application of arbuscular mycorrhizal fungi in the production of fruits and ornamental crops. In: Microbial Biotechnology in Horticulture (Ed) Ray RC and Ward OP, Science Publishers, USA, pp 201-231.

Sharma MP, Gupta S, Sharma SK, Vyas AK (2012) Effect of tillage and crop sequences on arbuscular mycorrhizal symbiosis and soil enzyme activities in soybean (*Glycine max*) rhizosphere. Indian J. Agric. Sci. 82 (1), 25-30.

Sharma MP and Sharma SK (2008). On-farm production of arbuscular mycorrhizal fungi. Biofertilizer Newsletter. 16 (1), 3-7.1.

Sharma MP, Sharma SK, Prasad RD, Pal KK, Dey R (2015) Application of arbuscular mycorrhizal fungi in production of annual oilseed crops. In: Mycorrhizal Fungi: Use in Sustainable Agriculture and Land Restoration, Soil Biology (Ed) Solaiman ZM Abbott LK and Varma A, Springer-Verlag, Berlin, Heidelberg, pp 119-148.

Singh S, Srivastava K, Badola JC, Sharma AK (2012) Aeroponic production of AMF inoculum and its application for sustainable agriculture. Wudpecker J Agric Res. 1(6), 186-190.

Sloane RJ, Patterson RP, Carter J TE (1990) Field drought tolerance of a soybean plant introduction. Crop Sci. 30(1), 118-123.

Sreenivassa MN (1992) Selection of an efficient vesicular arbuscular mycorrhizal fungus for Chili (*Capsicum annuum* L.). Sci. Hortic. 50, 53-58.

Stolf MR, Medri ME, Neumaier N, Lemos NG, Pimenta JA, Tobita S, Brogin RL, Marcelino GFC, Oliveira MCN, Farias JR, Abdelnoor RV, Nepomuceno AL (2010) Soybean physiology and gene expression during drought. Genet. Mol. Res. 9, 1946-1956.

Thompson JP (1986) Soilless culture of vesicular-arbuscular mycorrhizae of cereals: effects of nutrient concentration and nitrogen source. Can. J. Bot. 64, 2282-2294.

Tiwari P, Adholeya A (2003) Host dependent differential spread of Glomus intraradices on various Ri T-DNA transformed roots in vitro. Mycol. Prog. 2, 171-177.

Umezawa T, Nakashima K, Miyakawa T, Kuromori T, Tanokura M, Shinozaki K, Yamaguchi S K (2010) Molecular basis of the core regulatory network in ABA responses: sensing, signaling and transport. Plant Cell Physiol. 51(11), 1821-1839.

Valente MAS, Faria JAQA, Soares R JRL, Reis PAB, Pinheiro GL, Piovesan ND, Morais AT, Menezes CC, Cano MAO, Fietto LG (2009) The ER luminal binding protein (BiP) mediates an increase in drought tolerance in soybean and delays drought-induced leaf senescence in soybean and tobacco. J. Exp. Bot. 60(2), 533-546.

Voets L, de la Providencia IE, Fernandez K, IJdo M, Cranenbrouck S, Declerck S (2009) Extraradical mycelium network of arbuscular mycorrhizal fungi allows fast colonization of seedlings under *in vitro* conditions. Mycorrhiza. 19, 347-356.

Wang H, Zhou L, Fu Y, Cheung MY, Wong FL, Phang TH, Sun Z, Lam HM (2012) Expression of an apoplast localized BURP domain protein from soybean (GmRD22) enhances tolerance towards abiotic stress. Plant Cell Environ. 35(11), 1932-1947.

Wilkinson S, Davies WJ (2010) Drought ozone ABA and ethylene: new insights from cell to plant to community. Plant Cell Environ. 33(4), 510-525.

Wu Y, Cosgrove DJ (2000) Adaptation of roots to low water potentials by changes in cell wall extensibility and cell wall proteins. J Exp Bot. 51(350), 1543-1553.

Xiong L, Zhu JK (2002) Molecular and genetic aspects of plant responses to osmotic stress. Plant Cell Environ. 25(2), 131-139.

Xue HW, Chen X, Mei Y (2009) Function and regulation of phospholipid signalling in plants. Biochem. J. 421(2), 145.

Yamamoto E, Karakaya HC, Knap HT (2000) Molecular characterization of two soybean homologs of *Arabidopsis thaliana* CLAVATA1 from the wild type and fascination mutant. Biochem. Biophys. Acta. 1491(1), 333-340.

Yang J, Kloepper JW, Choong MR (2009) Rhizosphere bacteria help plants tolerate abiotic stress. Trends Plant Sci. 14(1), 1-4.

Yang L, Ji W, Gao P, Li Y, Cai H, Bai X, Chen Q, Zhu Y (2012) GsAPK, an ABA-activated and Calcium-Independent SnRK2-Type kinase from *Glycine soja*, mediates the regulation of plant tolerance to salinity and ABA stress. PLoS ONE 7(3), e33838.

Yoo JH, Park CY, Kim JC, Do Heo W, Cheong MS, Park HC, Kim MC, Moon BC, Choi MS, Kang YH (2005) Direct interaction of a divergent CaM isoform and the transcription factor, MYB2, enhances salt tolerance in Arabidopsis. J. Biol. Chem. 280 (5), 3697-3706.

Zhang H, Jiao H, Jiang CX, Wang SH, Wei ZJ, Luo JP, Jones RL (2010) Hydrogen sulfide protects soybean seedlings against drought-induced oxidative stress. Acta Physiol. Plant. 32(5), 849-857.

Zhang J, Jia W, Yang J, Ismail AM (2006) Role of ABA in integrating plant responses to drought and salt stresses. Field Crops Res. 97(1), 111-119.

Zhilei L, Yuanjing L, Lina M, Haichao W, Jianfeng Z, Xingyuan H, Chunjie T (2014) Coordinated regulation of arbuscular mycorrhizal fungi and soybean mapk pathway genes improved mycorrhizal soybean drought tolerance. Mol. Plant-Microbe Interact. 28, 408-419.

Zhou GA, Chang RZ, Qiu LJ (2010) Over expression of soybean ubiquitin-conjugating enzyme gene GmUBC2 confers enhanced drought and salt tolerance through modulating abiotic stress-responsive gene expression in *Arabidopsis*. Plant Mol. Biol. 72 (4-5), 357-367.

13

Microbial Bioinoculants for Quality Seedling Production in Forestry

Parkash Vipin, Saikia A.J. and Saikia M.

Abstract

Microorganisms inhabiting in the rhizosphere play a vital role in ecosystem functioning. Rhizosphere is an area around the root surface where pathogenic and beneficial microorganisms repose and influence the plant growth and health. These microbial groups of bioagents found in the rhizosphere include bacteria, fungi, nematodes, protozoa, algae and micro-arthropods. Because of their rich diversity, complexity of interactions and numerous metabolic pathways, microbes are an amazing resource for biological activity. These microorganisms require organic matter for their growth and activity in soil and provide valuable nutrients to the plants and hence maintaining the plant health. The most promising fields for the use of bioinoculants are agriculture, horticulture and forestry. There are many reports on the inoculation of bioagents in agriculture and horticulture crops but only a few reports are available in forestry. This article is an overview on various bioinoculants which are used during inoculation and a compilation of scattered reports of bioinoculants in forestry only.

Keywords: Bacteria, Bioinoculants, Bioinoculation, Cyanobacteria, Mycorrhizae, Synergism

1. Rhizospheric biota

Soil is one of the three major natural resources, alongside air and water. It is one of the marvellous products of nature and without which there would be no life. Soil is made up of three main components – minerals that come from rocks below or nearby, organic matter which is the remains of plants and animals and the living organisms that reside in the soil. Soil is a structured, heterogeneous and discontinuous system, generally poor in nutrients and energy sources with microorganisms living in discrete microhabitats (Nannipieri *et al.* 2003). A narrow zone of soil affected by the presence of plant roots is defined as rhizosphere (Hrynkiewicz and Baum 2011). The rhizosphere is an environment that the plant itself helps to create and where pathogenic and beneficial microorganisms constitute a major influential force on plant growth and health (Lynch 1990). Microbial groups comprising of bio-agents found in the rhizosphere include bacteria, fungi, nematodes, protozoa, algae and microarthropods (Lynch 1990; Raaijmakers 2001). Microorganisms that adversely affect plant growth and health are the pathogenic fungi, oomycetes, bacteria and nematodes, whereas microorganisms that are beneficial include nitrogen-fixing bacteria, endo- and ecto-mycorrhizal fungi, plant growth promoting rhizobacteria (PGPR) and fungi (Raaijmakers *et al.* 2009). Because of their rich diversity, complexity of interactions and numerous metabolic pathways, microbes are an amazing resource for biological activity (Emmert and Handelsman 1999; Alabouvette *et al.* 2006; Tejesvi *et al.* 2007). These microorganisms require organic matter for their growth and activity in soil and provide valuable nutrients to the plant (Rajendra *et al.* 1998).

2. Bioinoculants: The concept

Bioinoculants or biocontrol agents are the microorganisms that induce stimulatory effects on plant growth and/or suppressive effects on pests or pathogens through a variety of mechanisms when applied in an ecosystem. A large number of bioinoculants have been investigated to harness their beneficial effects on crop productivity. Although, the beneficial effects on legumes in improving soil fertility was known since ancient times and their role in biological nitrogen fixation was discovered more than a century ago, commercial exploitation of such biological processes is of recent interest and practice (Rajasekaran *et al.* 2012). The different definitions of bioinoculants, given by various scientists from time to time are cited in the following lines.

- The bioinoculant is a formulation with one or more strains/ species of microbes and appropriate as well as cost-effective carrier (Miransari 2014).

- Bioinoculants or biocontrol agents are the microorganisms that induce stimulatory effects on plant growth and/or suppressive effects on pests or pathogens through a variety of mechanisms when applied in an ecosystem (Khan and Anwar 2011).

- Bioinoculants are defined as the concoctions of microbial entities that are supplemented as biocontrol agents to induce or suppress both biotic and abiotic factors in promoting sustained growth (Gupta *et al.* 2007).

- Bioinoculants are carrier based microorganisms, which help to enhance productivity by biological nitrogen fixation and insoluble phosphate solublization or producing hormones, vitamins and other growth factors required for plant growth (Bhattacharya *et al.* 2000).

Microbial inoculants are biologically active products or of bacteria, algae and fungi (separately or in combination), which may help biological nitrogen fixation for the benefit of plants. Bioinoculants thus, include the following symbiotic nitrogen fixers *Rhizobium* Frank, asymbiotic free nitrogen fixers (*Azotobacter* Beijerinck, *Azospirillum* Tarrand etc.), algal biofertilizers (blue green algae or cyanobacteria in association with *Azolla* Lam.), phosphate solubilising bacteria, mycorrhizae and other organic fertilizers. The most striking relationship that these microbes have with plants is *'symbiosis'* in which the partners derive benefits from each other (Sujanya and Chandra 2011).

2.1 Biofertilizers and types of bioinoculants

Biofertilizer/s is/are ready to use live formulation/s of beneficial microorganism/s (bioinoculant/s), which on application to seed, root or soil mobilize the availability of nutrients by their biological activity and help in building up the micro-flora and thus, soil and plant health whereas fertilizers are any organic or inorganic material of natural or synthetic origin which is added to a soil in an attempt to provide plant nutrients. The main important functions of bioinoculants are to convert ambient nitrogen into forms that the plants can take (nitrate and ammonia) and take up of other elements like phosphorus; to increase soil porosity by gluing soil particles together; to defend plants against pathogens by out competing pathogens for food and saprophytic fungi in the soil break leaf litter down into usable nutrients. There are some of the following reasons as to why we should use biofertilizers;

Because they fix atmospheric nitrogen, increase availability or uptake of nutrients through solubilization or increased absorption and improves soil properties and sustaining soil fertility.

- Stimulate plant growth through hormonal or antibiotics action or by decomposing organic waste material.

- Lead to soil enrichment, compatible with long term sustainability and build up soil fertility in the long term.

- They are eco-friendly and pose no damage to the environment.

The various types of bioinoculants have been diagrammatically represented below.

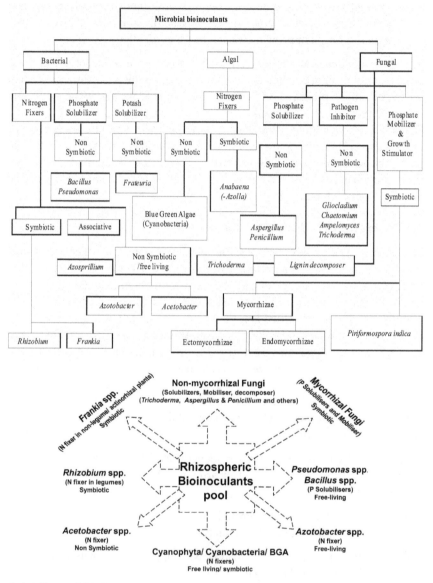

2.1.1 Fungal bioinoculants

Fungi are ubiquitous in nature and some are having beneficial effects on plants, while others may be detrimental (Anderson and Cairney 2004; Ipsilantis and Sylvia 2007; Park *et al.* 2005; Pereira *et al.* 2007; Shenoy *et al.* 2007; Soares and Barreto 2008; Than *et al.* 2008). The important genera of biocontrol fungi that have been tested against plant pathogenic fungi and nematodes include *Trichoderma* Persoon, *Aspergillus* Micheli, *Chaetomium* Kunze ex Fr., *Penicillium* Link, *Neurospora* Shear & Dodge, *Fusarium* (saprophytic) Link, *Rhizoctonia* DC., *Paecilomyces* Samson, *Pochonia* Zare and *Glomus* Tul. and Tul. Other kinds of bio-control agents such as plant growth-promoting organisms have also been evaluated for disease management (Papavizas 1985; Nair and Burke 1988). A number of fungi such as *Aspergillus* spp., *Penicillium* spp. and *Trichoderma* spp. are active phosphate-solubilizing microorganisms (PSM), which also suppress plant pathogens. Application of fungal PSMs can control soil-borne pathogens such as *Fusarium oxysporum* Schlecht. emend. Snyder and Hansen, *Macrophomina phaseolina* Tassi (Goid.), *Pythium aphanidermatum* (Edson) Fitzp., *Rhizoctonia solani* Kuhn, *Sclerotinia sclerotiorum* de Bary and nematodes such as *Meloidogyne incognita* Chit. (Khan and Anwer 2007, 2008; Khan *et al.* 2009).

Fungal bioinoculants basically work through parasitism (Papavizas 1985; Stirling 1993) against plant pathogenic fungi and nematodes (Khan 2005). Biocontrol agents and especially antagonistic fungi have been used to control plant diseases with 90% of applications being formulated using different strains of *Trichoderma e.g. T. harzianum* Rifai, *T. virens* Mill, Gidden & Foster, *T. viride* Pers. (Benítez *et al.* 2004). Many species of *Chaetomium e.g. Chaetomium globosum* Kunze, *C. cochlioides* Palliser, *C. cupreum* Ames can also be antagonistic against various soil microorganisms (Soytong *et al.* 2001; Kanokmedhakul *et al.* 2002, 2006). The following are some examples of fungal bioinoculants/agents.

a) **Trichoderma species:** *Trichoderma* species are common in soil and root ecosystems and are ubiquitous saprobes (Harman *et al.* 2004; Thormann and Rice 2007; Vinale *et al.* 2006; Kodsueb *et al.* 2008) and they are easily isolated from soil, decaying wood and other organic material (Howell 2003; Zeilinger and Omann 2007). There are several reports on the use of *Trichoderma* species as biological agents against plant pathogens (Harman *et al.* 2004; Zeilinger and Omann 2007). *Trichoderma* species *e.g. T. harzianum, T. viride, T. virens* have been used as biological control agents against a wide range of pathogenic fungi *e.g. Rhizoctonia* spp., *Pythium* spp., *Botrytis cinerea* Pers. and *Fusarium* spp. *Phytophthora palmivora* Butler, *P. parasitica* Dastur (Benítez *et al.* 2004; Sunantapongsuk *et al.* 2006; Zeilinger and Omann, 2007). Among them, *Trichoderma harzianum* is reported to be most widely used as an effective biological control agent (El-Katathy *et al.* 2001; Szekeres *et al.* 2004; Abdel-Fattah *et al.* 2007) and hence, maintains plant health.

Trichoderma species have been very successfully used as mycofungicides because they are fast growing, have high reproductive capacity, inhibit a broad spectrum of fungal diseases, have a diversity of control mechanisms, are excellent competitors in the rhizosphere, have a capacity to modify the rhizosphere, are tolerant or resistance to soil fungicides, have the ability to survive under unfavorable conditions, are efficient in utilizing soil nutrients, have strong aggressiveness against phytopathogenic fungi and also promote plant growth (Tang *et al.* 2001; Benítez *et al.* 2004; Vinale *et al.* 2006). Their ability to colonize and grow in association with plant roots is known as *'Rhizosphere Competence'* (Kaewchai *et al.* 2009).

b)   ***Ampelomyces* species:** *Ampelomyces quisqualis* Ces. is the mycoparasitic anamorphic Ascomycete that reduces the growth and kills powdery mildews. It can affect the pathogen through antibiosis and parasitism (Kiss 2003; Viterbo *et al.* 2007). The fungus *A. quisqualis* was the first organism reported to be a hyperparasite of powdery mildew and it can be easily found associated with powdery mildew colonies (Paulitz and Belanger 2001). Hyphae of *A. quisqualis* penetrate the hyphae of powdery mildews and grow internally then kill all the parasitized cells (Kiss 2003). *A. quisqualis* isolate M-10 has been formulated as AQ10 Biofungicide, developed by Ecogen, Inc., USA. This mycofungicide contains conidia of *A. quisqualis* and formulated as water-dispersible granules for the control of powdery mildew of carrot, cucumber and mango (Khetan 2001; Paulitz and Belanger 2001; Shishkoff and McGrath 2002; Kiss 2003; Viterbo *et al.* 2007). There are a few reports on *A. quisqualis* use in forest species (Markovic and Rajkovic 2011).

c)   ***Chaetomium* species:** *Chaetomium* species are normally found in soil and organic compost (Soytong *et al.* 2001). The genus, *Chaetomium* Kunze was first established in 1817 by Gustav Kunze (Soytong and Quimio 1989a). The application of *Chaetomim* as a biological control agent to control plant pathogens first commenced in about 1954 when *C. globosum* and *C. cochliodes* occurring on oat seeds and that these taxa provided some control of *Helminthosporium victoriae* Meehan & Murphy (Tviet and Moor 1954). *Chaetomium* species have been reported to be potential antagonists of various plant pathogens, especially soil-borne and seed borne pathogens (Soytong and Quimio 1989b; Dhingra *et al.* 2003; Aggarwal *et al.* 2004; Park *et al.* 2005). Many species of *Chaetomium* used as biological control agents have potential to suppress the growth of bacteria and fungi through competition (for substrate and nutrients), mycoparasitism, antibiosis, or various combinations of these and also plant growth stimulant (Marwah *et al.* 2007; Zhang and Yang 2007).

d) **Gliocladium species:** *Gliocladium* species are common soil saprobes and several species have been reported to be parasites of many plant pathogens (Viterbo *et al.* 2007). For example, *Gliocladium catenulatum* Gilman & Abbott parasitizes *Sporidesmium sclerotiorum* Uecker,. Adams et Ayers and *Fusarium* species. It destroys the fungal host by direct hyphal contact and forms *"pseudoappressoria"* (Punja and Utkhede 2004; Viterbo *et al.* 2007). *G. virens* (=*Trichoderma virens*) has been used as a biological control agent against a wide range of soil-borne pathogens such as, *Pythium* species and *Rhizoctonia* species under greenhouse and field conditions (Hebbar and Lumsden 1999; Viterbo *et al.* 2007). *G. virens* produces anti-biotic metabolites such as Gliotoxin which have anti-bacterial, anti-fungal, anti-viral and anti-tumor activities (Kaewchai *et al.* 2009).

e) **Piriformospora indica:** *Piriformospora indica* is a wide-host root colonizing endophytic fungus of the order Sebacinales which allows plants to grow under extreme physical and nutrient condition. It was discovered from orchid plants in the Thar desert, Rajasthan, India (Verma *et al.* 1998). It functions as a plant promoter and biofertilizer in nutrient deficient soils, as a bioprotector against biotic and abiotic stresses including root and leaf pathogens and insect invaders, inducing early flowering, enhanced seed production and stimulation of active ingredients in plants. Positive increments are established for many plants of medicinal and economic importance (Bagde *et al.* 2010).

2.1.2 Mycorrhizal bioinoculants

The symbiotic microbes belong to this group are called mycorrhizal fungi. The term *"Mycorrhiza"* was first coined by the German forest pathologist A.B. Frank in 1885, which literally means "Fungus Root" to denote the association between certain soil fungi and plant roots where the relationship is not pathogenic (Mohan *et al.* 2007). Mycorrhizae are fungi which form mutualistic relationships with roots of 90% of plants (Gaur and Adholeya 2004; Das *et al.* 2007; Rinaldi *et al.* 2008). Mycorrhizae promote absorption of nutrients and water, control plant diseases, and improve soil structure (Rola, 2000; Zhao *et al.* 2003; Chandanie *et al.* 2006; Rinaldi *et al.* 2008). Plants colonized by mycorrhizae grow better than those without them (Yeasmin *et al.* 2007; Singh *et al.* 2008) and are beneficial in natural and agricultural systems (Adholeya *et al.* 2005; Marin 2006). Mycorrhiza occurs in 90% of Angiosperms, 100% of Gymnosperms and 70% of Pteridophytes (Harley and Harley 1987). There are mainly two types of Mycorrhizal fungi namely Ectomycorrhizal (EcM) fungi and Endomycorrhizal (EdM) fungi. In this association, the fungus colonizes the host plant roots, either intracellular as in arbuscular mycorrhizal (AM) fungi or extracellularly as in ectomycorrhizal (EcM) fungi. These microbes are an important component of soil life and soil chemistry.

a) **Endophytic mycorrhizal bioinoculants:** In the symbiotic associations between plants and fungi, arbuscular mycorrhiza (AM), which is formed between plants and Glomeromycota fungi, has the widest distribution in the nature (Bagyaraj 2015). AM fungi inhabit a variety of ecosystems including agricultural lands, forests, grasslands and many stressed environments and colonize the roots of most plants, including Bryophytes, Pteridophytes, Gymnosperms and Angiosperms (Sadhana 2014). The importance of AM fungi to agricultural and forest plant species resides in its role in plant growth and nutrition. In tropical forests incidence of mycorrhizae profoundly influence soil fertility and thus, the growth and development of plants (Bagyaraj 1989).

The main advantage of mycorrhiza is its greater soil exploration and increasing uptake of N, P, K, Zn, Cu, S, Fe, Ca, Mg and Mn supply to the host roots (Li *et al.* 1991; Champawat and Pathak 1993; Marschner and Dell 1994; Smith *et al.*, 1994; Abdul Malik 2000). The *Glomus etunicatum* inoculated maize plants in sandy loam soil, under water stressed conditions, absorbed more phosphorus than non-mycorrhizal plants (Muller and Hofwer 1991). Michelsen and Rosendahl (1990) and Osonubi *et al.* (1992) observed that the AM fungi contribution to drought tolerance is minimal in *Acacia nilotica* (L.) Delile.

Bisleski (1973) reported that AM fungi may increase the effectiveness of absorbing capability of surface host root as much as ten times. Ions such as P, Zn and Cu do not diffuse readily through soil. Because of this poor diffusion, roots deplete these immobile soil nutrients from the zone immediately surrounding the root. The increase in plant growth resulting from AM symbiosis is usually associated with increased nutrient uptake by the hyphae from the soil (Harley and Smith 1983). It is widely accepted that a hyphal network associated with the roots of a living plant is capable of infecting the roots of other plants growing in its vicinity (Chiariello *et al.* 1983; Franschis and Read 1984; Newman 1988).

Arbuscular mycorrhizae are the important mutualistic symbionts of the soil edaphon in most agro-ecosystems. Microbial populations are key component of soil-plant system where they are immense in a network of interactions affecting plant development (Kumar *et al.* 2012). Mycorrhizal evolution is hypothesised to have progressed from endophytic to balanced symbiotic associations where both partners are interdependent due to the exchange of limiting resources (Brundrett 2002). About 90% of all terrestrial plant species are known to be forming this type of symbiosis (Smith and Read 1997). They play a role in shaping plant community structure by increasing the mineral supply to plants, improving water uptake and retention and thus drought tolerance (Lapointe and Molard 1997). The efficient utilization of AM fungal diversity is of crucial importance in sustainable plant production systems. The vesicular arbuscular mycorrhizal (VAM=AM) fungi are beneficial microorganisms that form symbiotic association with the fine roots of higher plants and can be utilized as biofertilizers in forestry and cropping systems (Parkash *et al.* 2005; Bagyaraj 2015). The mycorrhizal fungi play an important role in improving plant growth, nutrient uptake especially phosphorus (Mukerji and Dixon 1992; Gill *et al.* 2002; Khan and Uniyal 1992) and provide stress tolerance and disease resistance to plants (Mehrotra *et al.* 1995). Efforts are being made to improve the quality of seedlings of forest trees under nursery conditions through inoculation of suitable mycorrhizal strain alone or AM - *Rhizobium* combinations (Bagyaraj 1992; Sitaramaiah *et al.* 1998; Kaushik *et al.* 1992). Most host plants of non-arbuscular mycorrhizal fungi do not establish functional vesicular arbuscular mycorrhizal (VAM) symbiosis in nature (Giovannetti and Sbrana 1998). Among the known ectomycorrhizal hosts, *Eucalyptus* species is an exception as it also forms vesicular arbuscular mycorrhiza (Lapeyrie and Chilvers 1985) and/or mixed infection (Aggarwal *et al.* 2006; Parkash *et al.* 2011).

Parkash *et al.* (2005) studied the effect of single/alone inoculation of *Glomus mosseae* Gerd. & Trappe, *Glomus fasciculatum* Gerd. & Trappe emend. Walker & Koske, mixed VAM consortium (*Glomus, Acaulospora, Sclerocystis* and *Gigaspora* species mixed together) and *Trichoderma viride* on *Eucalyptus saligna* Sm. seedlings. The growth of inoculated seedlings improved significantly than non-inoculated seedlings. VAM spore number and mycorrhizal root colonization were found to be higher in *Glomus mosseae* treatment followed by mixed VAM consortium, *Glomus fasciculatum* and *Trichoderma viride* after 90 days of seedling growth than control seedlings in pot experiments.

The combined and synergistic effect of bioinoculants on *Eucalyptus saligna* Sm. seedlings was again tested by Parkash *et al.* (2011). Double inoculation (mixed VAM consortium + *Rhizobium* sp.) showed maximum increase in seedling height and shoot Phosphorus (P) content over other treatments including control. *Glomus mosseae* + *Rhizobium* sp. inoculated seedlings showed maximum root as well as shoot P content than other treatments. Triple inoculation also exhibited significant positive response on growth of *E. saligna* seedlings. Mixed VAM consortium + *Rhizobium* sp. + *Trichoderma viride* had more increase in height, VAM spore number and percentage mycorrhizal root colonization than other treatments including control; *G. mosseae* + *T. viride* + *Rhizobium* sp. treatment showed more root P content over other treatments while shoot P content was also more in *G. mosseae* + *Rhizobium* sp. + *T. viride*. However, P content of shoot and root in all double and triple inoculations was more over control treatment. Although, dual inoculation had good growth response but triple inoculation had more pronounced and significant response on growth and development on *E. saligna* seedlings.

b) **Ectomycorrhizal bioinoculants:** Ectomycorrhizae form symbiotic association between the terminal feeder roots of woody plant species and fungi. The great majority of ectomycorrhizal plants are woody perennials); although some sedges and herbaceous angiosperms are reported to exist in such associations (Fitter and Moyersoen 1996; Taylor and Alexander 2005). The ectomycorrhizal (EcM) fungi are characterized by the presence of complete sheath of fungal tissue covering the root epidermis called "*mantle*". The mantle serves as a barrier and prevents the entry of soil-borne pathogens and certain parasitic nematodes into the root systems of the host plants. There is an intercellular infection forming a network of fungal mycelium around the cortical cells called "*Hartig-net*". The Hartig-net structures serve as the centre for nutrient exchange between the host plant and the fungus. The fungi forming EcM mostly belong to the higher fungi (Basidiomycetes and Ascomycetes) primarily in the families like Amanitaceae, Boletaceae, Cortinariaceae, Russullaceae, Tricholo-mataceae, Rhizoponaceae, Sclerodermataceae, *etc.* some Ascomycetes especially truffles and one or two Phycomycetes (Trappe 1977). Among the Basidiomycetes, the fungi like *Amanita* Pers., *Boletus* L., *Cortinarius* Gray, *Laccaria* Berk. & Broome, *Lactarius* Pers., *Leccinum* Gray, *Russula* Pers., *Suillus* Gray, *Lycoperdon* Pers., *Pisolithus* Alb. & Schwein, *Rhizopogon* Fr., *Scleroderma* Pers., *etc.* are important (Smith 1971).

Scagel and Linderman (1996) studied the influence of ectomycorrhizal fungal inoculation on growth and root IAA concentrations of transplanted conifers. Douglas-fir (*Pseudotsuga menziesii* (Mirb.) Franco), lodgepole pine (*Pinus contorta* Dougl.) and ponderosa pine (*Pinus ponderosa* Dougl.) were inoculated at seeding level with ectomycorrhizal fungi and the seedlings were having a high, moderate or low capacity to produce either IAA or ethylene *in vitro*. Morphological responses to inoculation varied among different mycorrhizal fungi. Free IAA concentration of roots was increased in some inoculation treatments for all conifer species. Both IAA- and ethylene-producing capacity were significantly correlated with more morphological features in seedlings transplanted to a forest site than in seedlings transplanted to a nursery field. Some studies on their effect on different aspects are shown in a table under the subheading - bioinoculants application in forestry in this review.

2.1.2 Bacterial bioinoculants (plant growth promoting rhizobacteria)

The application of PGPR inoculants in agriculture can be traced back to the beginning of the past century, when a *Rhizobium* based product named *"Nitragin"* was patented by Nobbe and Hiltner during 1896 (Bashan 1998). Currently, the demand for microbial biofertilizers is increasing worldwide owing to a higher degree of environmental awareness, the increasing number of laws protecting the environment, and the ever-expanding demand for ecological products (Garcia-Fraile *et al.* 2015).

Biological Nitogen Fixation (BNF) occurs in the free living states, in association or in symbiosis with plants. From an ecological point of view, the most important N fixing systems are the symbiotic associations. Two groups of nitrogen-fixing bacteria, *e.g. Rhizobium* and *Frankia* Brunchorst have been found associated with plants. *Rhizobium* - a symbiotic Biofertilizer can be used for legume crop and trees (*e.g.Leucaena leucocephala* de Wit) and is a crop specific inoculant, for example, *Rhizobium leguminosarum* bv *trifoli* Frank for berseem; *Rhizobium melilotti* Dangeard (*=Sinorhizobium meliloti* De Lajude) for leucerne, *Rhizobium phaseoli* Dangeard for green gram, black gram, *Rhizobium japonicum* Buchanan (*=Bradyrhizobium japonicum* Jordan) for soyabean; *Rhizobium leguminosarum* Frank for pea, lentil; *Rhizobium lupini* Schroeter (*=Bradyrhizobium lupini* Peix) for chickpea. These inoculants are known for their ability to fix atmospheric N- in symbiotic association with plants forming nodules in roots (stem nodules in *Sesbania rostrata* Bremek. & Oberm.). *Rhizobium* is however limited by its specificity and only certain legumes are benefited by this symbiosis.

Frankia forms root nodules in more than 280 species of woody plants from 8 different families (Schwintzer and Tjepkema 1990). *Frankia* is a filamentous gram-positive nitrogen-fixing free-living bacterium (Normand *et al.* 1996) found in root nodules or in soil (Chaia *et al.* 2010) with actinorhizal plants (Handique and Parkash 2014). It has been classified in the order of Actinomycetales (Lechevalier 1994) and family Frankeniaceae, in which the capacity to fix nitrogen is restricted to *Frankia* (Wall 2000; Franche *et al.* 2009). The *Frankia*-actinorhizal plant symbiosis induces the formation of a perennial root organ called nodule, wherein bacteria is hosted and nitrogen is fixed (Perrine-Walker *et al*. 2011, Tromas *et al*. 2012); which exhibits variability with respect to forms and colours (Bargali 2011).

Nitrogen fixing bacteria of genus, *Azospirillum* Tarrand *et al*. is an important non-symbiotic associative microorganism, it fixes atmospheric nitrogen in soil (Krishnamoorthy 2002) and augments nitrogen fixation (Vijayakumari and Janardhanan 2003). *Azospirillum* promotes seedling growth, biomass and nutrient uptake (Sekar 1995; Rajendran *et al*. 2003; Kasthuri Rengamani *et al*. 2006). It also increases root biomass, root surface area, root diameter, density and length of root hairs (Okon and Kapulnik 1986). In agriculture, advantages of bioinoculants application are better known, but in tree species, the utility of biofertilizers is still in experimental stage (Kabi *et al*. 1982; Basu and Kabi 1987). Many root colonizing bacteria including the nitrogen fixing *Azospirillum* and phosphorus solubilizing *Pseudomonas* spp. are known to produce growth hormones which often lead to the increased root and shoot growth (Govindarajan and Thangaraju 2001).

Azotobacter species not only help in nitrogen fixation but also capable of producing antibacterial and antifungal compounds, hormones and siderophores (Tilak 1993). There are a wide no of nitrogen fixing becteria namely, *Rhizobium,* the obligate symbionts in leguminous plants and *Frankia* in non-leguminous trees and non-symbiotic (free-living, associative or endophytic) N_2- fixing forms such as Cyanobacteria*, Azospirillum, Azotobacter, Acetobacter diazotrophicus* Gillis *et al., Azoarcus* Reinhold-Hurek *et al*., etc. (Tilak *et al*. 2005). The family Azotobacteriaceae comprises of two genera namely: *Azomonas* (non-cyst forming) with three species (*A. agilis* Winogradsky*, A. insignis* Jensen and *A. macrocytogenes* New & Tchan) and *Azotobacter* (cyst forming) comprising of 6 species is generally regarded as free-living aerobic nitrogen fixer (Dash and Gupta 2011). Yield improvement is attributed more to the ability of *Azotobacter* to produce plant growth promoting substances such as phytohormones IAA and siderophore azotoactin, rather than to diazotrophic activity (Tchan 1984; Tchan and New 1984). Members of the genus *Azospirillum* fix nitrogen under microaerophilic conditions and are frequently associated with root and rhizosphere of large number of agriculturally important crops and cereals. Due to their frequent occurrence in the rhizosphere, these are known as associative diazotrophs (Okon 1985; Tilak and Subba 1987). *Azospirillum* plays a major role in osmoadaption through increase in osmotic stress (Basan and Holguin 1997).

The *Azotobacter* colonizing the roots not only remains on the root surface but also a sizable proportion of it penetrates into the root tissues and lives in harmony with the plants, belongs to family Azotobacteriaceae, aerobic, free living, and heterotrophic in nature. Azotobacters are present in neutral or alkaline soils and

A. chroococcum Beijerinck is the most commonly occurring species in arable soils. These are non-symbiotic free living aerobic bacteria possessing highest respiratory rate and can fix N up to 25 kg/ha under optimum conditions (Mazid and Khan 2014). They improve seed germination and plant growth by producing B-vitamins, NAA, GA and other chemicals (plant hormones) that are inhibitory to certain root pathogens (Mazid *et al.* 2011).

Azospirillium species (Family: Spirillaceae), when applied to rhizosphere it fixes atmospheric N (free living state) and makes it available to crop plants. This is also N-fixing microorganism, beneficial for non-leguminous plants, belongs to family Spirilaceae, heterotrophic and associative in nature. In addition to their N-fixing ability of about 20-40 kg/ha, they also produce growth regulating substances. Although, there are many species under this genus *i.e. A. amazonense* Magalhaes *et al., A. halopraeferens* Reinhold *et al., A. brasilense* Tarrand *et al.*, have worldwide distribution and benefits of inoculation have been proved mainly with the *A. lipoferum* Tarrand *et al.* and *A. brasilense* Tarrand *et al. Azospirillum* proved significantly beneficial in improving leaf area index and all yield attributing aspects (Mazid and Khan 2014).

Herbaspirillum species is an associative symbiont that is responsible for atmospheric N fixation in the roots of sugarcane. It has potential of enhancing the availability of N, promoting the uptake of nitrate, K, phosphate and production of growth promoting hormones (kinetin, gibberellic acid and auxin) (Khan *et al.* 2011b).

Several reports have examined the ability of different bacterial species to solubilize insoluble inorganic phosphate compounds such as tri-calcium phosphate, di-calcium phosphate, hydroxyapatite, and rock phosphate. Among the bacterial genera with this capacity are *Pseudomonas, Bacillus, Rhizobium, Burkholderia, Achromobacter, Agrobacterium, Microccocus, Aereobacter, Flavobacterium and Erwinia*. Phosphorus is both native in soil and when applied in inorganic fertilizers, becomes mostly unavailable to crops because of its low levels of mobility and solubility and its tendency to become fixed in soil. The phosphate solublizing bacteria (PSB) are life forms that can help in improving phosphate uptake of plants in different ways. The soil bacteria belonging to the genera *Pseudomonas* and *Bacillus* and fungi are more common. The major microbiological means by which insoluble-P compounds are mobilized is by the production of organic acids, accompanied by acidification of the medium. The PSB also have the possible potential to utilize India's abundant deposits of rock phosphate (Mazid and Khan 2014). For more information, one can see the review paper of Mohan (2015) who described in detail the various PGPR and their use in forestry.

2.1.3 Algal bioinoculants

Cyanobacteria or Blue green algae (BGA) are a diverse group of prokaryotes possessing oxygen evolving photosynthetic system (Prabina *et al.* 2004). They are known to possess the ability to form associations with vascular/non-vascular plants and produce growth-promoting substances (Nanjappan-Karthikeyan *et al.* 2007). Many, though not all, non-heterocystous cyanobacteria can fix N_2 and convert it into an available form of ammonia required for the plant growth. Nevertheless, these microorganisms may make a substantial contribution to the global nitrogen cycle (Bergman *et al.* 1997). Using of N_2-fixing cyanobacteria is the ultimate goal of N_2 fixation research which aims to decrease the dependence on chemical N fertilizers for food production. This is due to indiscriminate use of chemical fertilizers for a longer period drastically disturbed the natural ecological balance (Jha *et al.* 2001).

Cyanobacteria are photosynthetic nitrogen fixers and are free living. They are found in abundance in India. They too add growth promoting substances including vitamin B_{12}, improve the soil's aeration and water holding capacity and add to biomass when decomposed after life cycle. *Azolla* is an aquatic fern found in small and shallow water bodies and in rice fields. It has symbiotic relation with BGA.

Cyanobacteria such as *Nostoc muscorum* Agardh, *N. humifusum* Carmichael, *Anabaena oryzae* Fritsch, *Wollea* spp., *Phormedium* spp. and *Spirulina platensis* Turpin etc. are widely used in quality seed production. Cyanobacteria are one of the major components of the nitrogen fixing biomass in paddy fields, forestry and provides a potential source of nitrogen fixation at no cost. Due to the important characteristic of nitrogen fixation, cyanobacteria have a unique potential to enhance productivity in a variety of agricultural and ecological situations. Cyanobacteria play an important role to build-up soil fertility consequently increasing the yield. Biofertilizers being essential components of organic farming play vital role in maintaining long term soil fertility and sustainability by fixing atmospheric dinitrogen (N_2), mobilizing fixed macro and micro nutrients or convert insoluble phosphorus in the soil into forms available to plants, thereby increases their efficiency and availability. The blue green algae are capable of fixing the atmospheric nitrogen and convert it into an available form of ammonium required for plant growth. Dominant nitrogen-fixer blue-green algae are *Anabaena, Nostoc, Aulosira, Calothrix, Plectonema* etc. Blue-green algae have the abilities of photosynthesis as well as biological nitrogen fixation. The agricultural importance of cyanobacteria in rice cultivation is directly related with their ability to fix nitrogen and other positive effects for plants and soil. Biofertilizers are ecofriendly and have been proved to be effective and economical alternate of chemical fertilizers with lesser in put of capital and energy (Sahu *et al.* 2012).

The role of blue green alga as biofertilizers has been limited to its relevance and utilization in rice crops (Prasanna *et al.* 2008). Yanni and Abd El-Rahman (1993) stated that rice performance (as assessed by plant height, productive tillers, grain and straw yields and their N-contents and fertilizer N-use efficiency) was enhanced by inoculation with cyanobacteria along with urea fertilizer at 36 or 72kg N ha^{-1} rather than 108 kg N ha^{-1} without inoculation.

Cyanobacteria play an important role in maintenance and build up of soil fertility, consequently increasing rice growth and yield as a natural biofertilizer (Song *et al.* 2005). The acts of these algae include: (1) Increase in soil pores with having filamentous structure and production of adhesive substances. (2) Excretion of growth-promoting substances such as hormones (auxin, gibberellin), vitamins, amino acids (Roger and Reynaud 1982; Rodriguez *et al.* 2006). (3) Increase in water holding capacity through their jelly structure (Roger and Reynaud 1982). (4) Increase in soil biomass after their death and decomposition (Saadatnia and Riahi, 2009) (5) Decrease in soil salinity (Saadatnia and Riahi 2009). (6) Preventing weeds growth (Saadatnia and Riahi 2009). (7) Increase in soil phosphate by excretion of organic acids (Wilson 2006).

Hegazi *et al.* (2010) studied the influence of different cyanobacteria such as *Nostoc muscorum*, *N. humifusum*, *Anabaena oryzae*, *Wollea* sp., *Phormedium* and *Spirulina platensis* on growth and seed production of common beans (*Phaseolus vulgaris* L. cv. Nebraska) and reduction of utilization of mineral nitrogen fertilizer. They performed experiments to evaluate the influence of different application methods of Cyanobacteria mixture strains in the presence of different levels of nitrogen mineral fertilizer on performance of bean plants in clay loamy soil under surface irrigation. Most studied traits of bean such as vegetative growth, seed yield and its attributes, NPK and seed sugar (total and reducing) content showed positive significant effects when used in combination of two different applications (*viz.* - seed coating and soil drench). Cyanobacteria also enhance the soil biological activity in terms of increasing the total bacterial, total cyanobacterial counts, CO_2 evolution, dehydrogenase and nitrogenase activities.

Cyanobacteria *i.e. Chlorococcales, Mastigociadaceae, Nostocaceae, Oscikatoriaceae, Oscillatoriaceae, Rivulariaceae, Scytonematoceae, Stigonemataceae* these phototropic prokaryotic bacteria are effective only in submerged paddy in presence of bright sunlight by forming a bluish-green algae on standing water and by converting the insoluble P into soluble forms fixing N to the tune of 2-30 kg/ha thereby raising the crop yield by 10-15% when applied at 10kg/ha when BGA biomass decompose in soil and organic compounds are liberated and plant growth is regulated. These belong to eight different families, phototrophic in nature and produce indole acetic acid and gibberllic acid, fix 20-30 kg N/ha in submerged rice fields as they are abundant in paddy, so are also referred as 'paddy organisms'. N is the key input required in large quantities for low land rice production. Soil N and BNF by associated organisms are major sources of N for low land rice. The 50-60% N requirement is met through the combination of mineralization of soil organic N and BNF by free living and rice plant associated bacteria (Rahman *et al.* 2009). Although, the use of cyanobacteria is mostly limited to agriculture crops only but now-a -days its utilization in quality seedling production in forestry is also in vogue (Rai *et al.* 2004).

2.1.4 Synergistic effect of bioinoculants

All bioinoculants when applied alone showed promising results on growth and development of seedlings but when these are applied mixed together then their effects profound substantially. There are some research papers where such effects are reported. Mohan and Rajendran (2014) studied the effect of microorganisms on quality seed production of *Feronia elephantum* Corr. in Semi-Arid Region of Southern India. To improve the seedling quality, they performed nursery experiments to select the suitable bioinoculants for *F. elephantum* in tropical nursery conditions. The bioinoculants such as *Azospirillum* (*Azospirillum brasilense*), AM fungi (*Glomus fasciculatum*) and *Pseudomonas* (*Pseudomonas flurorescens* Migula) were isolated, mass multiplied and inoculated individually and in combinations. They found that the shoot length, root length, collar diameter and biomass were increased above 77.47 % in combined inoculation of *Azospirillum* + AM fungi + *Pseudomonas* in comparison to control. The combined inoculation of bioinoculants is beneficial for increasing growth, biomass and good quality seedling production.

Kumar (2012) evaluated the influence of bioinoculants on growth and mycorrhizal occurrence in the rhizosphere of *Mentha spicata* Linn. In this investigation, he made an attempt to evaluate the potential of two AM fungi (*Glomus mosseae* Gerd. & Trappe and *Acaulospora laevis* Gerd. & Trappe) along with *Trichoderma viride* on growth attributes of *M. spicata*. He found that all the inoculated seedlings showed significant results over control after 45 and 90 days of inoculation in polyhouse pot experiment. Dual inoculation of *G. mosseae* and *A. laevis* showed the maximum mycorrhizal inoculation effect.

Meenakshisundaram *et al.* (2011) conducted an experiment to select suitable bioinoculants and their combination to improve quality seedling production of *Delonix regia* (Hook.) Raf. and also studied its effect. Seedlings were inoculated individually and in combinationwith *Azospirillum, Azotobacter* and AM fungi (*Glomus fasciculatum*). They found that bioinoculants caused a significant increase the growth and biomass, total nitrogen, chlorophyll, protein content in *D. regia* when compared to uninoculated control plants.

Sujanya and Chandra (2011) conducted an experiment to study the effect of bioinoculants in replacement with chemical fertilizer on Groundnut, *Arachis hypogea* L. Different treatment combinations were employed, 75%, 50% or 25% of NPK and organic farm yard manure (FYM) were tested for partial replacement. Three biological agents, BGA, *Azotobacter* and *Azospirillum*, were studied along with NPK and FYM combinations, to analyze the use of biofertilizers with organic and chemical fertilizers. The use of biological agents, as a part replacement for chemical fertilizers has revealed an overall improvement in crop parameters studied.

Revathi *et al.* (2013) applied integrated nutrient management on the growth enhancement of *Dalbergia sissoo* Roxb. seedlings. An attempt was made to determine the effect of inorganic fertilizers *i.e.,* macro (N, P and K) and micro nutrients (Fe, Zn, B and Mo) with biofertilizers such as N fixing bacteria (*Rhizobium*) and P mobilizing symbiotic fungi (AM fungi) and bio-manures (leaf manures) in normal and alkaline soils on seedling growth of *D. sissoo.* They observed that dual inoculation with biofertilizers (*Rhizobium* and AM) was impressive in improving the growth and biomass of Shisham under normal soil whereas in alkaline soil, blending of micronutrients with biofertilizers (*Rhizobium* + AM) had better growth and biomass. Integrated Nutrient Management (INM) takes into account the optimizing performance of soil through augmenting chemical and biological properties of soil.

Kumar *et al.* (2009) studied the Influence of AM fungi and *Trichoderma viride* on growth performance of *Salvia officinalis* L. (Sage), a popular kitchen herb, member of mint family (Lamiaceae), has been cultivated for its wide range of medicinal values. They analyzed the effect of two AM fungi (*Acaulospora laevis* and *Glomus mosseae*) along with *Trichoderma viride*, alone and in combination, on different growth parameters of *S. officinalis* in a green house pot experiment with sterilized soil. AM inoculum and *T. viride* showed significant increase of different growth parameters after 45 and 90 days of inoculation. Among all treatments, dual combination of *A. laevis* plus *T. viride* was most effective in increasing shoot length, leaf area, root length, root weight, AM spore number and percent root colonization. Moreover, maximum increase in shoot biomass was found in plants treated with *T. viride*.

Kumar *et al.* (2013) conducted an experiment with six treatments comprised of *Dalbergia sissoo-Rhizobium* inoculation, 2 levels of N- fertilization alone and along with each other including control. The results of the experiment revealed that root inoculation of *D. sissoo* seedlings with *D. sissoo - Rhizobium* inoculum was found significantly effective in improving growth, biomass production and nodulation over only inorganic N application and control. A positive response was found when N in less level was given along with *Rhizobium* inoculation.

The PGPR influence the root colonization, penetration and nutrient translocation activity of VAM fungi (Azcon-Aguilar and Barea 1978, 1979). A synergistic interaction of VAM fungi with saprobic fungus- *Trichoderma* also resulted in enhanced growth responses in host plants (Parkash and Aggarwal 2011). VAM fungi have been found to improve growth, nodulation and nitrogen fixation in legume-*Rhizobium* symbiosis in *Acacia nilotica* (Khan *et al.* 2001). The mycorrhizal interaction with *Frankia* sp. resulted in greater dry weight, number of nodules and nodule weight and increased nitrogenase activity on *Alnus nepalensis* D. Don (Tiwari 1995). Synergism of *Azotobacter* with VAM fungi have been resulted in more nitrogen fixation and plant growth (Bagyaraj and Menge 1978). It has been reported that the effect of VAM fungi is increased when they were co-inoculated with other microflora of rhizosphere of plants (Parkash *et al.* 2011).

There are some more examples of inoculation where both PGPR and ectomycorrhizae were tested on plants but alone (not in combined treatments) and significant results were found. Asif *et al.* (2013) conducted an experiment to study the field performance of blue pine (*Pinus wallichiana* Jackson) seedlings inoculated with selected species of bioinoculants under nursery conditions. The experiment comprised of seven inoculants (*Azotobacter* sp., *Azospirillum* sp., *Pseudomonas fluorescens, Bacillus subtilis* Cohn, *Pisolithus tinctorius* Coker & Couch, *Laccaria laccata* Cooke and control). Various growth characters *viz.*, shoot height, collar diameter, root length and seedling survival at various intervals responded significantly to all the microbial inoculants. Among microbial inoculants the two ectomycorrhizae *viz.*, *P. tinctorius* and *L. laccata* proved beneficial for all growth parameters than rest of the inoculants. It was followed by *Azotobacter* sp., *Azospirillum* sp., *P. fluorescens* and *B. subtilis*. For root length, *P. fluorescens* and *B. subtilis* had best results than *Azotobacter* sp. and *Azospirillum* sp. Microbial inoculation of *P. tinctorius* and *L. laccata* revealed best results with respect to percent decrease in seedling mortality rate of the species. Thus, the two ectomycorrhizal species *viz.*, *Pisolithus tinctorius* and *Laccaria laccata* proved superior for all the studied growth parameters.

Umashankar *et al.* (2012) studied the effect of microbial inoculants on the growth of silver oak (*Grevillea robusta* A. Cunn.) in nursery condition. Experiment was conducted to evaluate the performance of microbial inoculants *viz.,* *Trichoderma*, P-solubilizer and N fixer for growth promoting activity of

Silver oak. The study was aimed to assess the effects of *Bacillus coagulans, Azotobacter* and *Trichoderma* isolates on plant growth of silver oak in the nursery condition. Maximum growth was observed in *Trichoderma* inoculated plants.

2.2 Mechanism of bioinoculants action

Bioinoculants suppress plant pathogens by direct parasitism, lysis, competition for food, direct antibiosis or indirect antibiosis through production of volatile sub-stances, *viz.*- ethylene, hydrogen cyanide, alcohols, monoterpenes and aldehydes (Juan *et al.* 2005). Activity of bioinoculants mainly depends on the physicochemical environmental conditions to which they are subjected. These mechanisms are complex and what has been defined as biocontrol is the final result of varied mechanisms acting antagonistically to achieve disease control.

Ectomycorrhizal (EcM) fungi play an important role in tree growth by increment in root absorption surface due to extensive network of fungal mycelia (Colinas *et al.* 1994); soil binding to create favourable soil structures (Borchers and Perry 1992); facilitation of soil nutrient transfer (Simard *et al.* 1997) and production along with supply of plant growth regulators (Heinrich *et al.* 1989).

Endomycorrhizae especially AM fungi interact with a wide range of other soil microorganisms in the rhizosphere of plants. These interactions may be either stimulatory or inhibitory. These are stimulatory when they increase the growth response of the host in presence of other microorganisms and inhibitory when they control soil-borne pathogens. AM fungi play a dominant role in increasing phosphorus solubilization and uptake of P, N, Ca, S, K, Mg, Mn, Cl by plants. The AM hyphae growing through soil pore spaces can affect phosphate absorption beyond the depleted zone. The fungal hyphae transport phosphate over large distance into the root cortical cells. The phosphate absorbed by AM fungi from soil solution is accumulated in the vacuoles of the fungus as polyphosphate (Poly-P) granules. The (Poly P) granules in fine branches of arbuscules are broken down by enzymatic activities and releasing phosphorus (inorganic) in the cytoplasm.

The various mechanisms of action of mycorrhizal effect on plant health and disease control put forward from time to time are shown below.

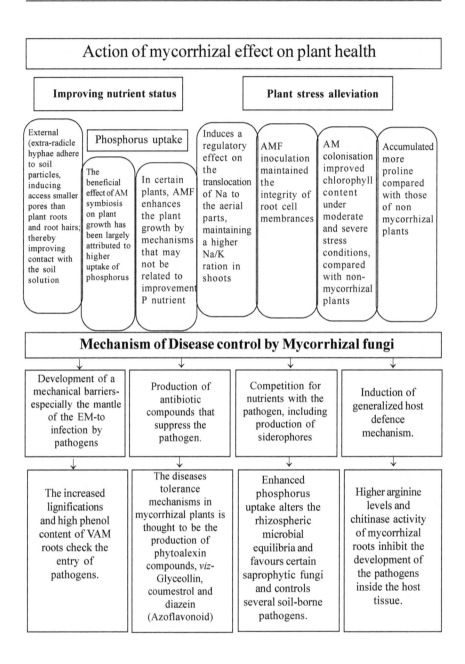

Action of mycorrhizal effect on plant health

Improving nutrient status

Plant stress alleviation

External (extra-radicle hyphae adhere to soil particles, inducing access smaller pores than plant roots and root hairs; thereby improving contact with the soil solution

Phosphorus uptake

The beneficial effect of AM symbiosis on plant growth has been largely attributed to higher uptake of phosphorus

In certain plants, AMF enhances the plant growth by mechanisms that may not be related to improvement P nutrient

Induces a regulatory effect on the translocation of Na to the aerial parts, maintaining a higher Na/K ration in shoots

AMF inoculation maintained the integrity of root cell membrances

AM colonisation improved chlorophyll content under moderate and severe stress conditions, compared with non-mycorrhizal plants

Accumulated more proline compared with those of non mycorrhizal plants

Mechanism of Disease control by Mycorrhizal fungi

Development of a mechanical barriers-especially the mantle of the EM-to infection by pathogens

Production of antibiotic compounds that suppress the pathogen.

Competition for nutrients with the pathogen, including production of siderophores

Induction of generalized host defence mechanism.

The increased lignifications and high phenol content of VAM roots check the entry of pathogens.

The diseases tolerance mechanisms in mycorrhizal plants is thought to be the production of phytoalexin compounds, viz-Glyceollin, coumestrol and diazein (Azoflavonoid)

Enhanced phosphorus uptake alters the rhizospheric microbial equilibria and favours certain saprophytic fungi and controls several soil-borne pathogens.

Higher arginine levels and chitinase activity of mycorrhizal roots inhibit the development of the pathogens inside the host tissue.

2.3 Bioinoculants Application

The following table shows some of selected and important effects of bioinoculants on various forestry plant species.

Name of plant species	Effects	References
Fungal (Non-mycorrhizal) Inoculation		
Trichoderma species		
Prunus cerasus L._P. dulcis_ (Mill.) D.A. Webb _Ziziphus jujuba_ Mill.	Bio-control of root and butt disease Enhancement in above- and below-ground biomass	Asef _et al._ (2008) Roohbakhsh and Davarynejad (2013)
Hevea brasiliensis (Willd. ex A. Juss.) Müll. Arg.	Increase in morphometric indices and leaf Phosphorus content	Promwee _et al._ (2014)
Ampelomyces species		
Salix cinerea L., _Betula pendula_ Roth., _Fagus sylvatica_ L., _Acer campestre_ L., _A. ginnala_ Maxim., _A. negundo_ L., _A. platanoides_ L.	Bio-control of tree mildews	Sucharzewska _et al._ (2012)
Piriformospora indica		
Azadirachta indica A. Juss._Adhatoda vasica_ L._Cassia angustifolia_ L._Terminalia arjuna_ L.	Plant promoter, plant protector, resistance against heavy metals, bioherbicide, immune-modulator, resistance against temperature, salt and stress tolerance as bio fertilizer	Bagde _et al._ (2010)
Chaetomium species		
Populus species	Growth parameters and bio-control efficacy of _Valsa_ canker (caused by _Valsa sordida_)	XinYan _et al._ (2009)
Hevea brasiliensis (Willd. ex A. Juss.) Müll. Arg.	Bio-control of white root disease (caused by _Rigidoporus microporus_)	Kaewchai and Soytong (2010)
Arbuscular Mycorrhizal Inoculation		
Echinacea purpurea L.	On secondary metabolites production and biomass	Araim _et al._ (2009)

(Contd.)

Name of plant species	Effects	References
Artemisia annua L.	Effects on shoot biomass, essential oil content and pharmaceutically active compound	Chaudhary *et al.* (2008)
Ocimum sanctum L. *Foeniculum vulgare* L.	Accumulation of secondary compounds	Copetta *et al.* (2006) Gupta *et al.* (2002) Kapoor *et al.* (2004) Toussaint *et al.* (2007)
Valeriana officinalis L., *Salvia officinalis* L., *Trifolium pretense* L., *Origanum vulgare* L.	On biomass production and concentration of pharmaceutically active compounds	Nell (2009)
Brassica oleracea L. var. *italica*	Enhancement of growth parameters and yield	Tanwar *et al.* (2014)
Eucalyptus saligna Sm.	Effect on growth and P uptake Growth and biomass production (Synergistic effect of AM fungi)	Parkash *et al.* (2005) Parkash and Aggarwal (2011)
Ruta graveolens L.	Growth and biomass production	Parkash *et al.* (2011a)
Rauwolfia serpentina Benth ex. Kurtz.	Growth and physical parameters	Kaushish *et al.* (2012)
Dendrocalamus strictus L.	Synergistic effect on growth	Parkash *et al.* (2011b)
Acacia catechu Willd.	Synergistic effect on growth	Parkash and Aggarwal (2009)
Eleusine corocana Gaertn.	Enzyme production and Grain yield	Parkash and Sharma (2013)
Acacia nilotica Delile and *Calliandra calothyrsus* Meisn.	Effect of VAM fungi on growth	Bagyaraj and Menge (1978)
Acacia nilotica Lamk. *Prosopis cineraria* L.	Growth effect and development	Rani *et al.* (1998a, 1998b)
Hevea brasiliensis (Willd. ex A. Juss.) Müll. Arg.	Enhancement of growth parameters, plastochron interval index and above- & below-ground biomass	Basumatary *et al.* (2014)
Aquilaria malaccensis Lamk.	Growth effect and plastochron interval index	Parkash and Biswas (2015)
Mesua ferrea L.	Growth effect and plastochron interval index	Parkash (2014)
Abroma augusta L.	Growth effect, conservation and secondary metabolite production	Parkash *et al.* (2014)
Ectomycorrhizal inoculants		
Pinus kesiya Royle ex Gordon	Growth performance	Tiwari and Mishra (1995)
Pinus kesiya Royle ex Gordon	Forest and degraded soils	Rao *et al.* (1996)
Eucalyptus tereticornis L.	Biocontrol of damping off disease in eucalyptus	Bhat *et al.* (1997) (Contd.)

Name of plant species	Effects	References
Pseudotsuga menziesii(Mirb.) Franco (Douglas-fir seedlings)	Growth performance	Mortier *et al.* (1998)
Eucalyptus tereticornis L.	Growth performance and colonization	Badshah *et al.* (2006)
Pinus patula Schiede ex Schlecht. & Cham.	Phosphorus uptake Growth performance	Ajungla *et al.* (2006)
		Ajungla *et al.* (2010)
Bacterial inoculants (Plant Growth Promoting Rhizobacteria)		
Pinus taeda L.	Enhancement of plant growth and biomass yield	Enebak *et al.* (1998)
Acacia farnesiana (L.) Willd.	Improvement of seedling vigour	Ceccon *et al.* (2012)
Cynobacterial inoculants		
Lupinus termis Forssk. var. Balady	Growth, Protein Pattern and Some Metabolic Activities	Haroun and Hussein (2003)
Prosopis juliflora L.	Enhancements in plant biomass, photosynthetic pigments, protein content and *in vivo* nitrate reductase activity (Synergistic effect)	Rai *et al.* (2004)\

3 Acknowledgements

Senior author (Vipin Prakash) is thankful to Rain Forest Research Institute (Indian Council of Forestry Research & Education), Jorhat, Assam for providing the all possible laboratory facilities for carrying out the research review work.

References

Abdel-Fattah MG, Shabana MY, Ismail EA, Rashad MY (2007) *Trichoderma harzianum*: a biocontrol agent against *Bipolaris oryzae*. Mycopathologia 164, 81-89.

Abdul Mallik M (2000) Association of arbuscular mycorrhizae with some varieties of tobacco (*Nicotiana tobacum* L.) and its effect on their growth, nutrition and certain soilborne diseases, Ph.D. Thesis, Bharathidasan University, Tiruchirapalli, India

Adholeya A, Tiwari P, Singh R (2005) Large scale inoculum production of arbuscular mycorrhizal fungi on root organs and inoculation strategies. In: *In Vitro* Culture of Mycorrhizae - Soil Biology, Volume 4 (Ed) Declerck S, Strullu DG and Fortin A, Springer-Verlag Berlin Heidelberg, pp 315-338.

Aggarwal A, Parkash V, Mehrotra RS (2006) VAM fungi as biocontrol agents for soil borne plant pathogens. In: Plant Protection in new Millennium (Ed) Gadewar AV and Singh BP, Satish Serial Pub. House, New Delhi, India, pp 131-147.

Aggarwal R, Tewari AK, Srivastava KD, Singh DV (2004) Role of antibiosis in the biological control of spot blotch (*Cochliobolus sativus*) of wheat by *Chaetomium globosum*. Mycopathologia. 15, 369-377.

Ajungla T, Imliyanger, Tzudir (2010) Effects of ectomycorrhizal fungi on the growth and performance of *Pinus patula* (Schiede ex Schlecht. & Cham.). Environ. Biol. Conserv. 15, 29-31.

Ajungla T, Sharma GD, Dkhar MS (2006) Uptake of phosphorus by ectomycorrhizal pine seedlings in the metal polluted soil. J. Environ. Biol. Conserv. 11, 15-18.

Alabouvette C, Olivain C, Steinberg C (2006) Biological control of plant diseases: the European situation. Eur. J. Plant Pathol. 114, 329-341.

Anderson CI, Cairney WGJ (2004) Diversity and ecology of soil fungal communities: increased understanding through the application of molecular techniques. Environ. Microbiol. 6, 769-779.

Araim G, Saleem A, Arnason JT, Charest C (2009) Root colonization by an arbuscular mycorrhizal (AM) fungus increases growth and secondary metabolism of purple coneflower, *Echinacea purpurea* (L.) Moench. J. Agric. Food Chem. 57, 2255-2258.

Asef MR, Goltapeh EM, Danesh YR (2008) Antagonistic effects of *Trichoderma* species in biocontrol of *Armillaria mellea* in fruit trees in Iran. J. Plant Prot. Res. 48(2), 213-222.

Asif M, Lone S, Lone FA, Hamid A (2013) Field performance of Blue Pine (*Pinus wallichiana*) seedling inoculated with selected species of Bio-Inoculants under nursery conditions. Int. J. Pharma. Bio. Sci. 4, 632 – 640.

Azcon-Aguilar C, Azcon R, Barea JM (1979) Endomycorrhizal fungi and *Rhizobium* biological fertilizers for *Medicago sativa* in normal cultivation. Nature 279, 325-327.

Azcon-Aguilar C, Barea JM (1978) Effects of interaction between different culture fractions of phosphobacteria and *Rhizobium* on mycorrhizal infection, growth and nodulation of *Medicago sativa*. Can. J. Bot. 24, 520-524.

Badshah N K U, Naik ST (2006) Growth and colonization in *Eucalyptus tereticornis* (Hybrid) seedlings inoculated with different inoculum formulations of *Pisolithus tinctorius*. Ind. For. 132, 575-580.

Bagde US, Prasad R, Varma A (2010) Interaction of Mycobiont: *Piriformospora indica* with medicinal plants and plants of economic importance. Afr. J. Biotechnol. 9, 9214-9226.

Bagyaraj D J (1989) Mycorrhizae, In: Tropical rain forest ecosystem (Ed.) Lieth H and Werger MJA, Elsevier Science Pub., Amsterdam, pp 537-545.

Bagyaraj DJ (1992) Vesicular arbuscular mycorrhiza: Application in agriculture. Methods Microbiol. 24, 360-373.

Bagyaraj DJ (2015) Status Paper on Arbuscular Mycorrhizal Fungi, In: Advances in Mycorrhiza & Useful Microbes in Forestry (Ed.) Harsh NSK and Kumar A, Greenfields Publishers & ICFRE, Dehradun, India, pp 21-37.

Bagyaraj DJ, Menge JA (1978) Interaction between a VA mycorrhiza and *Azotobacter* and their effects on the rhizosphere microflora and plant growth. New Phytol. 80, 567-573

Bargali K (2011) Actinorhizal plants of Kumaun Himalaya and their ecological significance. Afr. J. Plant Sci. 5, 401–406.

Basan Y, Holguin G (1997). *Azospirillum*-plant relations: Environmental and physiological advances. Can. J. Microbiol. 43, 103-121.

Bashan Y (1998) Inoculants of plant growth-promoting bacteria for use in agriculture. Biotechnol. Adv. 16, 729-770.

Basu PK, Kabi MC (1987) Effect of application of biofertilizers on the growth and nodulation of seven forest legumes. Ind. For. 113 (4), 249-257.

Basumatary N, Parkash V, Tamuli AK, Saikia AJ, Teron R (2015) Distribution and diversity of arbuscular mycorrhizal fungi along with soil nutrient availability decline with plantation age of *Hevea brasiliensis* (Willd. ex A. Juss.) Müll. Arg. J. Biodivers. 115, 401-412.

Benítez T, Rincón MA, Limón MC, Codón CA (2004) Biocontrol mechanisms of *Trichoderma* strains. Int. Microbiol. 7, 249-260.

Bergman B, Gallon JR, Rai AN, Stal LJ (1997) N2 Fixation by non-heterocystous cyanobacteria. FEMS Microbiol. Rev. 19 (3), 139-185.

Bhat M, Jeyrajan R, Ramaraj B (1997) Bio-control of damping off of *Eucalyptus tereticornis* using ectomycorrhizae. Ind. For. 123, 307-311.

Bhattacharya P, Jain RK, Paliwal MK (2000) Biofertilizers for Vegetables. Ind. Hort. 40, 12-13.

Bhattacharya PM, Misra D, Saha J, Chaudhari S (2000) Arbuscular mycorrhizal dependency of *Eucalyptus tereticornis* Sm.: How real is it? Mycorrhiza News 12, 11-15.

Bisleski RL (1973) Phosphate transport and phosphate availability. Ann. Rev. Pl. Physiol. 24, 225-252.

Borchers SL, Perry DA (1992) The influence of soil texture and aggregation on carbon and nitrogen dynamics in Southwest Oregon forest and clear-cuts. Can. J. For. Res. 21, 198-305.

Brundrett MC (2002) Coevolution of roots and mycorrhizas of land plants. New Phytol. 154, 275–304

Ceccon E, Almazo-Rogel A, Martinez-Romero E, Toledo I (2012) The effect of inoculation of an indigenous bacteria on the early growth of *Acacia farnesiana* in a degraded area. *CERNE.* 18(1), 49-57.

Chaia EE, Wall LG, Huss-Danell K (2010) Life in soil by the actinorhizal root nodule endophyte Frankia-A review. Symbiosis 51(3), 201–226.

Champawat RS, Pathak VN (1993) Effect of Vesicular-arbuscular mycorrhizal fungi on growth and nutrition uptake of pearl millet. Indian J. Mycol. Pl. Pathol. 23(1), 30-34.

Chandanie WA, Kubota M, Hyakumachi M (2006). Interactions between plant growth promoting fungi and arbuscular mycorrhizal fungus *Glomus mosseae* and induction of systemic resistance to anthracnose disease in cucumber. Plant Soil 286, 209-217.

Chaudhary V, Kapoor R, Bhatnagar AK (2008) Effectiveness of two arbuscular mycorrhizal fungi on concentrations of essential oil and artemisinin in three accessions of *Artemisia annua* L. Appl. Soil Ecol. 40, 174-181.

Chiariello N, Hickman JC, Mooney HA (1983) Endomycorrhizal role for interspecific transfer of phosphorus in a community of annual plants. Soil. Sci. 21, 941- 943.

Colinas C, Molina R, Trappe J, Perry D (1994) Ectomycorrhizas and rhizosphere microorganisms of seedlings of *Pseudotsuga menzeisii* (Mirb.) Franco planted on a degraded site and inoculated with forest soils pretreated with selective biocides. New Phytol. 127, 529-537.

Copetta A, Lingua G, Berta G (2006) Effects of three AM fungi on growth, distribution of glandular hairs, and essential oil production in *Ocimum basilicum* L. var. *genovese*. Mycorrhiza. 16, 485-494.

Das A, Prasad R, Srivastava A, Giang HP, Bhatnagar K, Varma A (2007) Fungal siderophores: structure, functions and regulation. In: Soil Biology Volume 12 Microbial Siderophores (Ed) Varma A and Chincholkar SB, Springer-Verlag Berlin Heidelberg, pp 1-42.

Dash S, Gupta N (2011) Microbial bioinoculants and their role in plant growth and development. Int. J. Biotechnol. Mol. Biol. Res. 2(13), 232-251.

Dhingra OD, Mizubuti ES, Santana FM (2003) *Chaetomium globosum* for reducing primary inoculum of *Diaporthe phaseolorum* f. sp. *meridionalis* in soil-surface soybean stable in field condition. Biol. Control. 2, 302-310.

El-Katathy MH, Gudelj M, Robra K-H, Elnaghy MA, Gübitz GM (2001) Characterization of a chitinase and an endo-β-1,3-glucanase from *Trichoderma harzianum* Rifai T24 involved in control of the phytopathogen *Sclerotium rolfsii*. Appl. Microbiol. Biotechnol. 5, 137-143.

Emmert EAB, Handelsman J (1999) Biocontrol of plant disease: a (Gram-) positive perspective. FEMS Microbiol. Lett. 17, 1-9.

Enebak SA, Wei G, Kloepper JW (1998) Effects of plant growthpromoting rhizobacteria on Loblolly and Slash pine seedlings. For. Sci. 44, 139-143.

Fitter AH, Moyersoen B (1996) Evolutionary trends in root-microbe symbioses. Phil. Trans. Royal Soc. London, Series B – Biol. Sci. 351, 1367-1375.

Franche C, Lindström K, Elmerich C (2009) Nitrogen-fixing bacteria associated with leguminous and non-leguminous plants. Plant Soil 321, 35–59.

Francis R, Read D J (1984) Direct transfer of carbon between plants connected by arbuscular mycorrhizal mycelium. Nature 307, 53-56.

García-Fraile P, Menéndez E, Rivas R (2015) Role of bacterial biofertilizers in agriculture and forestry. AIMS Bioeng. 2, 183-205.

Gaur A, Adholeya A (2004) Prospects of arbuscular mycorrhizal fungi in phytoremediation of heavy metal contaminated soils. Curr. Sci. 8, 528-534.

Gill TS, Singh RS, Kaur J (2002) Comparison of four arbuscular mycorrhizal fungi for root colonization, spore population and plant growth response in chickpea. Indian Phytopath. 55(2), 210-213.

Giovannetti M, Sbrana C (1998) Meeting a non-host: The behaviour of AM fungi. Mycorrhiza. 8(3), 123-130.

Govindarajan K, Thangaraju M (2001) Use of biofertilizers in quality seed production. In: Recent Techniques and Participatory Approaches on Quality Seed Production, Department of Seed Science and Technology, Coimbatore, pp 127-130.

Gupta ML, Prasad A, Ram M, Kumar S (2002) Effect of the vesicular-arbuscular mycorrhizal (VAM) fungus *Glomus fasciculatum* on the essential oil yield related characters and nutrient acquisition in the crops of different cultivars of menthol mint (*Mentha arvensis*) under field conditions. Bioresour. Technol. 81, 77-79.

Gupta RP, Kalia A, Kapoor S (2007) Bioinoculants: a step towards sustainable agriculture. New India Publishing Agency, New Delhi.

Handique L, Parkash V (2014) Nodular micro-endosymbiotic and endomycorrhizal colonization in *Elaeagnus latifolia* L.: A noble host of symbiotic actinorhizal and endomycorrhizal association. IJSR. 3, 1713-1714.

Harley JL, Smith SE (1983) Mycorrhizal Symbiosis. Academic Press, New York.

Harley JL, Harley EL (1987) A check list of mycorrhiza in the British flora. New Phytol. 105(2), 1–102.

Harman GE, Howell CR, Viterbo A, Chet I, Lorito M (2004) *Trichoderma* species opportunistic, avirulent plant symbionts. Nat. Rev. Microbiol. 2, 43-56.

Haroun SA, Hussein MH (2003) The promotive effect of algal biofertilizers on growth, protein pattern and some metabolic activities of *Lupinus termis* plants grown in siliceous soil. Asian J. Plant Sci. 2, 944-951.

Hebbar PK, Lumsden R D (1999) Biological control of seedling diseases. In: Methods in Biotechnology Vol. 5: Biopesticides: Use and Delivery (Ed) Franklin RH and Julius JM, Humana Press, Inc. Totowa, NJ, pp 103-116.

Hegazi AZ, Mostafa SSM, Ahmed HMI (2010) Influence of different Cyanobacterial application methods on growth and seed production of common bean under various levels of mineral nitrogen fertilization. Nat. Sci. 8, 183-194.

Heinrich PA, Muligen DR, Patrick JF (1989) The effect of ectomycorrhizal on the phosphorus and dry weight acquisition of Eucalyptus seedlings. Plant Soil 108, 147-149.

Howell RC (2003) Mechanisms employed by *Trichoderma* species in the biological control of plant diseases: the history and evolution of current concepts. Plant Dis. 87, 4-10.

Hrynkiewicz K, Baum, C (2011) The Potential of Rhizosphere Microorganisms to Promote the Plant Growth in Disturbed Soils. In: Environmental Protection Strategies for Sustainable Development Strategies for Sustainability (Ed.) Malik A and Grohmann E, Springer Science, Business Media, Netherlands, pp 35-64.

Ipsilantis I, Sylvia DM (2007) Abundance of fungi and bacteria in a nutrient-impacted Florida wetland. Appl. Soil Ecol. 35, 272-280.

Jha MN, Prasad AN, Sharma SG, Bharati RC (2001) Effects of fertilization rate and crop rotation on diazotrophic cyanobacteria in paddy field. World J. Microbiol. Biotechnol. 17(5), 463-468.

Juan AVN, Luis S, Angela B, Francisca V, Santiago GR, Rosa MH, Enrique M (2005) Screening of antimicrobial activities in *Trichoderma* isolates representing three *Trichoderma* sections. Mycol. Res. 109, 1397–1406.

Kabi MC, Poni SC, Bhaduru PN (1982) Potential of nitrogen nutrition of leguminous crops through rhizobial inoculation in West Bengal soils. Trans. of 12th Int. Cong. Soil Sci. 6, 1-52.

Kaewchai S, Soytong K, Hyde KD (2009) Mycofungicides and fungal biofertilizers. Fungal Divers. 3, 25-50.

Kaewchai S, Soytong K (2010) Application of biofungicides against *Rigidoporus microporus* causing white root disease of rubber trees. Int. J. Agric. Technol. 6, 349-363.

Kanokmedhakul S, Kanokmedhakul K, Nasomjai P, Louangsysouphanh S, Soytong K, Isobe M, Kongsaeree P, Prabpai S, Suksamran A (2006) Antifungal azaphilones from *Chaetomium cupreum* CC3003. J. Nat. Prod. 69, 891-895.

Kanokmedhakul S, Kanokmedhakul K, Phonkerd N, Soytong K, Kongsaree P, Suksamrarn A (2002). Antimycobacterial anthraquinonechromanone compound and diketopiperazine alkaloid from the fungus *Chaetomium globosum* KMITL-N0802. Planta Med. 6, 834-836.

Kapoor R, Giri B, Mukerji KG (2004) Improved growth and essential oil yield and quality in *Foeniculum vulgare* Mill. on mycorrhizal inoculation supplemented with P-fertilizer. Bioresour. Technol. 93, 307-311.

Karthikeyan N, Prasanna R, Nain L, Kaushik BD (2007) Evaluating the potential of plant growth promoting cyanobacteria as inoculants for wheat. Eur. J. Soil Biol. 43, 23-30.

Kasthuri Rengamani S, Jothibasu M, Rajendran K (2006) Effect of bioinoculants on quality seedlings production of Drumstick (*Moringa oleifera* L.). J. Non-Timber For. Pro. 13 (1), 41-46.

Kaushik A, Dixon RK, Mukerji KG (1992) Vesicular asbuscular mycorrhizal relationships of *Prosopis julifera* and *Zizipus jujuba*. Phytomorphol. 42, 133-147.

Kaushik S, Kumar A, Aggarwal A, Parkash V (2012) Influence of Inoculation with the endomycorrhizal fungi and *Trichoderma viride* on morphological and physiological growth parameters of *Rauwolfia serpentina* Benth. Ex. Kurtz. Ind. J. Microbiol. 52, 295-299.

Khan MR (2005) Biological control of Fusarial wilts and root-knot of legumes. New Delhi: Department of Biotechnology, Ministry of Science and Technology. pp 1-56.

Khan MR, Anwar MA (2011) Fungal Bioinoculants for Plant Disease Management. In: Microbes and Microbial Technology: Agricultural and Environmental Applications. (Eds) Ahmad I, Ahmad F, Pichtel J, Springer, New York, pp 447-488.

Khan MR, Anwer MA (2007) Molecular and biochemical characterization of soil isolates of *Aspergillus niger* and assessment of antagonism against *Rhizoctonia solani*. Phytopathol. Mediterr. 46, 304–315.

Khan MR, Anwer MA (2008) DNA and some laboratory tests of nematode suppressing efficient soil isolates of *Aspergillus niger*. Indian Phytopathol. 61, 212–225.

Khan MR, Khan N, Khan SM (2001) Evaluation of agricultural materials as substrate for mass culture of fungal biocontrol agents of fusarial wilt and root-knot nematode disease. Ann. Appl. Biol. (TAC-21 Suppl.) 22, 50–51.

Khan MR, Kousar K, Hamid A (2002) Effect of certain rhizobacteria and antagonistic fungi on root-nodulation and root-knot nematode disease of green gram. Nematol. Mediterr. 30, 85–89.

Khan S, Uniyal K (1999) Growth response of two species to VAM and *Rhizobium* inoculations. Ind. Forester 125, 1125-1128.

Khan TA, Mazid M, Mohammad F (2011) Ascorbic acid: an enigmatic molecule to developmental and environmental stress in plant. Int. J. Appl. Biol. Pharm. Technol. 2(33), 468-483.

Khetan SK (2001) Microbial pest control. New York, Basel, Marcel Dekker, Inc, pp. 300.

Kiss L (2003) A review of fungal antagonists of powdery mildews and their potential as biocontrol agents. Pest Manage. Sci. 5, 475-483.

Kodsueb R, McKenzie EHC, Lumyong S, Hyde KD (2008) Diversity of saprobic fungi on Magnoliaceae. Fungal Divers. 3, 37-53.

Krishnamoorthy G (Ed.) (2002) Agrobook, Issue April-June, Usha printers, New Delhi, pp. 22-24.

Kumar A, (2012) The Influence of bioinoculants on growth and mycorrhizal occurrence in the rhizosphere of *Mentha spicata* Linn. Bulletin of Environment, Pharmacology and Life Sciences . Environ. Pharmacol. Life Sci. 1, 60 – 65.

Kumar A, Aggarwal A, Kaushish S (2009) Influence of arbuscular mycorrhizal fungi and *Trichoderma viride* on growth performance of *Salvia officinalis* Linn. J. Appl. Nat. Sci. 1, 13-17.

Kumar A, Dash D, Jhariya M K (2013) Impact of *Rhizobium* on growth, biomass accumulation and nodulation in *Dalberia sisoo* seedlings . Bioscan. 8, 553-560.

Kumar A, Bhatti SK , Aggarwal A, (2012) Biodiversity of endophytic mycorrhiza in some ornamental flowering plants of Solan, Himachal Pradesh. BFAIJ. 4: 45-51

Lapeyrie FF, Chilvers GA (1985) An endomycorrhiza-ectomycorrhiza succession associated with enhanced growth by *Eucalyptus dumosa* seedlings planted in calcareous soils. New Phytol. 100, 93-104.

Lapointe L, Molard J (1997) Costs and benefits of mycorrhizal infection in a spring ephemeral, *Erythronium americanum*. New Phytol. 135, 491-500.

Lechevalier MP (1994) Taxonomy of the genus *Frankia* (Actinomycetales). Int. J. Syst. Bacteriol. 44, 1–8.

Li XL, Marschner H, George E (1991) Acquisition of phosphorus and copper by arbuscular mycorrhizal hyphae and root to shoot transport in white clover. Plant Soil 136, 49-57.

Lynch, J (1990) The rhizosphere. Wiley, London, UK.

Marin M (2006). Arbuscular mycorrhizal inoculation in nursery practice. In: Handbook of Microbial Biofertilizers (Ed) Rai MK, Food Products Press, pp 289-324.

Markovic M, Rajkovic N (2011) *Ampelomyces quisqualis* Ces. ex Schlecht. As an alternative measure of protection. In: Microbes in Applied Research Current Advances and Challenges (Ed) Mendez-Vilas A, World scientific Publishing Company Pvt. Ltd., Singapore, pp 8-12.

Marschner H, Dell B (1994) Nutrient uptake in mycorrhizal symbiosis. Plant Soil. 159, 89-102.

Marwah RG, Fatope MO, Deadman ML, Al- Maqbali YM, Husband J (2007) Musanahol: a new aureonitol-related metabolite from a *Chaetomium* sp. Tetrahedron Lett. 6, 8174- 8180.

Mazid M, Khan TA (2014) Future of bio-fertilizers in Indian agriculture: An overview. Int. J. Agric. Food Res. 3, 10-23.

Mazid M, Khan TA, Mohammad F (2011) Cytokinins, A classical multifaceted hormone in plant system. J. Stress Physiol. Biochem. 7, 347-368.

Meenakshisundaram K, Santhaguru, Rajenderan K (2011) Effects of bioinoculants on quality seedlings production of *Delonix regia* in tropical nursery conditions. Asian J. Biochem. Pharma. Res. 1, 2231-2560.

Mehrotra VS, Baijal U, Mishra SD, Panday DP, Mathews T (1995) Movement of 32P in sunflower plants insulated with single and dual inocula of VAM fungi. Curr. Sci. 68, 751-753.

Michelson A, Rosendahl S (1990) The effect of VA mycorrhizal fungi, phosphorus and drought stress on the growth of *Acacia nilotica* and *Leucaena leucocephala* seedlings. Plant Soil 124, 7-13.

Miransari M, Abrishamchi A, Khoshbakht K, Niknam V (2014) Plant hormones as signals in arbuscular mycorrhizal symbiosis. Crit. Rev. Biotechnol. 34, 123-133.

Mohan E, Rajendran K (2014) Effect of Plant growth-promoting Microorganisms on Quality Seedling Production of *Feronia elephantum* (Corr.) in Semi-Arid Region of Southern India. Int. J. Curr. Microbiol. App. Sci. 3, 103-116.

Mohan V (2015) Plant Growth Promoting Rhizobacteria (PGPR) in Forestry, In: Advances in Mycorrhiza & Useful Microbes in Forestry (Ed) Harsh NSK and Kumar A, Greenfields Publishers & ICFRE, Dehradun, India. pp 67-89.

Mohan V, Manimekalai M, Manokaran P (2007) Status of arbuscular mycorrhizal (AM) associations in seedlings of important forest tree species in Tamil Nadu. Kavaka. 35, 45-50.

Mortier F, LeTecon, Garbaye (1998) Effect of inoculums type and inoculation dose on ectomycorrhizal development, root nurseries and growth of Douglas-fir seedlings inoculated with *Laccaria laccata* in a nursery. Ann. Sci. For. 45, 301-310.

Mukerji KG, Dixon RK (1992) Mycorrhizae in reforestation. In: Proceedings International Symposium on Rehabilitation of Tropical Rainforest Ecosystems, Research and Development (Ed) Said AB, Sarawak, Malaysia, pp 66-82.

Nair MG, Burke BA (1988) A few fatty acid methyl ester and other biologically active compounds from *Aspergillus niger*. Phytochemistry. 27, 3169–3173

Nannipieri P, Ascher J, Ceccherini MT, Landi L, Pietramellara G, Renella G (2003) Microbial diversity and soil functions. Eur. J Soil Sci. 54, 655-670.

Nell, M. (2009). Effect of the Arbuscular Mycorrhiza on Biomass Production and Accumulation of Pharmacologically Active Compounds in Medicinal Plants. Ph.D. Thesis, University of Natural Resources and Applied Life Sciences, Vienna.

Newman EI (1988) Mycorrhizal links between plants: their functioning and ecological significance. Adv. Eco. Res. 18, 243-270.

Okon Y (1985) *Azosprillum* as a potential inoculants for agriculture. Trend. Biotechnol. 3, 223-228.

Okon Y, Kapulnik Y (1986) Development and Function of *Azospirillum* inoculated roots. Plant Soil 90, 3-16.

Osonubi O, Bakare ON, Mulongy K (1992) Interactions between drought stress and vesicular-arbuacular mycorrhiza on the growth of *Faidherbia albida* (Syn. *Acacia albida* and *Acacia nilotica* in sterile and non-sterile soils. Biol. Fertil. Soils. 14, 159-165.

Papavizas GC (1985) Biological control of soil borne diseases. Summa Phytopathol. 11, 173-179.

Park JH, Choi GJ, Jang SK, Lim KH, Kim TH, Cho YK, Kim JC (2005) Antifungal activity against plant pathogenic fungi of chaetoviridins isolated from *Chaetomium globosum*. FEMS Microbiol. Lett. 252, 309-313.

Parkash V (2014) Biotization with Arbuscular Mycorrhizal fungi improves Plastochrome, biovolume and quality indices in nursery grown *Mesua ferrea* L. seedlings. In: Souvenir cum Abstract Book of National Seminar on "Emerging Bio-inputs in Biotechnology for a Green Environment", Department of Bioengineering and Technology, Gauhati University Institute of Science and Technology, Guwahati – 014, Assam in collaboration with North-East Biotechnological Association (NEBA) Department of Biotechnology, Gauhati University Guwhati-781 014, Assam. pp 49.

Parkash V, Aggarwal A (2009) Diversity of endomycorrhizal fungi and its synergistic effect on growth of *Acacia catechu* Willd. J. For. Sci. 55, 461-468.

Parkash V, Aggarwal A (2011) Interaction of VAM fungi with *Rhizobium* sp. and *Trichoderma viride* on establishment and growth of *Eucalyptus saligna* Sm. seedlings. E-Int. Sci. Res. J. 3, 200-209.

Parkash V, Aggarwal A, Bipasha (2011b) Rhizospheric effect of vesicular arbuscular mycorrhizal inoculation on biomass production of *Ruta graveolens* L.: A potential medicinal and aromatic herb. J. Plant Nutr. 34, 1386-96.

Parkash V, Aggarwal A, Sharma S, Sharma D (2005) Effect of endophytic mycorrhizae and fungal bioagent on the development and growth of *Eucalyptus saligna* seedlings. Bull. Nat. Inst. Ecol. 1, 127-131.

Parkash V, Biswas SC (2015) Arbuscular Mycorrhizal biotization improves morphometric indices in *Aquilaria malaccensis* Lamk. In: Souvenir cum Abstract Book of National Seminar on Recent advances on Agarwood Research in India (Ed) Parkash V, Das DJ, Dutta D, Hazarika P, Kumar R and Pathak KC,, Rain Forest Research Institute (ICFRE). pp 42.

Parkash V, Biswas SC, Saikia AJ (2014) Locational variability in rhizospheric endomycorrhizae associated with *Abroma augusta* L. and their synergistic role in accumulation of some phytochemicals. In: Souvenir cum Abstract Book of National Seminar on "Unravelling Plant Microbe Interaction for supporting plant health", Department of Botany, Gauhati University, Guwahati. pp 71.

Parkash V, Sharma S (2013) Endomycorrhizal inoculation affects APase and ALPase enzymes in rhizosphere and biomass yield of *Eleusine coracana* (Linn.) Gaertn. Adv. Crop Sci. 3, 811-819.

Parkash V, Sharma S, Aggarwal A (2011a) Symbiotic and synergistic efficacy of endomycorrhizae with *Dendrocalamus strictus* L. Plant Soil Environ., 57: 447-451.

Paulitz TC, Belanger RR (2001) Biological control in greenhouse system. Ann. Rev. Phytopathol. 39, 103-133.

Pereira OL, Barreto RW, Cavallazzi JRP, Braun U (2007) The mycobiota of the cactus weed *Pereskia aculeata* in Brazil, with comments on the life-cycle of *Uromyces pereskiae*. Fungal Divers. 25, 127-140.

Perrine-Walker F, Gherbi H, Imanishi L, *et al.* (2011) Symbiotic signaling in actinorhizal symbioses. Curr. Protein Pept. Sci. 12, 156–164.

Prabina BJ, Kumar K, Kannaiyan S (2004) Growth pattern and chlorophyll content of the cyanobacterial strains for their utilization in the quality control of cyanobacterial biofertilizers. In: Biofertilizers Technology Coimbatore (Ed) Kannaiyan S, Kumar K and Govindarajan K, India, pp 446-450.

Prasanna R, Jaiswal P, Singh YV, Singh PK (2008) Influence of biofertilizers and organic amendments on nitrogenase activity and phototrophic biomass of soil under wheat. Acta Agron. Hung. 56(2), 149-159.

Promwee A, Issarakraisila M, Intana W, Chamswarng C, Yenjit P (2014) Phosphate solubilization and growth promotion of rubber tree (*Hevea brasiliensis* Muell. Arg.) by *Trichoderma* Strain, J. Agric. Sci. 6, 8-12.

Punja ZK, Utkhede RS (2004) Biological control of fungal diseases on vegetable crops with fungi and yeasts. In: (ed.) Arora DK, Fungal Biotechnology in Agricultural, Food, and Environmental Applications, New York, Basel, pp 157-171.

Raaijmakers JM (2001) Rhizosphere and rhizosphere competence (Ed) Maloy OC and Murray TD, Encyclopedia of Plant Pathology, Wiley, USA, pp 859–860.

Raaijmakers JM, Paulitz, TC, Steinberg C, Alabouvette C, Moënne-Loccoz Y (2009) The rhizosphere: a playground and battlefield for soilborne pathogens and beneficial microorganisms. Plant Soil. 321, 341–36.

Rahman MM, Amano T, Shiraiwa T (2009) Nitrogen use efficiency and recovery from N fertilizer under rice-based cropping systems. Aust. J. Crop Sci. 3, 336-351.

Rai UN, Pandey K, Sinha S, Singh A, Saxena R, Gupta DK (2004) Revegetating fly ash landfills with *Prosopis juliflora* L.: impact of different amendments and *Rhizobium* inoculation. Environ. Int. 30, 293-300.

Rajasekaran S, Sankar Ganesh K, Jayakumar K, Bhaaskaran MRC, Sundaramoorthy P (2012) Biofertilizers-Current Status of Indian Agriculture. Int. J. Environ. Bioenergy 4, 176-195.

Rajendra P, Singh S, Sharma S N (1998) Interrelationship of fertilizers use and other agricultural inputs for higher crop yields. Fert. News. 43, 35-40.

Rajendran K, Sugavanam V, Devaraj P (2003) Effect of biofertilizers on quality seedling production of *Casuarina equisetifolia*, J. Trop. For. Sci. 15, 82 96.

Rani P, Aggarwal A, Mehrotra RS (1998a) Establishment of nursery technology through *Glomus mosseae*, *Rhizobium* sp. and *Trichoderma harzianum* on better biomass yield of *Prosopis cineraria* Linn. Proc. Nat. Acad. Sci. Sec. B 68, 301-305.

Rani P, Agarwal A, Mehrotra RS (1998b) Growth Responses in *Acacia nilotica* inoculated with VAM fungi (*Glomus fasciculatum), Rhizobium* sp. and *Trichoderma harzianum*. J. Mycopath. Res. 36, 13-16.

Rao CS, Sharma GD, Shukla AK (1996) Ectomycorrhizal efficiency of various mycobionts with pinus kesiya seedlings in forest and degraded soils. Proc. Ind. Nat. Sci. Acad. B62(5), 427-434.

Revathi R, Mohan V, Jha MN (2013) Integrated nutrient management on the growth enhancement of *Dalbergia sissoo* Roxb. seedlings. J. Acad. Indus. Res. 1(9), 225-252.

Rinaldi AC, Comandini O, Kuyper TW (2008) Ectomycorrhizal fungal diversity: separating the wheat from the chaff. Fungal Divers. 33, 1-45.

Rodriguez A A, Stella A A, Storni M M, Zulpa G and Zaccaro MC (2006) Effects of cyanobacterial extracelular products and gibberellic acid on salinity tolerance in *Oryza sativa* L. Saline System. 2, 7.

Roger PA, Reynaud PA (1982) Free-living Blue-green Algae in Tropical Soils. Martinus Nijhoff Publisher, The Hague.

Rola CA (2000) Economic perspective for agricultural biotechnology research planning. Philippine institute for development studies, Discussion paper No. 2000-10, April 2000: pp 28.

Roohbakhsh H, Davarynejad GH (2013) How addition of *Trichoderma* would affect further growth of jujube cuttings? Intl. J. Agri. Crop Sci. 6, 905-912.

Saadatnia R (2009). Cyanobacteria from paddy fields in Iran as a biofertilizer in rice plants. Plant Soil Environ. 55, 207-212.

Sadhana B (2014) Arbuscular mycorrhizal fungi (AMF) as a Biofertilizer- a Review. Int. J. Curr. Microbiol. App. Sci. 3, 384-400.

Sahu D, Priyadarshani I, Rath B (2012) Cyanobacteria - As potential biofertlizers. J. Microbiol. 1, 20-26.

Scagel C F, Linderman R G (1996) Influence of ectomycorrhizal fungal inoculation on growth and root IAA concentrations of transplanted conifer. Tree Physiol. 18, 73-747.

Schwintzer R, Tjepkema JD (1990) The biology of *Frankia* and *Actinorrhizal* Plants, Academic Press inc. San Diego, USA. pp 99.

Sekar I, Vanangamudi K, Suresh K, Suresh KK (1995) Effects of biofertilizers on the seedling biomass VAM colonization, enzyme activity and phosphrous uptake in the shoal tree species, My Forest. 31, 21- 26.

Shenoy BD, Jeewon R, Hyde K D (2007) Impact of DNA sequence-data on the taxonomy of anamorphic fungi. Fungal Divers. 26, 1-54.

Shishkoff N, McGrath MT (2002) AQ10 biofungicide combined with chemical fungicides or AddQ spray adjuvant for control of cucurbit powdery mildew in detached leaf culture. Plant Dis. 86, 915-918.

Simard SW, Perry DA, Jones MD, Durall DM, Molina R (1997) Net transfer of carbon between tree species with shared ectomycorrhizal fungi. Nature. 388, 579-582.

Singh S, Pandey A, Palni LMS (2008) Screening of arbuscular mycorrhizal fungal consortia developed from the rhizospheres of natural and cultivated tea plants for growth promotion in tea (*Camellia sinensis* (L.) O. Kuntze). Pedobiologia. 52, 119-125.

Sitaramaiah K, Khanna R, Trimurtulu N (1998) Effect of *Glomus fasciculatum* on growth and chemical composition of maize. J. Mycol. Plant Pathol. 28, 38-41.

Smith AH (1971) Taxonomy of ectomycorrhiza-forming fungi. In: Mycorrhizae. Miscellaneous Publication 1189, (Ed) Hacskaylo E, Washington, DC: US Department of Agriculture, Forest Service, pp 1–8.

Smith SA, Read DJ (1997) Mycorrhizal Symbiosis. 2nd Edn., Academic Press, Cambridge.

Smith SE, Dickson S, Morris C, Smith EA (1994)Transfer of phosphate from fungus to plant in arbuscular mycorrhizas in *Allium porrum* L. New Phytol. 127, 93-99.

Soares DJ, Barreto RW (2008) Fungal pathogens of the invasive riparian weed *Hedychium coronarium* from Brazil and their potential for biological control. Fungal Divers. 28, 85-96.

Song T, Martensson L, Eriksson T, Zheng W, Rasmussen U (2005) Biodiversity and seasonal variation of the cyanobacterial assemblage in a rice paddy field in Fujian, China. The Federation of European Materials Societies Microbiol. Ecol. 54, 131–140.

Soytong K, Kanokmadhakul S, Kukongviriyapa V, Isobe M (2001) Application of *Chaetomium* species (Ketomium®) as a new broad spectrum biological fungicide for plant disease control: A review article. Fungal Divers. 7, 1-15.

Soytong K, Quimio TH (1989a) A taxonomic study on the Philippines species of *Chaetomium*. Philipp. Agric. Sci. 72, 59-72.

Soytong K, Quimio TH (1989b) Antagonism of *Chaetomium globosum* to the rice blast pathogen, *Pyricularia oryzae*. Kasesart J. (Nat. Sci.) 23, 198-203.

Stirling GR (1993) Biocontrol of plant pathogenic nematode and fungus. Phytopathol. 83, 1525–1532

Sucharzewska E, Dynowska M, Kubiak D, Ejdys E, Biedunkiewicz A (2012) *Ampelomyces hyperparasites* – occurrence and effect on the development of ascomata of *Erysiphales* species under conditions of anthropopressure. Acta Soc. Bot. Pol. 81, 147–152.

Sujanya S, Chandra S (2011) Effect of part replacement of chemical fertilizers with organic and bio-organic agents in ground nut, *Arachis hypogea*. J. Algal Biomass Utln. 2, 38– 41.

Sunantapongsuk V, Nakapraves P, Piriyaprin S, Manoch L (2006) Protease production and phosphate solubilization from potential biological control agants *Trichoderma viride* and *Azomonas agilis* from Vetiver rhizosphere. International Workshop on Sustained Managament of Soil- Rhizosphere System for Efficient Crop Production and Fertilizer Use. Land Development Department, Bangkok, Thailand, pp 1-4.

Szekeres A, Kredics L, Antal Z, Kevei F, Manczinger, L (2004). Isolation and characterization of protease overproducing mutants of *Trichoderma harzianum*. Microbiol. Lett. 23, 215-222.

Tang W, Yang H, Ryder M (2001) Research and application of *Trichoderma* spp. in biological control of plant pathogen. In: Bio-Exploitation of Filamentous Fungi Fungal diversity Research Series 6 (Ed) Pointing SB and Hyde KD, Bangkok, Thailand, pp 403- 435.

Tanwar A, Aggarwal A, Parkash V (2014) Effect of bioinoculants supplemented with phosphate fertilizers on the growth and yield of Broccoli (*Brassica oleracea* L. var. *Italica* Plenck). New Zeal. J. Crop Hort., DOI:10.1080/01140671.2014.924537.

Taylor AES, Alexander I (2005) The Ectomycorrhizal symbiosis: life in the real world. Mycologist. 19, 102-112.

Tchan YT (1984) Family II. Azotobacteriaceae. In: Bergey's Manual of Systematic Bacteriology (Ed) Krieg NR and Holt JG, Williams and Wikins, Baltimore, Vol. 1, pp. 219.

Tchan YT, New PT (1984) Genus I. *Azotobacter beijerinck.* In: Bergey's Manual of Systematic Bacteriology (Ed) Krieg NR and Holt JG, Williams and Wikins, Baltimore, Vol. 1, pp. 220.

Tejesvi MV, Kini KR, Prakash HS, Ven Subbiah, Shetty HS (2007) Genetic diversity and antifungal activity of species of *Pestalotiopsis* isolated as endophytes from medicinal plants. Fungal Divers. 24, 37-54.

Than PP, Prihastuti H, Phoulivong S, Taylor PWJ, Hyde KD (2008a) Chilli anthracnose disease caused by *Colletotrichum* species. J. Zhejiang Univ. Sci. B. 9, 764-778.

Than PP, Shivas RG, Jeewon R, Pongsupasamit S, Marney TS, Taylor PWJ, Hyde KD (2008b) Epitypification and phylogeny of *Colletotrichum acutatum* J.H. Simmonds. Fungal Divers. 28, 97-108.

Thormann MN, Rice AV (2007) Fungal from peatlands. Fungal Divers. 24, 241-299

Tilak KVBR (1993) Association effects of vesicular arbuscular mycorrhizae with nitrogen fixers. *Proc.* Indian. Natl. Sci. Acad. B 59, 325-332.

Tilak KVBR, Ranganayaki N, Pal KK, De R, Saxena AK, Shekhar NC, Mittal S, Tripathi AK, Johri BN (2005). Diversity of plant growth and soil health supporting bacteria. Curr. Sci. 89, 136-150.

Tilak KVBR, Subba RNS (1987) Association of *Azospirillum brasilense* with pearl millet (*Pennisetum americanum* (L.) Leeke). Biol. Fertil. Soil. 4, 97-102.

Tiwari SC (1995) *Alnus nepalensis* D. Do biomass production and growth response to inoculation with *Frankia* and vesicular arbuscular mycorrhiza. In: Mycorrhizae: Biofertilizers for The Future, (Ed.) Adholeya A and Singh S, TERI, Lodhi Road, New Delhi, India, pp 184-188

Tiwari SC, Mishra RR (1995) Effects of *Boletus edulis*, *Laccaria laccata*, *Pisolithus* and *Rhizopogon luteolus* on the growth performance of *P. kesiya* (Royle ex Gordon) in NE India. Ind. J. For. 1, 293-300.

Toussaint, J.P, Kraml, M., Nell, M., Smith, S.E., Smith, F.A., Steinkellner, S., Schmiderer, C., Vierheilig, Novak, J. (2008) Effect of *Glomus mosseae* on concentrations of rosmarinic and caffeic acids and essential oil compounds in basil inoculated with *Fusarium oxysporum* f. sp. *basilici*. Plant Pathol. 57: 1109-1116.

Trappe JM (1977) Selection of fungi for ectomycorrhizal inoculation in nurseries. Annu. Rev. Phytopathol. 15, 203-222.

Tromas A, Parizot B, Diagne N, Champion A, Hocher V, Cissoko M, Crabos A, Prodjinoto H, Lahouze B, Bogusz D, Laplaze L, Svistoonoff S (2012) Heart of endosymbioses: transcriptome analysis reveals conserved genetic programme between arbuscular mycorrhizal, actinorhizal and legume-rhizobial symbioses. PloS ONE 7(9), e44742. doi:10.1371/journal.pone.0044742.

Tviet M, Moor MB (1954) Isolates of Chaetomium that protect oats from *Helminthosporium victoriae*. Phytopathology 44, 686-689.

Umashankar N, Venkateshamurthy P, Krishnamurthy R, Raveendra HR, Satish KM (2012) Effect of microbial inoculants on the growth of silver oak (*Grevillea robusta*) in nursery Condition. IJESD. 3, 71-76.

Verma S, Varma A, Rexer K-H, Hassel A, Kost G, Sarbhoy A, Bisen P, Bütehorn B, Franken P (1998) *Piriformospora indica*, gen. *et* sp. nov., a new root-colonizing fungus. Mycologia 90, 896–903.

Vijayakumari B, Janardhanan K (2003) Effect of biofertilizers on seed germination, seedling growth and biochemical changes in silk cotton [*Ceibapentanndra* (Linn.) Gaertn.] Crop Res. 25, 328-332.

Vinale F, Marra R, Scala F, Ghisalbert EL, Lorito M, Sivasithamparam K (2006) Major secondary metabolotes produced by two commercial *Trichoderma* strains active against different phytopathogens. Lett. Appl. Microbiol. 43, 143-148.

Vinale F, Sivasithamparam K, Ghisalberti EL, Marra R, Woo SL, Lorito M (2008). *Trichoderma* plant pathogen interactions. Soil Biol. Biochem. 40, 1-10.

Viterbo A, Inbar J, Hadar Y, Chet I (2007) Plant disease biocontrol and induced resistance via fungal mycoparasites. In: Environmental and Microbial Relationships, 2nd edn. (Ed) Kubicek CP and Druzhinina IS, The Mycota IV. Springer-Verlag Berlin Heidelberg, pp 127-146.

Wall L G (2000) The actinorhizal symbiosis, J. Plant Growth Regul. 19(2), 167-182.

Wilson LT (2006). Cyanobacteria: A potential nitrogen source in rice fields. Texas Rice 6, 9–10.

XinYan Y, QingGuo M, SiDong R, ZhenYe L, MaoYun W, KeXiang G (2009) Effect of *Chaetomium globosum* ND35 on plant growth and preliminary study of its biocontrol efficacy. Hunan Agric. Sci. Technol. 10, 126-129.

Yanni YG, Abd El-Rahman AAM (1993) Assessing phosphorus fertilization of rice in the Nile delta involving nitrogen and cyanobacteria. Soil Biol. Biochem. 25, 289-293.

Yeasmin T, Zaman P, Rahman A, Absar N, Khanum NS (2007) Arbuscular mycorrhizal fungus inoculum production in rice plants. Afr. J. Agric. Res. 2, 463- 467.

Zeilinger S, Omann M (2007) *Trichoderma* biocontrol: signal transduction pathways involved 50 in host sensing and mycoparasitism. Gene Regul. Syst. Bio. 1, 227-234.

Zhang HY, Yang Q (2007) Expressed sequence tags-based identification of genes in the biocontrol agent *Chaetomium cupreum*. Appl. Microbiol. Biotechnol. 74, 650-658.

Zhao ZW, Wang GH, Yang L (2003) Biodiversity of arbuscular mycorrhizal fungi in tropical rainforests of Xishuangbanna, southwest China. Fungal Divers. 13, 233-242.

14

Microbial Synthesis of Nanoparticles for Use in Agriculture Ecosystem

Tarafdar J.C. and Indira Rathore

Abstract

Microbiological synthesis of nanoparticles has emerged as rapidly developing research area in nanotechnology across the globe with various biological entities being employed in production of nanoparticles constantly forming an impute alternative for conventional methods. Simple prokaryotes to complex eukaryotic organisms including higher plants are used for the fabrication of nanoparticles. In the present article, the synthesis of important plant nutrients like Mg, Zn, Ag, Au, Ti, P and Fe nanoparticles from different fungal species, their characterization and application potential of nanonutrients in agriculture is elaborately discussed.

Keywords: Agriculture, Microbial nanoparticles, Nanonutrients.

1. Introduction

Nanotechnology is emerging as the sixth revolutionary technology in the current era after the Industrial Revolution of Mid 1700s, Nuclear Energy Revolution of the 1940s, The Green Revolution of 1960s, Information Technology Revolution of 1980s and Biotechnology Revolution of the 1990s. It is an emerging and fast growing field of science which is being exploited over a wide spectrum of disciplines such as physics, chemistry, biology, material science, electronics, medicine, energy, environment and health sectors. Nano-technology deals with the matter considered at nanoscale (1-100 nm) and their implication for the welfare of human beings. Materials reduced to the nanoscale show some unusual properties which are different from what they exhibit on a macro scale, enabling unique systematic applications.

Synthesis of nanoparticles can be performed using a number of routinely used chemical, physical and aerosol techniques (Tarafdar and Raliya 2011). However, altogether these methods are energy and capital intensive, and they employ toxic chemicals and nonpolar solvents in the synthesis procedure and later on synthetic additive or capping agents, thus precluding their application in clinical, agriculture and biomedical fields. Therefore, the need for the development of a clean, reliable, biocompatible, benign and eco-friendly process to synthesize nanoparticles leads to turning researchers towards 'green' chemistry and bioprocesses (Tarafdar *et al.* 2013). In recent years, microbial synthesis of nanoparticles has emerged as a promising field of research in nanotechnology. Microorganisms such as bacteria, fungi, actinobacteria, yeasts, and viruses are reported to have the innate potential to produce metal nanoparticles either intra- or extra- cellularly and are considered as potential biofactories for nanoparticle synthesis (Narayanan and Sakthivel, 2010). A great deal of effort has been devoted towards the biosynthesis of metal nanoparticles using bacteria (Klaus *et al.* 1999; Husseiny *et al.* 2007), fungi (Tarafdar *et al.* 2012; Raliya and Tarafdar 2012), actinobacteria (Ahmad *et al.* 2003; Shastry *et al.* 2003), yeast (Kowshik *et al.* 2003) and viruses (Lee *et al.* 2002; Merzlyak and Lee 2006). In addition to the above mentioned synthesis methods, biogenic synthesis that utilizes parts of whole plants as biological factories to synthesize metallic nanoparticles is under exploitation and is an advantageous and profitable approach (Akhtar *et al.* 2013).

2. Biosynthesis of nanoparticles

Biosynthesis of nanoparticles is a kind of bottom up approach where the main reaction occurring is reduction/oxidation. The microbial enzymes or the plant phytochemicals with antioxidant or reducing properties are usually responsible for reduction of metal compounds into their respective nanoparticles.

For preparation of nanoparticles, micro-organisms were forced to grow in the respective salt solutions to see their compatibility. Selected organisms (those who are growing well) were kept for fungal ball preparation which was then allowed to release for enzymes to breakdown of salts into nano form. The biosynthesized nanoparticles are more stable due to natural encapsulation by mother protein.

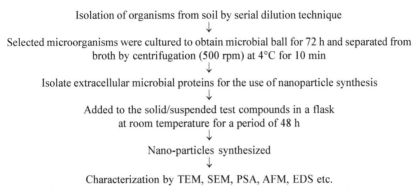

Isolation of organisms from soil by serial dilution technique
↓
Selected microorganisms were cultured to obtain microbial ball for 72 h and separated from broth by centrifugation (500 rpm) at 4°C for 10 min
↓
Isolate extracellular microbial proteins for the use of nanoparticle synthesis
↓
Added to the solid/suspended test compounds in a flask at room temperature for a period of 48 h
↓
Nano-particles synthesized
↓
Characterization by TEM, SEM, PSA, AFM, EDS etc.

Fig. 1: Methodology for nanoparticle synthesis from microorganisms

A flowchart for biosynthesis of nano particles is shown in Fig. 1. Several organisms have been developed and identified for different nanonutrient production in our laboratory (Table 1). Fungi were found to be more efficient in biosynthesis of nanonutrients. Some well known examples of bacteria synthesizing inorganic materials include magetotactic bacteria (synthesizing magnetic nano-particles) and S layer bacteria which produce gypsum and calcium carbonate layers (Shankar *et al.* 2004).

Table 1: List of some microorganisms possessing properties of nano-particle formation

Organism	Type	NCBI Accession no.	Nano particle production capacity
Ascomycota clone	CZF-6	KC412873	Zn and Mg
Aspergillus aeneus	NJP12	HM222934	Zn
Aspergillus brasiliensis	TFR-23	JX999490	Mg
Aspergillus flavus	CZR-2	JF681301	Zn, Mg, Ti, Au and Ag
Aspergillus fumigatus	TFR-8	JQ675291	Mg, Zn, P
Aspergillus japonicus	AJP01	JF770435	Fe
Aspergillus niger	TFR-4	JQ675305	Mg, Zn and Ti
Aspergillus ochraceus	TFR-23	KC806053	K
Aspergillus oryzae	TFR-9	JQ675292	Zn, P, Au, Ag, Fe and Ti
Aspergillus terreus	CZR-1	JF681300	Zn, Mg, P and Fe
Aspergillus tubingensis	TFR- 3	JN126255	Mg
Bacillus megaterium	JCT 13	JX442240	P
Emericella nidulans	TFR-14	KC175549	Zn and Mg
Emericella quadrilineata	TFR-25	KC806055	N
Emericella variecolor	TFR-16	KC175551	Mg and P
Fusarium solani	CZF-4	KC142125	Zn and Mg
Pantoea tarafdar	JCT14	KC806057	N
Penicillium janthinellum	CZF-5	KC412872	Zn and Mg
Rhizoctonia bataticola	TFR-6	JQ675307	Ag, Au and Zn

Some microorganisms can survive and grow even at higher metal ion concentration due to their resistance to the metal. The mechanisms involve: efflux systems, alteration of solubility and toxicity *via* reduction or oxidation, biosorption, bioaccumulation, extra cellular complexation or precipitation of metals and lack of specific metal transport systems (Husseiny *et al.* 2007). For e.g. *Pseudomonas stutzeri* AG 259 isolated from silver mines has been shown to produce silver nanoparticles (Mohanpuria *et al.* 2008).

In the microbial synthesis of metal nanoparticles by fungus, extracellular secreting enzymes are produced which reduces the metal salt of macro and micro scale into nanoscale diameter through catalytic effect (Jain *et al.* 2011). Extracellular secretion of enzymes offers the advantage to obtain pure, monodisperse nanoparticles, which are free from cellular components, associated with downstream processing. A possible mechanism for phosphorus nanoparticle biosynthesis is hypothesized (Fig. 2). The microbial synthesis of phosphorus nanoparticles involves an enzyme mediated process which is present in extracellular secrets and another protein mentioned as capping protein play role in the further encapsulation of phosphorus nanoparticles and increase stability. We found that cationic proteins with a molecular weight of 32-33 kDa were mainly responsible for the synthesis of the nanonutrients. Jain *et al.* (2011) indicated that silver nanoparticle synthesis by *A. flavus* occurs initially by "33 kDa" protein followed by a protein (cystein and free amine groups) electrostatic attraction which stabilizes the nanoparticle by forming a capping agent. In general, biosynthesis of nanoparticles is not fully understood but microbial cell wall surface electrostatic attraction, reduction and accumulation has been proposed.

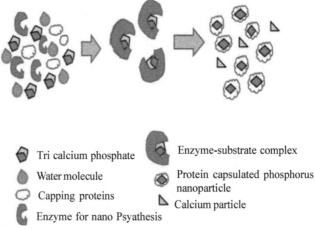

Tri calcium phosphate	Enzyme-substrate complex
Water molecule	Protein capsulated phosphorus nanoparticle
Capping proteins	Calcium particle
Enzyme for nano Psyathesis	

Fig. 2: Hypothetical mechanism for microbial synthesis of phosphorus nanoparticles using *Aspergillus tubingensis* TFR-5 from tri calcium phosphate.

3. Characterisation of nanoparticles

Particles size analyzer (PSA) with zeta potential, transmission electron microscopy (TEM), scanning electron microscopy (SEM), energy dispersive X-ray spectroscopy (EDS), atomic force microscopy (AFM), X-ray diffraction, UV-VIS spectroscopy, lithography, and fourier transform infrared spectroscopy (FTIR) are used to characterize different aspects of nanoparticles. Particle size analyzer will provide the distribution of the particle. It also gives the polydispersity index (PDI) value which indicates whether particles are monodisperse or not. Zeta potential will indicate the stability of the particles. SEM, TEM and AFM can be used to visualize the location, size, and morphology of the nanoparticles, while UV-VIS spectroscopy can be used to confirm the metallic nature, size and aggregation level. Energy dispersive analysis of X-ray is used to determine elemental composition, and X-ray diffraction is used to determine chemical composition and crystallographic structure. FTIR will provide the indication of bonding mechanism. By using lithographic technique we may prepare the prototype and pattern of distribution. Characterization biosynthesized Mg and Zn nanoparticles are shown as Figs. 3 and 4.

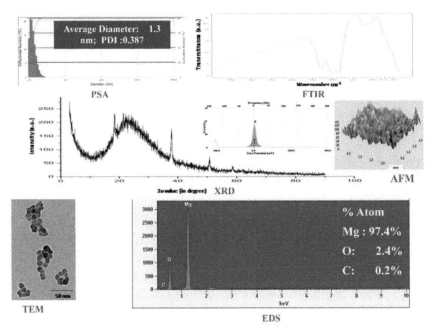

Fig. 3: Characterization of Biosynthesized Mg Nanoparticles

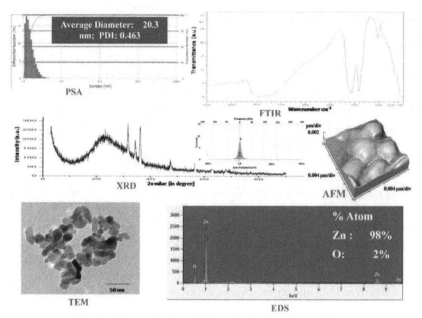

Fig. 4: Characterization of Biosynthesized Zn Nanoparticles

4. Application of nanoparticles

Application of nanonutrients to plants and microorganisms, aerosol spray (with the help of nebulizer) was found to be superior than traditional spray. In general, only 15% loss of nanoparticles was found during aerosol spray as compared to 33% loss by natural sprayer (Tarafdar *et al.* 2012). Lower concentrations (5 ppm or less) penetrates better through plant tissues. Use of particles of size 20 nm or less may be more beneficial (Tarafdar *et al.* 2012). Cube shaped nanoparticles is found to be the better shape for more penetration both in plants and microorganisms as compared to the other shapes (plate, wire, cage etc.) of the particles.

Nanoparticles are adsorbed on the plant surfaces and taken up through natural nano- or micrometer-scale plant tissue openings. Several pathways exist or are predicted for nanoparticle association and uptake in plants (Fig. 5).

Uptake rates will depend on the size and surface properties of the nanoparticles. Smaller size nanoparticles can penetrate through cuticle but larger nanoparticles through cuticle-free areas, such as hydathodes, the stigma of flowers and stomatas. The nanoparticles after entering may travel through sugar solution or cell sap (Fig. 6) and with time becomes mega particle due to agglomeration. TEM images shows many Mg nanoparticles were accumulated in vacuole.

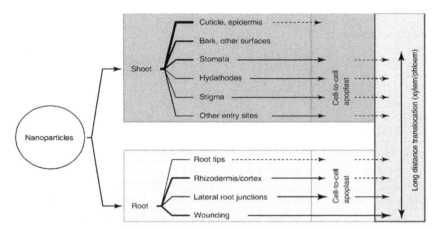

Fig. 5: Tentative pathways for nano particle translocation in plants

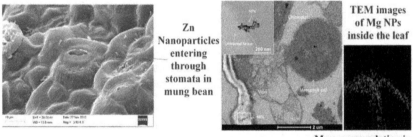

Fig. 6: Penetration and movement of Zn nanoparticles through mung bean leaves

5. Mitigation of problems by biosynthesized nanonutrients

Some of the chronic agricultural problems can be solved by the application of nutrients in nano form. For example, phosphorus (P) particles in nano form may reduce the absorption of P by Ca, Fe, Al, clay and organic matter as noticed with the mega particles and nullify the competition with silica (Si) in soil due to size difference, thus enhancing P use efficiency from 15% (P fertilizer) to 58% (nano-P application). The nano-P can be encapsulated with oleic acid for preparation of nano-P fertilizer which has tremendous impact on crop yield (Table 2). The P was applied as foliar spray on two weeks old plants.

Table 2: Effect of nano-P fertilizer (40 ppm P concentration @ 16L ha^{-1}) on clusterbean and pearl millet yield under arid field condition

Treatment	Grain yield (kg ha^{-1})	
	Clusterbean	Pearl millet
Control	625	910
Mega-P as SSP	650 (4.0)*	950 (4.4)
Nano-P (20 nm average size)	900 (44.0)	1346 (47.9)

*percent increase over control; SSP – single super phosphate

Significant improvement (44-253%) of phosphatase and phytase secretion by plants was observed with the foliar application of 10 ppm Zn nanoparticles. Nano-P, Mg and Zn also enhance root length and area of arid crops (clusterbean, moth bean, mung bean and pearl millet) by 11-32%. Nano Zn and Fe spray is associated with high protein content and low semi-oxide dismutase (SOD) activity of the plants resulted in more stress tolerance. Zn and Fe nanoparticles were also found to be essential for prevention of membrane damage at lower concentration (2-5 ppm). In general 17-33% more sunlight absorption was noticed by plant leaves with the application of 20 ppm Mg nanoparticles which enhanced 16-22% chlorophyll content. The nodulation (45-92%) and root nitrate reductase activity (13-19%) was enhanced with the application of 4 ppm Zn nanoparticles.

Soils of the arid region are generally coarse-textured, contain low soil organic carbon and have poor structure. As a result, these soils dry rapidly and are extremely prone to wind erosion. These twin factors of wind erosion and rapid drying adversely affect the survival of microorganisms in arid soils. Exo-polysaccharides produced by microorganisms can absorb moisture and can also bind the soil particles together to improve soil aggregation. Thus increasing production of exo-polysaccharides can help in reducing erodibility of soil, increasing soil moisture retention, survivability of microorganisms and C build up into the system. Nanoparticles Zn (10 ppm) and Fe (30 ppm) may help to increase more polysaccharide release by microorganisms (Fig. 7) and C as well as microbial build up in the soil.

In general, 10-14% more moisture retention was observed with the application of nano-induced polysaccharide powder in arid soils (Table 3).

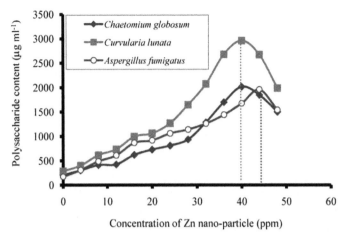

Fig. 7: Effect of Zn nanoparticles on polysaccharide production by fungi

Table 3: Effect of nano-induced polysaccharide powder on improvement in moisture retention in aridisol

Polysaccharide percentage	Improvement in moisture retention (%)
1	10.7
2	12.2
3	12.5
4	12.8
5	13.6
6	14.2
LSD (p = 0.05)	2.3

Advantages of biosynthesized nanonutrients are:

- Ecofriendly approach
- Three fold increase in Nutrient Use Efficiency (NUE)
- 80-100 times less requirement than chemical fertilizer
- Complete bio-source so environment friendly
- 10 times more stress tolerant by the crops
- 30% more nutrient mobilization in the rhizosphere
- 17-54% improvement in crop yield
- Improvement in soil aggregation (33-82%), moisture retention (10-14%) and C build up (2-5%) in the soil.

6. Other applications in agriculture

Khodakovaskaya *et al.* (2009) at the University of Arkanasas, USA, have reported the use of carbon nano-tube for improving the germination of tomato seeds through permeation of moisture. Their data has clearly demonstrated that carbon nanotubes (CNTs) serve as new pores for water permeation by penetration of seed coat and act as a gate to channelize the water from the substrate in to the seeds. These processes facilitate germination which can be exploited in rainfed agricultural system.

Nanofertilizer has the opportunity to profoundly impact energy, the economy, and the environment by reducing nitrogen loss due to leaching, emission, and long-term incorporation by soil microorganisms (De Rosa *et al.* 2010). Nano-fertilizers that utilize natural materials for coating and cementing granules of soluble fertilizer have the advantage of being less expensive to produce than those fertilizers that rely upon manufactured coating materials.

Developing a target specific herbicide molecule encapsulated with nanoparticle is aimed for specific receptor in the roots to target weeds, which enter into system and translocated to parts that inhibit glycosis of food reserve in the root system. This will make the specific weed plant to starve for food and get killed (Chinnamuthu and Kokiladevei 2007). In this way, less herbicide is required as well as if the active ingredient is combined with a smart delivery system, herbicide will be applied only when necessary according to the conditions prevailing in the field. Nano-pesticides will reduce the rate of application because the quantity of product actually being effective is at least 10-15 times smaller than that applied with classical formulation; hence a much smaller amount is required to have much better and prolonged management.

Nanoscale carrier can be utilized for the efficient delivery of fertilizers, pesticides, herbicides, plant growth regulators etc. The mechanisms involved in the efficient delivery, better storage and controlled release induce: encapsulation and entrapment, polymers and dendrimers, surface ionic and weak bond attachments among others. These mechanisms help to improve stability against degradation in the environment and ultimately reduce the amount to be applied, which reduces chemical runoff and alleviates environmental problems.

Biosensors provide high performance capabilities for use in detecting contaminants in food or environmental media. They offer high specificity and sensitivity, rapid response, user-friendly operation, and compact size at low cost (Amine *et al.* 2006). Several nanobased biosensors on direct enzyme inhibition have been developed to detect contaminants. They may prove useful as a screening tool making them useful to those in the field. Moreover, nanotechnology has shown its ability in modifying the genetic constitution of the crop plants, thereby, helping in further improvement of crop plants (Jones, 2006; DeRosa *et al.* 2010).

Nanoscale devices are envisioned that would have the capability to detect and treat diseases, nutrient deficiencies or any other maladies in crops long before symptoms are visually exhibited. "Smart Delivery System" for agriculture can possess timely controlled, spatially targeted, self-regulated, remotely regulated, pre-programmed, or multi-functional characteristics to avoid biological barriers to successful targeting. The nano-scale monitors may be linked to the recording and the tracking devices to improve identify preservation (IP) of food and agricultural products. The IP system is highly useful to discriminate organic *versus* conventional agricultural products.

Nanobarcodes have been used as ID tags for multiplexed analysis of gene expression and intracelluar histopathology. In the near future, more effective identification and utilization of plant gene trait resources is expected to introduce rapid and cost effective capability through advances in nanotechnology based gene sequencing.

Crop growth and field conditions like moisture level, soil fertility, temperature, crop nutrient status, insects, plant diseases, weeds etc. can be monitored through advancement in nanotechnology. Such real-time monitoring is done by employing networks of wireless nanosensors across the cultivated fields, providing essential data for agronomic processes like optimal time of planting and harvesting of the crop. It is also helpful for monitoring the time and level of water application, fertilizers, pesticides and herbicides and other treatments. These processes are needed to be administered for a given specific plant physiology, pathology and environmental conditions, which ultimately ensure the judicious use of the resource input and yield maximization.

Nanoparticles have been widely used to improve various reactions as reductant and/or catalysis due to their large surface areas and specific characteristics (Hilderbrand *et al.* 2008). Magnetic nanoparticle has been used to improve the microbiological reaction rates. In fact, magnetic nanoparticles were utilized not only for their catalytic function but also for their good ability to disperse. Shen *et al.* made use of the coated microbial cells of *Pseudomonas delafieldii* with magnetic Fe_3O_4 nanoparticles to fulfill desulfurization of dibenzothiophene (Shan *et al.* 2005). The high surface energies of nanoparticles resulted in their strong adsorption on the cells. The application of an external magnetic field ensured that the cells were well diffused in the solution even without mixing and enhanced the possibility to collect cells for reuse.

7. Biosafety issues

The toxicity tests for the nanoparticle should be prescribed on a case-by-case basis, based on the nanoparticles and their dose, the crop and the economic products. Clear guidelines should be formulated regarding the parameters and methodology of monitoring and evaluation. In the context of the environmental exposure to nanotechnology-based material and products, thereby increasing the risk to human (since water resources are particularly vulnerable to direct and indirect contamination with nanomaterials), the potential toxicity and environmental implications of nano-materials to aquatic organisms need to be evaluated. The nano-toxicity studies in agriculture are very limited, and need to be strengthened.

8. Conclusion and future prospects

Biosynthesized nanonutrients is in its infancy stage but have tremendous potential. Much work is needed to improve the synthesis efficiency and control of particle size and morphology. The synthesis process is quite slow, need a few days, compared to physical and chemical approaches. Reduction of synthesis time will make this biosynthesis process much more attractive. Particle size, monodispersity and stability are the three important issues in the evaluation of nanoparticle synthesis. Therefore, effective control of the particle size, monodispersity and stability must be extensively investigated. Our experiences have shown that the nanoparticles formed by microorganisms may be decomposed after a certain period of time. Thus the stability of nanoparticles produced by biological methods deserve further study. Since the control of particle shape in chemical and physical synthesis of nanoparticles is still an ongoing area of research, biological process with the ability to strictly control particle morphology would therefore offer considerable advantage. By varying parameters like microorganisms type, growth stage (phase) of microbial cells, growth medium, synthesis conditions, pH, substrate concentration, source compound of target nanoparticle, temperature, reaction time, and addition of non-target ions; it might be possible to obtain sufficient control of particle size and monodispersity. Biosynthesis methods are also advantageous because nanoparticles are sometimes coated with a lipid layer that confers physiological solubility and stability, which is critical for biomedical applications and is the bottleneck of other synthetic methods. Research is currently carried out manipulating cells at the genomic and proteomic levels. With a better understanding of the synthesis mechanisms on a cellular and molecular level, including isolation and identification of the compounds responsible for the reduction of nanoparticles, it is expected that short reaction time and high efficiency can be obtained. With the recent progress and the ongoing efforts in improving particle synthesis efficiency and exploring their biomedical and agricultural applications, it is hoped that the implementation of these approaches on a large scale and their commercial application will take place in coming years.

References

Ahmad A, Senapati S, Khan MI, Kumar R, Sastry M (2003) Extracellular biosynthesis of monodisperse gold nanoparticles by a novel extremophilic actinomycete, *Thermomonospora* sp. Langmuir 19, 3550-3553.

Akhtar MS, Panwar J and Yun YS (2013) Biogenic synthesis of metallic nanoparticles by plant extracts. ACS Sustain. Chem. Eng. 1, 591-602.

Amine A, Mohammadi H, Bourais I. and Palleschi G (2006) Enzyme inhibition-based biosensors for food safety and environmental monitoring. Biosens. Bioelectron. 21, 1405-1423.

Chinnamuthu CR and Kokiladevi E (2007) Weed management through nanoherbicides. In: Application of Nanotechnology in Agriculture (Ed) Chinnamuthu CR, Chandrasekaran B and Ramasamy C, Tamil Nadu Agricultural University, Coimbatore, India, pp 23-36.

DeRosa MC, Monreal C, Schnitzer M, Walsh R and Sultan Y (2010) Nanotechnology in fertilizers. Nature Nanotechnol. 5, 91.

Hilderbrand H, Mackenzie K and Kopinke FD (2008) Novel nano-catalysts for waste water treatment. Global NEST Journal 10, 47-53.

Husseiny MI, El-Azia MA, Badr Y. and Mahmoud MA (2007) Biosynthesis of gold nanoparticles using *Pseudomonas aeruginosa*. Spectrochim. Acta Part A 67, 1003-1006.

Jain N, Bhargava A, Majumdar S, Tarafdar J, Panwar J (2011) Extracellular biosynthesis and characterization of silver nanoparticles using *Aspergillus flavus* NJP08 a mechanism perspective. Nanoscale 3: 635-641.

Jones P (2006) A Nanotech revolution in agriculture and the food industry. Information Systems for Biotechnology. http://www.isb.vt.edu/articles/jun0605.htm.

Khodakovskaya M, Dervishi E, Mahmood M, Yang Xu, Zhongrui Li, Watanabe F and Biris AS (2009) Carbon nanotubes are able to penetrate plant seed coat and dramatically affect seed germination and plant growth. ACS Nano 3, 3221–3227.

Klaus T, Joerger R, Olsson E and Granqvist CG (1999) Silver based crystalline nanoparticles, microbically fabricated. Proc. Natl Acad. Sci. 96, 13611-13614.

Kowshik M, Arhtaputre S, Kharrazi S, Vogel W, Urban J, Kulkarni S K and Paknikar KM (2003) Extracellular synthesis of silver nanoparticles by a silver-tolerant yeast strain MKY3. Nanotechnology 14, 95-100.

Lee SW, Mao C, Flynn C and Belcher AM (2002) Ordering of quantum dots using genetically engineered viruses. Science 296, 892-895.

Merzlyak A and Lee S W (2006) Phage as template for hybrid materials and mediators for nanomaterials synthesis. Curr. Opin. Chem. Biol. 10, 246-252.

Mohanpuria P, Rana NK and Yadav SK (2008) Biosynthesis of nanoparticles: technological concepts and future applications. J. Nanoparticle Res. 10, 507-517.

Narayanan KB and Sakthivel N (2010) Biological synthesis of metal nanoparticles by microbes. Adv. Colloid Interface Sci. 156, 1-13.

Raliya R and Tarafdar JC (2012) Novel approach for silver nanoparticles synthesis using *Aspergillus terreus* CZR-1: Mechanism perspective. J. Bionanoscience 6, 12-16.

Sastry M, Ahmad A, Khan MI and Kumar R (2003) Biosynthesis of metal nanoparticles using fungi and actinomycete. Curr. Sci. 85, 162-170.

Shan G, Xingh J, Zhang H and Liu H (2005) Biodesulfurization of dibenzothiophene by microbial cells coated with magnetite nanoparticles. Appl. Environ. Microbiol. 71, 4497-4502.

Shankar SS, Rai A, Ankamwar B, Singh A, Ahmad A and Sastry M. (2004) Biological synthesis of triangular gold nanoprism. Nat. Mater. 3, 482-488.

Tarafdar JC and Raliya R (2011) The Nanotechnology, Scientific Publisher (India) pp215.

Tarafdar JC, Xiang Y, Wang WN, Dong Q and Biswas P (2012). Standardization of size, shape and concentration of nanoparticle for plant application. Appl. Biol. Res. 14, 138-144.

Tarafdar JC, Raliya R and Rathore I (2012) Microbial synthesis of phosphorus nanoparticles from Tri-calcium phosphate using *Aspergillus tubingensis* TFR-5. J. Bionanoscience 6, 84-89.

Tarafdar JC, Agrawal A, Raliya R., Kumar P, Burman U and Kaul RK (2012) ZnO nanoparticles induced synthesis of polysaccharides and phosphatases by *Aspergillus* fungi. Adv. Sci. Eng. Med. 4, 1-5.

Tarafdar JC, Sharma S and Raliya, R. (2013) Nanotechnology: Interdisciplinary science of application. Afr. J. Biotech. 12, 219-226.

CPSIA information can be obtained
at www.ICGtesting.com
Printed in the USA
LVHW081844080320
649338LV00008B/98

9 780367 140717